Emerging Scientific and Technical Challenges and Developments in Key Power Electronics and Mechanical Engineering

Emerging Scientific and Technical Challenges and Developments in Key Power Electronics and Mechanical Engineering

Editors

Jichao Hong
Wei Wang
Xinxi Zeng
Xiaoming Xu

MDPI

Basel • Beijing • Wuhan • Barcelona • Belgrade • Novi Sad • Cluj • Manchester

Editors
Jichao Hong
University of Science and Technology Beijing
Beijing
China

Wei Wang
University of Science and Technology Beijing
Beijing
China

Xinxi Zeng
University of Science and Technology Beijing
Beijing
China

Xiaoming Xu
University of Science and Technology Beijing
Beijing
China

Editorial Office
MDPI
St. Alban-Anlage 66
4052 Basel, Switzerland

This is a reprint of articles from the Special Issue published online in the open access journal *Electronics* (ISSN 2079-9292) (available at: https://www.mdpi.com/journal/electronics/special_issues/ustb_70th_anniversary).

For citation purposes, cite each article independently as indicated on the article page online and as indicated below:

Lastname, A.A.; Lastname, B.B. Article Title. *Journal Name* **Year**, *Volume Number*, Page Range.

ISBN 978-3-0365-8704-2 (Hbk)
ISBN 978-3-0365-8705-9 (PDF)
doi.org/10.3390/books978-3-0365-8705-9

Contents

About the Editors

Jichao Hong

Jichao Hong obtained his Ph.D. in Mechanical Engineering from the Beijing Institute of Technology in 2020. He is currently an Associate Professor at the School of Mechanical Engineering at the University of Science and Technology Beijing. His current research interests include advanced energy storage systems, big data mining and analysis, power system integration and intelligent control technology. As a technical backbone, he participated in the National Natural Science Foundations of China, National Key Research and Development Programs, special projects of the Beijing Municipal Science and Technology Commission, etc. He is a reviewer for over 40 SCI journals and has published over 60 EI/SCI papers, including the journals Applied Energy, Energy, IEEE Transactions on Transportation Electrification, Journal of Cleaner Production, International Journal of Energy Research, Journal of Energy Research, and International Journal of Energy Research. He has also applied for more than 40 invention patents. Some of the research results have been promoted and applied in the national monitoring and management platform for new energy vehicles of China and served for millions of new energy vehicles.

Wei Wang

Wei Wang received his Ph.D. from the University of Science and Technology Beijing and was jointly trained by the University of Cambridge in the UK in January 2016. He currently works at the School of Metallurgy and Ecological Engineering in Beijing Institute of Technology. So far, he has published more than 100 SCI academic papers and applied for more than 20 authorized patents. His work has been published in a range of journal including Advanced Materials, Advanced Energy Materials, Advanced Functional Materials, Energy Environmental Science, Nano Today, ACS Nano, and fourteen of his publications are highly cited, having been cited more than 7000 times.

Xinxi Zeng

Xinxi Zeng received his Ph.D. from Tsinghua University in 2019. Dr. Zeng's research interests are in the broad areas of metamaterials, shape-controlled 3D/4D printing, information functional materials and devices, robotics and intelligent structures. He was supported by three funds including the Postdoctoral Innovative Talent Support Program and funds for basic scientific research in central universities. He has participated in the National Key Research and Development Program of China and the National Natural Science Foundation of China. He has published more than 20 papers in ACS Applied Materials & Interfaces, Applied Physics Letters, Optics Express and other journals.

Xiaoming Xu

Xiaoming Xu received his Ph.D. from Nanjing University of Aeronautics and Astronautics in 2012. He is currently a Professor of the School of Mechanical Engineering in Beijing Institute of Technology. He is mainly engaged in the design and safety technology research of power battery systems/fuel cell systems for new energy vehicles and has hosted seven related projects including the National Natural Science Foundation of China, National Key R&D Program Tasks, Jiangsu Provincial Natural Science Foundation, China Postdoctoral Fund, Tsinghua University Key Laboratory Fund, and Jilin University Key Laboratory Fund. He has published 40 related academic papers as the first author/corresponding author, including 29 SCI index papers and 11 EI index papers. In addition, Xiaoming Xu has also applied for more than 20 invention patents and has published an academic monograph. Professor Xu Xiaoming pays special attention to the combination of international academic frontiers and major national needs. He won second prize for the Jiangsu Science and Technology Progress Award, first prize for the Innovation Achievement Award of the China Industry–University–Research Cooperation Promotion Association, and first prize for the Chinese Invention Association's invention and entrepreneurship achievement (all ranked first).

electronics

MDPI

Editorial

Emerging Scientific and Technical Challenges and Developments in Key Power Electronics and Mechanical Engineering

Jichao Hong [1,2]

[1] School of Mechanical Engineering, University of Science and Technology Beijing, Beijing 100083, China; hongjichao@ustb.edu.cn
[2] Shunde Innovation School of University of Science and Technology Beijing, Foshan 528000, China

1. Introduction

In celebration of the 70th anniversary of the University of Science and Technology Beijing (USTB), this Special Issue presents the electrical and mechanical engineering research of the USTB, with the aim of providing timely solutions to emerging scientific and technical challenges in key power electronics and mechanical engineering at the frontier of modern industrial development. High-quality original technical papers and advanced review papers are included herein.

This Special Issue contains thirteen articles, of which twelve are research articles and one is a review. The next section will provide a brief introduction to each article.

2. Brief Description of the Published Articles

Lu et al. [1] proposed a multi-signal vehicle speed prediction model based on the long short-term memory (LSTM) network, improving the accuracy of vehicle speed prediction by considering multiple signals. First, various signals were collected by simulating the vehicle model, and a Pearson correlation analysis was performed on the collected multiple signals in order to improve the model's prediction accuracy, and the appropriate signal was selected as the input of the prediction model. The experimental results indicated that the prediction method greatly improves the predictive effect compared with the support vector machine (SVM) vehicle speed prediction method. Secondly, the method was combined with the model predictive control–equivalent consumption strategy (MPC-ECMS) to form a control strategy suitable for power maintenance conditions, enabling the equivalent factor to be adjusted adaptively in real time and the target state of charge (SoC) value to be set. Pontryagin's minimum principle (PMP) enables the battery to calculate the range extender output power at each moment. PMP, as the core algorithm of the ECMS, is a common real-time optimal control algorithm. Then, taking into account the engine's operating characteristics, the calculated range extender power was filtered to make the engine run smoothly. Finally, hardware-in-the-loop simulation (HIL) was used to verify the model.

Wu et al. [2] proposed a skin pathological mirror classification method based on discrete wavelet down-sampling feature reconstruction. The wavelet down-sampling method was introduced first, and the multichannel attention mechanism was then introduced to realize the pathological feature reconstruction of high-frequency and low-frequency components, which reduces the loss of pathological feature information due to down-sampling and effectively utilizes the channel information. A skin cancer classification model is presented, using a combination of depth-separable convolution and 3 × 3 standard convolution and wavelet down-sampling as the input backbone of the model to ensure the perceptual field while reducing the number of parameters; the residual module of the model was optimized using wavelet down-sampling and the Hard-Swish activation function to enhance the feature representation capability of the model. The network weight

Citation: Hong, J. Emerging Scientific and Technical Challenges and Developments in Key Power Electronics and Mechanical Engineering. *Electronics* **2023**, *12*, 2958. https://doi.org/10.3390/electronics12132958

Received: 29 June 2023
Accepted: 4 July 2023
Published: 5 July 2023

parameters were initialized on ImageNet using transfer learning and then debugged on the augmentation HAM10000 dataset. The experimental results showed that the accuracy of the proposed method for skin cancer pathological mirror classification significantly improved, reaching 95.84%.

Xu et al. [3] simulated the thermal runaway triggering process of Li-ion batteries and analyzed the relationship between the local heating of the cathode collector surface and the change in the high-temperature area distribution of the diaphragm layer. The thermal runaway mechanism is further revealed. Based on the simulation results, the following conclusions can be drawn: phosphonitene compounds can delay the decomposition of the solid electrolyte interphase membrane and reduce the energy yield of battery-side reactions. Compared with the phosphonitene compound, the optimized structure of adding phosphonitene has little effect on the thermal stability of the battery.

Zhao et al. [4] developed an integrated vibration elimination system (IVES) containing a dynamic vibration-absorbing structure between the IWM and the suspension. It also includes an active suspension system based on a delay-dependent H∞ controller. Furthermore, a novel frequency-compatible tire (FCT) model was constructed to improve the IVES's accuracy. The mechanical–electrical–magnetic coupling effects of IWMD EVs were theoretically analyzed. A virtual prototype for the IVES was created by combining CATIA, ADAMS, and MatLab/Simulink, resulting in a high-fidelity multi-body model, validating the IVES's accuracy and practicability. Simulations for the IVES considering three different suspension structure types and time delay considerations were performed. Analyses in frequency and time domains for the simulation results showed that the root mean square of sprung mass acceleration and the eccentricity were significantly reduced via the IVES, indicating an improvement in ride comfort and IWM vibration suppression.

Zhou et al. [5] proposed a wavelet transform longitudinal denoising method, combined with a genetic algorithm (GA-WT), which was proposed to handle the big noise of the measured data from each signal channel of the flatness meter, and Legendre orthogonal polynomial fitting was employed to extract the effective flatness features. Based on the preprocessed actual production data, the adaptive moment estimation (Adam) optimization algorithm was applied to intelligently identify the flatness control efficiency. This paper takes the actual production data of a 1420 mm tandem cold mill as an example to verify the performance of the new method. Compared with the control efficiency determined by the empirical method, the flatness residual MSE 0.035 was 5.4% lower. The test results indicated that the GA-WT–Legendre–Adam method can effectively reduce the noise, extract the flatness features, and achieve the intelligent determination of the flatness control efficiency.

Li et al. [6] proposed a novel liquid cooling plate with mini-channels and improved it with disturbance structures. First, an accurate battery heat generation model was established and verified by experiments. The error was less than 4%, indicating that the heat generation power is reliable. Then, five designs are proposed first to determine the suitable number of disturbance structures, and plan 3 with five disturbance structures showed satisfactory performance in heat dissipation and flow field. Moreover, four layout plans were proposed, namely uniform, interlaced, thinning, and gradually denser distribution. The results showed that plan 5 (uniform) achieved the best performance: the maximum average temperature was 36.33 °C and the maximum average temperature difference was 0.16 °C. Last, an orthogonal experiment and range analysis were adopted to optimize the structure parameters. The results showed that the best combination for the space between the adjacent disturbance structures was d1 = 20 mm, length d2 = 5 mm, width d3 = 1.5 mm and tilt angle $\beta = 60°$.

Zhang et al. [7] studied the spiral climbing motion of a snake-like robot on the outer surface of a cylindrical object based on the three-dimensional motion of a biological snake and then carried out the analysis and optimization of the motion-influencing factors. First the spiral climbing motion of the snake-like robot was implemented by the angle control method, and the target motion was studied and analyzed by combining numerical and environmental simulations. We integrated the influence of kinematics and dynamics factors on the spiral climbing motion. Based on this, we established a multi-objective optimization function that utilized the influence factors to optimize the joint module. In addition, through dynamics simulation analysis, the change in the general clamping force of the snake-like robot's spiral climbing motion was transformed into the analysis of the contact force between the joint module and the cylinder. On the basis of the results, the effect of the control strategy adopted in this paper on the motion and change rule of the spiral climbing motion was analyzed. This paper presents the analysis of the spiral climbing motion, which is of great theoretical significance and engineering value for the realization of the three-dimensional motion of the snake-like robot.

Zhang et al. [8] proposed a strain rate-dependent material model for accurately evaluating the dynamic response of CFRTP laminates with different stacking sequences. The model was composed of three components: a strain-rate-dependent constitute model, a strain-rate-related damage initiation model, and an energy-based damage evolution model. The strain rate effect of modulus and strength was described by a stacking-sequence-related matrix, and the damage initiation model could describe the matrix, fiber, and delamination damage of CFRTP laminates without introducing cohesive elements. The material model was implemented into finite element software ABAQUS by user-defined subroutine VUMAT. The low-velocity impact tests of CFRTP laminates with quasi-isotropic and angle-ply stacking sequences were used to provide validation data. The dynamic response of CFRTP laminates from the numerical results was highly consistent with the experimental results. The mechanical response of CFRTP laminates was affected by the stacking sequence and impact energy, and the numerical error of the proposed material model significantly decreased with the increasing impact energy, especially for the laminae where the damage had occurred.

Yang et al. [9] proposed a collaborative multi-vehicle localization scheme based on GNSS and vehicle networks. The vehicle first estimates the location based on GNSS positioning information and then shares this information with the environmental vehicles through vehicle network communication. The vehicle further integrates the relative position of the ambient vehicle observed by the radar with the ambient vehicle position information obtained by communication. A smaller error estimate of the position of self-vehicle and environmental vehicles is obtained by correcting the positioning of self-vehicle and environmental vehicles. The proposed method is validated by simulating multi-vehicle motion scenarios in both lane change and straight-ahead scenarios. The root-mean-square error of the co-location method is below 0.5 m. The results demonstrate that the combined vehicle network communication approach has higher accuracy than single GNSS positioning in both scenarios.

Xiang et al. [10] investigated the fatigue fracture of bilateral drive drum shafts in casting bridge cranes including their fracture morphology and other factors, such as materials, manufacturing processes, and loads. Seven conditions were designed to test the effects of changes in the speed and torque of the drum shafts during start-up, commissioning, and braking under different loads. A dynamic model was developed for the structure and control system of the hoisting mechanism. Changes in the speed and torque of the motor and drum shafts were simulated under common operating conditions such as the speed and load changes of the motor, control asynchrony, and single-motor towing. The results showed that asynchronous motor starting and braking, motor dragging, and other behaviors led the left and right drum shafts to undergo oscillated torque with a value reaching 2×10^5 N·m in a period of approximately 13 s, and a residual torque of about 3×10^4 N·m was retained after braking. The torques on the drum shafts changed suddenly during the processes of starting, shifting, and braking. Dynamic loading was the root cause of the fatigue fracture of the drum shafts.

Zhang et al. [11] designed a new, three-dimensional honeycomb with a negative Poisson's ratio. A honeycomb cell was first designed by out-of-plane stretching of a re-entrant honeycomb, and the honeycomb was built by spatially combining the cells. The in-plane response and energy absorption characteristics of the honeycomb were studied through the finite element method (FEM). Some important characteristics were studied and are listed as follows: (1) the effects of cell angle and impact velocity on the dynamic response were tested. The results show that the honeycomb exhibits an obvious negative Poisson's ratio and unique platform stress enhancement effect under the conditions of low and medium velocity. An obvious necking phenomenon appears when the cell angle parameter is 75°. (2) Based on the one-dimensional shock wave theory, the empirical formula of the platform stress was proposed to predict the dynamic bearing capacity of the honeycomb. (3) The energy absorption under different conditions was investigated. The results showed that as the impact velocity increases, the energy absorption efficiency gradually decreases. In addition, with the increase in the cell angle, the energy absorption efficiency is gradually improved. The above study shows that the honeycomb studied has good potential for use in the automotive industry as an energy absorption material. It also provides a new strategy for the multi-objective optimization of mechanical structure design.

Lu et al. [12] proposed a 3D-based breast ultrasound system, which can automatically diagnose ultrasound images of the breasts and generate a representative 3D breast lesion model through typical ultrasonography. In this system, the authors used a weighted ensemble method to combine three different neural networks and explore different combinations of the neural networks. On this basis, a breast locator was designed to measure and transform the spatial position of lesions. The breast ultrasound software generates a 3D visualization report through the selection and geometric transformation of the nodular model. The ensemble neural network improved in all metrics compared with the classical neural networks (DenseNet, AlexNet, GoogLeNet, etc.). The results proved that the ensemble neural network proposed in this work can be used for the intelligent diagnosis of breast ultrasound images. For 3D visualization, magnetic resonance imaging (MRI) scans were performed to achieve their 3D reconstructions. By comparing two types of visualized results (MRI and the 3D model), the authors determined that the models generated by the 3D-based breast ultrasound system have similar nodule characteristics and spatial relationships with MRI. In summary, this system implements the automatic diagnosis of ultrasound images and presents lesions through 3D models, which can obtain complete and accurate ultrasound image information. Thus, the system has clinical potential.

Wu et al. [13] described the EMI sources and coupling paths of EMI in third-generation semiconductor devices used in power electronic converters. The modeling methods of EMI are summarized from the perspectives of power devices and coupling paths. The suppression methods of conducted noise are summarized by suppressing EMI sources and improving the coupling path characteristics. This paper provides a reference for the electromagnetic compatibility design of power electronic converters for third-generation semiconductor devices.

3. Future Directions

Electrical and mechanical engineering has developed rapidly in recent years. One obvious trend is digitization and intelligence. This is reflected in the emergence and development of technologies such as artificial intelligence, big data, and digital twins. These technologies can increase the efficiency of equipment operation, extend operation times, and provide more information services. As a practitioner in the relevant industries, it is of positive significance to understand and observe the development of the field.

Funding: This research was funded by (National Natural Science Foundation of China) grant number (52107220). And The APC was funded by (Jichao Hong).

Data Availability Statement: The data presented in this study are available on request from the corresponding author.

Acknowledgments: I would like to thank of all the authors for the papers submitted to this Special Issue. We would also like to acknowledge all of the reviewers for their careful and timely reviews to help improve the quality of this Special Issue. Last but not least, we would like to thank the editorial team of the *Electronics* journal for all of the support provided during the publication of this Special Issue.

Conflicts of Interest: The author declares no conflict of interest.

References

1. Lu, L.; Zhao, H.; Liu, X.; Sun, C.; Zhang, X.; Yang, H. MPC-ECMS Energy Management of Extended-Range Vehicles Based on LSTM Multi-Signal Speed Prediction. *Electronics* **2023**, *12*, 2642. [CrossRef]
2. Wu, Q.-E.; Yu, Y.; Zhang, X. A Skin Cancer Classification Method Based on Discrete Wavelet Down-Sampling Feature Reconstruction. *Electronics* **2023**, *12*, 2103. [CrossRef]
3. Xu, X.; Zhang, X.; Hong, J. Research on the Effects of Different Electrolyte Ratios on Heat Loss Control in Lithium-Ion Batteries. *Electronics* **2023**, *12*, 1876. [CrossRef]
4. Zhao, Z.; Gu, L.; Wu, J.; Zhang, X.; Yang, H. An Integrated Vibration Elimination System with Mechanical-Electrical-Magnetic Coupling Effects for In-Wheel-Motor-Driven Electric Vehicles. *Electronics* **2023**, *12*, 1117. [CrossRef]
5. Zhou, X.; Li, L.; Ma, X.; Xu, T. Data-Driven Intelligent Recognition of Flatness Control Efficiency for Cold Rolling Mills. *Electronics* **2023**, *12*, 875. [CrossRef]
6. Li, R.; Yang, Y.; Liang, F.; Liu, J.; Chen, X. Investigation on Battery Thermal Management Based on Enhanced Heat Transfer Disturbance Structure within Mini-Channel Liquid Cooling Plate. *Electronics* **2023**, *12*, 832. [CrossRef]
7. Zhang, P.; Zang, Y.; Guan, B.; Wu, Z.; Gao, Z. Analysis and Optimization Based on Factors Affecting the Spiral Climbing Locomotion of Snake-like Robot. *Electronics* **2022**, *11*, 4002. [CrossRef]
8. Zhang, Y.; Liu, B. A Strain Rate Dependent Damage Model for Evaluating the Dynamic Response of CFRTP Laminates with Different Stacking Sequence. *Electronics* **2022**, *11*, 3728. [CrossRef]
9. Yang, H.; Hong, J.; Wei, L.; Gong, X.; Xu, X. Collaborative Accurate Vehicle Positioning Based on Global Navigation Satellite System and Vehicle Network Communication. *Electronics* **2022**, *11*, 3247. [CrossRef]
10. Xiang, D.; Li, Y.; Zhang, Y.; Xu, F. Experimental Study, Simulation and Analysis of the Fracture Failure of the Drum Shaft of a Casting Bridge Crane. *Electronics* **2022**, *11*, 3043. [CrossRef]
11. Zhang, J.; Shi, B.; Han, T. Dynamic Response and Energy Absorption Characteristics of a Three-Dimensional Re-Entrant Honeycomb. *Electronics* **2022**, *11*, 2725. [CrossRef]
12. Lu, Y.; Chen, Y.; Chen, C.; Li, J.; He, K.; Xiao, R. An Intelligent Breast Ultrasound System for Diagnosis and 3D Visualization. *Electronics* **2022**, *11*, 2116. [CrossRef]
13. Wu, X.; Gao, X.; Wang, J.; Li, Z.; Du, S.; Gao, S.; Li, F.; Du, J.; Shchurov, N.I.; Zhang, X. Advances in Modeling and Suppression Methods of EMI in Power Electronic Converters of Third-Generation Semiconductor Devices. *Electronics* **2023**, *12*, 2348. [CrossRef]

electronics

MDPI

Article

MPC-ECMS Energy Management of Extended-Range Vehicles Based on LSTM Multi-Signal Speed Prediction

Laiwei Lu [1], Hong Zhao [1,*], Xiaotong Liu [1], Chuanlong Sun [1], Xinyang Zhang [2] and Haixu Yang [2]

[1] College of Mechanical and Electrical Engineering, Qingdao University, Qingdao 266071, China; 2021023552@qdu.edu.cn (L.L.); 2020025567@qdu.edu.cn (X.L.); 2020020440@qdu.edu.cn (C.S.)

[2] School of Mechanical Engineering, University of Science and Technology Beijing, Beijing 100083, China; m202120637@xs.ustb.edu.cn (X.Z.); haixuyang@xs.ustb.edu.cn (H.Y.)

* Correspondence: zhaohong@qdu.edu.cn

Abstract: Rule-based energy management strategies not only make little use of the efficient area of engines and generators but also need to perform better planning in the time domain. This paper proposed a multi-signal vehicle speed prediction model based on the long short-term memory (LSTM) network, improving the accuracy of vehicle speed prediction by considering multiple signals. First, various signals were collected by simulating the vehicle model, and a Pearson correlation analysis was performed on the collected multiple signals in order to improve the model's prediction accurate, and the appropriate signal was selected as the input to the prediction model. The experimental results indicate that the prediction method greatly improves the predictive effect compared with the support vector machine (SVM) vehicle speed prediction method. Secondly, the method was combined with the model predictive control-equivalent consumption strategy (MPC-ECMS) to form a control strategy suitable for power maintenance conditions enabling the equivalent factor to be adjusted adaptively in real-time and the target state of charge (SoC) value to be set. Portryagin minimum principle (PMP) enables the battery to calculate the range extender output power at each moment. PMP, as the core algorithm of ECMS, is a common real-time optimal control algorithm. Then, taking into account the engine's operating characteristics, the calculated range extender power was filtered to make the engine run smoothly. Finally, hardware-in-the-loop simulation (HIL) was used to verify the model. The simulation results demonstrate that this method uses less fuel than the equivalent fuel consumption minimum strategy (ECMS) by 1.32%, 9.47% when compared to the power-following control strategy, 15.66% when compared to the SVM-MPC-ECMS, and only 3.58% different from the fuel consumption of the dynamic programming (DP) control algorithm. This shows that this energy management approach can significantly improve the overall vehicle fuel economy.

Keywords: extended-range bus; LSTM vehicle speed prediction; model prediction control (MPC); equivalent factor adaptive (ECMS)

Citation: Lu, L.; Zhao, H.; Liu, X.; Sun, C.; Zhang, X.; Yang, H. MPC-ECMS Energy Management of Extended-Range Vehicles Based on LSTM Multi-Signal Speed Prediction. *Electronics* **2023**, *12*, 2642. https://doi.org/10.3390/electronics12122642

Academic Editor: Carlos Andrés García-Vázquez

Received: 16 May 2023
Revised: 6 June 2023
Accepted: 7 June 2023
Published: 12 June 2023

1. Introduction

New energy vehicles have drawn more and more attention in recent years as energy and environmental issues have become more severe. Zhao et al. [1] studied the impact of unregulated emissions from fuel-powered cars on the environment and promoted Qingdao's "Oil-to-Gas" plan. Although pure electric vehicles have many advantages, such as low emissions and high efficiency, due to the limitations of small battery capacity, long charging time, and the small number of charging piles, the driving range of pure electric vehicles is short, which greatly affects the popularity of pure electric vehicles. Extended-range electric vehicle has become a research hotspot today because of a range extender that can continuously charge the battery, greatly increasing its driving range. The control strategy implements energy management by distributing power between the range extender and the power battery at various periods. In order to better distribute the range extender and battery output power, the rule-based control strategy was developed. Single-point and multi-point control are two rule-based control methods. Although the rule-based control strategy solves the problem of power allocation, it does not achieve optimal planning in most scenarios, so many scholars have proposed an optimization-based control strategy. The optimization-based control strategies, such as DP and PMP [2,3], adaptive equivalent fuel consumption minimum strategies (A-ECMS) [4], etc., typically optimize the original control strategy by using algorithms or calculate the global optimal control quantity from the global perspective.

Numerous academics have examined dynamic programming in recent years. Yuan et al. [5] designed a dynamic programming management strategy and applied it to automobiles, obtaining better experimental results and good vehicle economy. Zhang et al. [6] studied the control strategy for the optimal curve to improve the traditional power-following control strategy fuel economy. Chen et al. [7] designed a range-extender car using the DP control strategy. Pan et al. [8] applied the DP to supercapacitor hybrid electric vehicles for optimization, aiming to make the vehicle more economical. It is still difficult to achieve global optimization because of the uncertainty of future operating conditions. ECMS is a real-time control algorithm that changes from the Pontryagin minimum principle (PMP), which can equivalently convert electricity to fuel consumption by equivalent factors to calculate the control quantity and reduce fuel consumption. However, faced with constantly changing working conditions during driving, it is necessary to change the equivalent factor to maintain a constant SoC. Researchers have recently proposed more and more factor-adaptive approaches. Rezai et al. [9] proposed an ECMS strategy that adapts the equivalent factors according to the driving conditions. A hierarchical energy management strategy optimized in real time was proposed in [10], and heuristic DP was used to obtain optimal control and significantly improved fuel economy. Many researchers [11–13] adopted different speed prediction methods and combined MPC to optimize energy management strategies, such as the long short-term memory (LSTM), support vector machine(SVM), etc. Chen et al. [14] proposed an energy management method for driving condition recognition based on LSTM, which coordinated and optimized the engine speed according to different working conditions. Han et al. [15] studied the influence of electric motor (EM) thermodynamics on energy utilization efficiency based on vehicle speed prediction and MPC energy management. Lin et al. proposed an integrated-learning speed prediction-based energy management approach that considers adaptive reference SoC for driving modes [16].

Ritter et al. [17] proposed a novel method for successfully integrating long prediction ranges influenced by uncertainty in a stochastic MPC framework for hybrid electric car energy management. To increase the energy economy of metropolitan buses, Li et al. [18] proposed a control strategy that can manage energy based on driver behaviour types. A new driving pattern recognition method was designed to improve fuel economy and optimize power distribution by classifying the drive blocks by the K-means clustering algorithm [19]. An MPC based on an energy management strategy (EMS) and double-Q learning (DQL) was proposed for power distribution among multiple power sources in PHEVs [20]. Lin et al. [21] proposed an approach that can adjust the ECMS equivalent factor on time based on the unfinished mileage. The parameter identification method was used to identify the "black box" HEV model, and an EMS based on a two-layer MPC was proposed to optimize and control the operation process of HEV [22]. This method significantly reduced fuel consumption. Wang et al. [23,24] used improved control strategies to save fuel according to the driving patterns. Turker et al. [25] used neural networks for the energy management of cars, improving fuel economy. Al-Saadi et al. [26] improved vehicle economy by combining adaptive cruise control and MPC while ensuring safety. Wang et al. [27] proposed a power allocation strategy based on car operating conditions, driving mode prediction, and MPC, saving fuel while maintaining a constant SoC based on the tracking framework. Mu et al. [28] established a traffic flow statistical framework for long-term statistics, improving statistical efficiency and accuracy. Various random factors were considered to model the thermal runaway problem of overcharging batteries with multiple factors, and the reliability of this method was verified [29]. Hong et al. [30] established a new battery temperature prediction model through a new clustering data partitioning method and self-attention mechanism, providing accurate predictions throughout the seasons, reliably guaranteeing battery safety and greatly reducing the risk of battery runaway. Sun et al. [31] investigated LSTM speed prediction and improved MPC to improve the fuel economy.

Although many people have proposed MPC-EMS energy management strategies based on different neural network-structured vehicle speed prediction models, few studies have focused on the effect of multi-vehicle signals on speed prediction accuracy. Based on these studies, this paper proposes a combination of multiple signals LSTM prediction models and MPC-ECMS to adjust the fuel economy. The rest of this article is organized as follows. Section 2 introduces the optimal economic curve of the vehicle powertrain and the range extender. In Section 3, we carry out Pearson's correlation analysis of multiple signals during vehicle driving. Multiple signals are screened out to facilitate vehicle speed prediction, and the multi-signal LSTM vehicle and SVM vehicle velocity prediction are introduced. The predictive effects of the two methods and subsequently compared and analysed. Section 4 introduces the MPC-ECMS and compares the fuel consumption of various control strategies, including multi-signal-LSTM-MPC-ECMS, SVM-MPC-ECMS, DP-ECMS, and power-following control strategies.

The research in this paper provides a good reference for the energy management of various current models. In addition, this control strategy has good applicability to a wide range of vehicles. Multiple signal inputs can improve the accuracy of vehicle speed prediction, and the neural network has more support compared with a single signal. Furthermore, the prediction accuracy can be improved more effectively by using the LSTM memory function. Combining MPC with ECMS can effectively utilize the real-time performance of ECMS.

2. Powertrain Modelling

This article takes a four-wheel drive bus powered by a range extender and a battery as the research object; the engine and generator are mechanically connected to form the range extender. A DC/DC converter connects the battery and other power components, and the automotive structure used in this article is shown in Figure 1. The parameters of each major component are shown in Table 1.

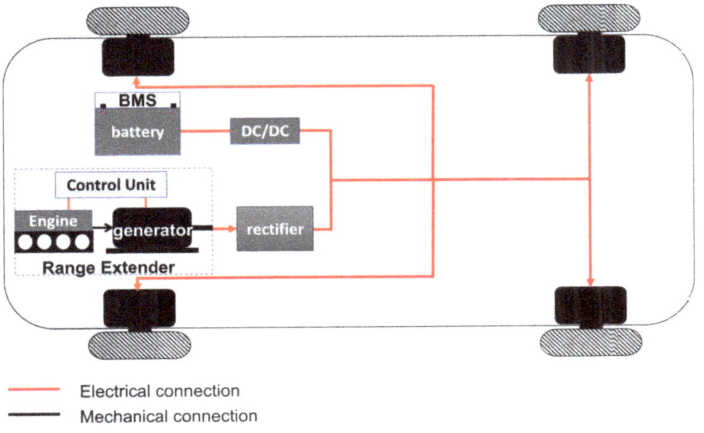

Figure 1. Range-extended vehicle structure.

Table 1. Main component parameters of range-extended vehicles.

Name	Parameters	Value	Parameters	Value
Vehicle	Total mass Aerodynamic drag coefficient	7000 kg 0.6	Wheel radius Front area	0.60 m 9 m^2
Engine	Engine type	Diesel engine	Maximum torque	2200 Nm
	Maximum speed	2000 r/min		
Electric motor	Maximum torque Type	1800 Nm Permanent magnet synchronous	Maximum speed	2650 r/min
Generator	Maximum torque	2200 Nm	Maximum speed	2000 r/min
Battery pack	Voltage	580 V	Capacity	200 Ah

2.1. Longitudinal Dynamics Model of the Vehicle

The longitudinal dynamics model is used to evaluate the performance of the vehicle under different operating conditions. It mainly considers the driving resistance according to the longitudinal dynamics principle of the whole vehicle; the driving resistance can be expressed by Equation (1)

$$F = Gfcos\alpha + \frac{C_D A}{21.15}u^2 + Gsin\alpha + \delta m\frac{du}{dt} \tag{1}$$

where G is the gravity acting on the vehicle, N; α is the slope of the road; C_D is the air resistance coefficient; A is the windward area, m^2; δ is the rotational mass conversion factor; m is the mass, kg; and $\frac{du}{dt}$ is the driving acceleration, m/s^2.

2.2. Engine Model

The engine is the main power component of the model in this paper. In this study, the engine, generator and drive motor all adopt the quasi-static model. The quasi-static model is a simplified model that facilitates the description of static operating characteristics. The quasi-static model of the engine is shown in Figure 2. The quasi-static engine model demonstrate the changes in the engine fuel consumption rate with engine speed and torque, and is beneficial for the control strategy to find the optimal working state of the engine. The model can calculate the instantaneous fuel consumption in the following way:

$$m_f = T_{eng} \times n_{eng} \times be(T_{eng}, n_{eng}) \tag{2}$$

where m_f represents the engine fuel consumption rate, kg/h; T_{eng} is the engine torque, Nm; n_{eng} represents the speed, r/min; and be represents the equivalent fuel consumption rate, kg/KWh.

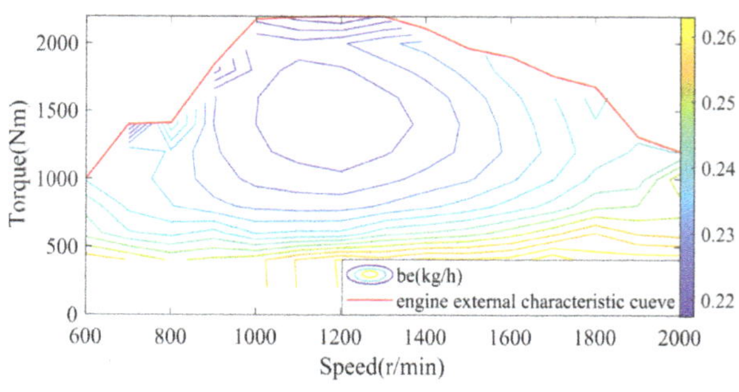

Figure 2. Engine BSFC.

2.3. Generator and Drive Motor Models

Generator and motor have similar characteristics and are used to achieve the conversion between mechanical and electrical energy. The quasi-static model of the generator and drive motor is shown in Figure 3. Because the generator and engine are mechanically connected, the optimal fuel consumption curve of the range-extender can be calculated by combining the drive generator and engine models. The optimal fuel consumption curve of the range-extender is shown in Figure 4a. The method for calculating the vehicle economy is as follows:

$$P_{Gen} = P_{eng}\eta(T_{eng}, n_{eng}) \tag{3}$$

$$feul_{eng} = f(T_{eng}, n_{eng}) \tag{4}$$

where P_{Gen} represents the generator output power; $\eta(T_{eng}, n_{eng})$ represents the generation efficiency, obtained according to the quasi-static model; P_{eng} represents the engine output power; $feul_{eng}$ represents the range-extender fuel consumption rate; T_{eng} represents the engine torque; and n_{eng} represents the engine speed.

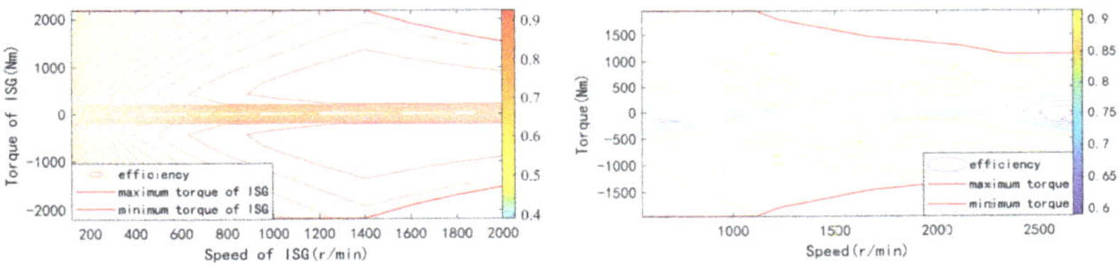

(a) (b)

Figure 3. Generator and motor efficiency. (**a**) Generator efficiency; (**b**) motor efficiency.

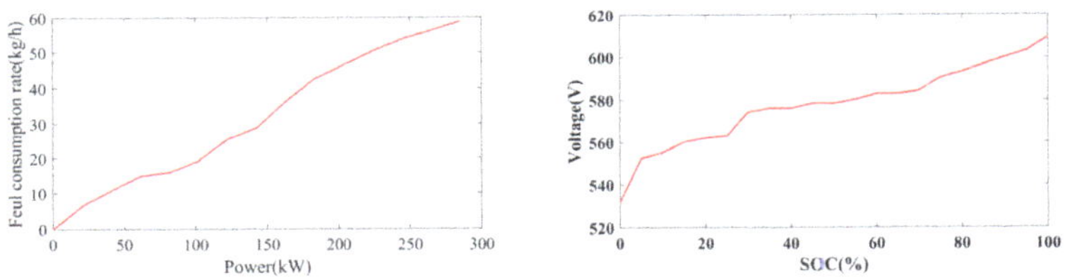

(a) (b)

Figure 4. Range-extender optimal fuel consumption curve and battery open circuit voltage. (**a**) Range-extender optimal fuel consumption curve; (**b**) battery open circuit voltage.

2.4. Power Battery Model

Using a battery as the energy storage element is an essential component of electric vehicles. The battery pack model adopts an equivalent circuit model, meaning the battery is equivalent to a voltage source and resistance. The voltage varies with the SoC, as shown in Figure 4b.

The modelling principle of power batteries is expressed as follows:

$$I(t) = \frac{U_{VOC}}{2R} - \sqrt{\frac{U_{VOC}^2}{2R^2}\frac{P_b}{R}} \tag{5}$$

$$SOC(t) = SOC_{init} + \int_{t_0}^{t} \frac{U_{VOC} - \sqrt{U_{VOC}^2 - 4RP_b}}{2RQ} dt \tag{6}$$

where SOC_{init} represents the initial SoC value; U_{voc} represents the battery open circuit voltage; R represents the internal resistance; P_b represents output power; and Q represents the capacity.

The SoC and current are also within the battery's tolerance range.

3. Speed Prediction

This paper first collects various vehicle signals through simulation, then conducts Pearson's correlation analysis to select four signals with high correlation in the historical horizon, combined with vehicle speed to form the input layer data. Then, we input different signals in the historical time domain into the trained model to predict the vehicle speed. Finally, we propose a control strategy combining the MPC and ECMS to obtain adaptable equivalent factors and control quantities in the control time domain. The combination of MPC and ECMS enables real-time adjustment of the range-extender output power while facilitating short-term planning. In this section, we compare two machine learning-based prediction methods: (1) vehicle speed prediction by multi-signal LSTM; (2) vehicle speed prediction based on SVM. The test set adopts world transient vehicle cycle (WTVC), while the training set consists of six working conditions: NEDC, WLTC, UDDS, HEFET, FTP75, and EUDC. The speed changes in the training set are shown in Figure 5. The workflow of this study is shown in Figure 6.

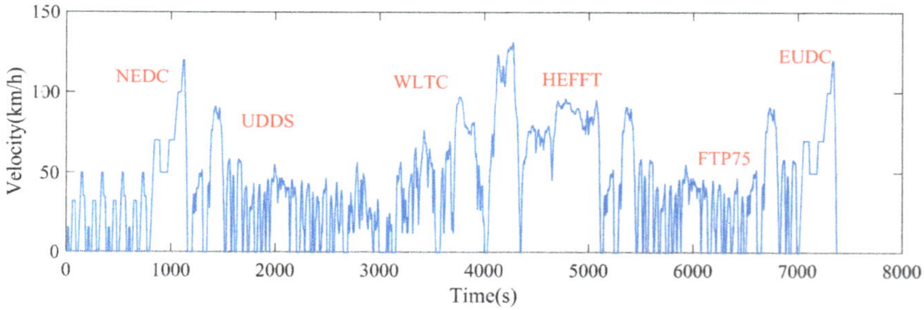

Figure 5. Vehicle speed prediction training set data.

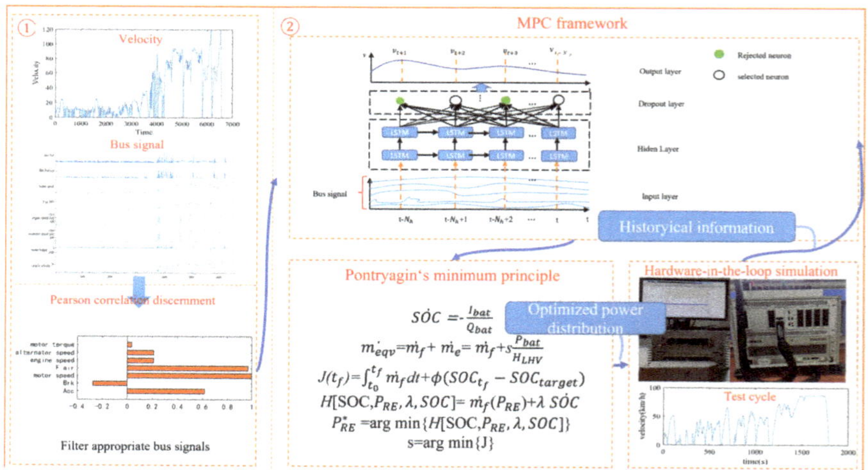

Figure 6. Flow chart of the velocity forecasting and energy management research.

3.1. Data Processing Based on Pearson's Correlation

In this paper, each vehicle signal is detected by simulation, and the correlation analysis is carried out with vehicle speed. Finally, four signals are selected together with the speed in the historical time domain to form the input data of the prediction model. The Pearson's correlation was developed by the British statistician Carl Smith and proposed by Pearson in the 20th century. The calculation formula for Pearson's correlation coefficient is as follows:

$$\rho_{X,Y} = \frac{\mathrm{cov}(X,Y)}{\sigma_X \sigma_Y} = \frac{E[(X-\mu_X)(Y-\mu_Y)]}{\sigma_X \sigma_Y} \tag{7}$$

where $\rho_{X,Y}$ is Pearson's correlation coefficient, representing the bus signal and vehicle speed; cov represents the covariance; σ indicates the standard deviation; E denotes the expectation; X represents each bus signal; and Y represents the actual vehicle speed. Table 2 shows the correlation coefficients between different bus signals and the actual speed.

Table 2. Pearson's correlation coefficients of each vehicle signal.

CAN Bus Signal	Pearson's Correlation Coefficient
Accelerator pedal opening	0.6192
Brake pedal opening	−0.2737
Motor speed	1
Air resistance	0.9687
Engine speed	0.2085
Alternator speed	0.2085
Motor torque	0.0356

According to Table 2, the accelerator pedal position, brake pedal position, motor speed, air resistance, and vehicle velocity are used as inputs for velocity prediction in the prediction domain.

3.2. Vehicle Speed Prediction Based on SVM

SVM is a machine learning algorithm commonly used for classification, which is extended to support vector regression (SVR) when applied to regression problems. SVR is a regression algorithm based on SVM, which can handle non-linear data and is usually used for time-series prediction. The process of SVR speed prediction is to collect the vehicle speed in the past n seconds to form the input and take the future speed in m seconds as the output. Figure 7 is a schematic diagram of the SVR to solve the linear regression problem. In the figure, the white circles indicate the vehicle speed in the past time, and the red circles indicate the vehicle speed predicted by the model.

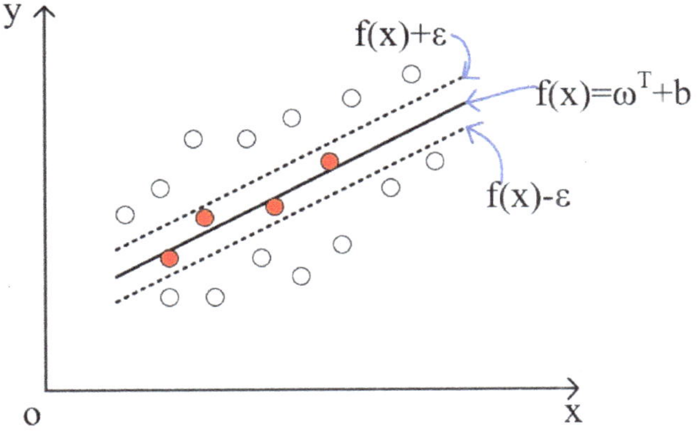

Figure 7. Schematic diagram of the support vector regression.

In contrast to traditional regression models, SVR assumes a minimum value ϵ exists such that ϵ is greater than or equal to the absolute difference between any predicted value $f(x)$ and the actual value y. Losses are calculated when the absolute value of

$y - f(x)$ exceeds ϵ, and the prediction is considered accurate when the absolute value of the difference from y is less than ϵ. Equations (8) and (9) show the expression of SVR and the loss function.

$$\min_{\omega,b} = \frac{1}{2}\|\omega\|^2 + C\sum_{j=1}^{m} l_\varepsilon(f(x_j) - y_j) \tag{8}$$

$$l_\varepsilon(z) = \begin{cases} 0, & if|z| \leq \varepsilon; \\ |z| - \varepsilon, & otherwise; \end{cases} \tag{9}$$

where ω represents the normal vector; b represents the threshold; C represents the penalty parameter; and $\text{ł}_{epsilon}$ represents the insensitive loss of ϵ.

The slack variables ξ_j and $\xi_j{}^*$ are introduced and used to rewrite Equation (9), as shown in Equation (10).

$$\begin{gathered} \min_{\omega,b,\xi_j,\xi_j{}^*} \tfrac{1}{2}\|\omega\|^2 + C\sum_{j=1}^{m}(\xi_j + \xi_j{}^*) \\ s.t. f(x_j) - y_j \leq \varepsilon + \xi_j, \\ y_j - f(x_j) \leq \varepsilon + \xi_j{}^*, \\ \xi_j, \xi_j{}^* \geq 0, j = 1, 2, \cdots, m \end{gathered} \tag{10}$$

The Lagrange multiplier is introduced to the original problem $\alpha_i, \alpha_i{}^*, \mu_i, \mu_i{}^*$ to construct the Lagrange function in Equation (11).

$$\begin{gathered} L(\omega, b, \alpha, \alpha^*, \mu, \mu^*, \xi, \xi^*) = \frac{1}{2}\|\omega\|^2 + C\sum_{j=1}^{m}(\xi_j + \xi_j{}^*) - \sum_{j=1}^{m}\mu_j\xi_j - \\ \sum_{j=1}^{m}\mu_j{}^*\xi_j{}^* + \sum_{j=1}^{m}\alpha_j(f(x_j) - y_j - \varepsilon - \xi_j) + \sum_{j=1}^{m}\alpha_j{}^*(f(x_j) - y_j - \varepsilon - \xi_j{}^*) \end{gathered} \tag{11}$$

Then, according to the dual condition and the KKT condition, the dual SVR problem is shown in Equation (12).

$$\begin{gathered} \max_{\alpha,\alpha^*} \sum_{j=1}^{m} y_j(\alpha_j{}^* - \alpha_j) - \varepsilon(\alpha_j{}^* + \alpha_j) - \tfrac{1}{2}\sum_{j=1}^{m}\sum_{j=1}^{m}(\alpha_i{}^* - \alpha_j)(\alpha_j{}^* - \alpha_j)x_i{}^T x_j \\ s.t. \sum_{j=1}^{m}(\alpha_j{}^* - \alpha_j) = 0, \\ 0 \leq \alpha_j, \alpha_j{}^* \leq C. \end{gathered} \tag{12}$$

Then, since the partial derivatives of ω, b,ξ_i and $\xi_i{}^*$ with respect to L are all 0, we obtain:

$$\omega = \sum_{j=1}^{m}(\alpha_j{}^* - \alpha_j)x_j \tag{13}$$

Then, the solution to the above regression problem is as follows:

$$f(x) = \sum_{i=1}^{m}(\alpha_i{}^* - \alpha_i)x_i{}^T x_i + b \tag{14}$$

where x is then mapped to a feature vector representation in a high-dimensional space denoted as $\phi(x)$, and then introduced to the kernel function. The above equation can be represented as:

$$f(x) = \sum_{j=1}^{m}(\alpha_j{}^* - \alpha_j)\kappa(x, x_i) + b \tag{15}$$

The kernel function used in this article is the Gaussian radial basis function, as shown in Equation (16):

$$\kappa(x, x_j) = \exp(-g\|x - x_j\|^2) \tag{16}$$

From the above derivation, we know that the penalty factor C and kernel function bandwidth g directly impact the accuracy of the SVM speed predictions. The forecast accuracy can be increased by choosing the right C and g values.

The process of using SVR to solve the vehicle speed prediction problem is shown in Figure 8.In the figure, the arrows indicate that as time changes, '...' indicates the sliding window cycle calculation. It adopts a rolling prediction method to advance every second, using the vehicle speed in the historical time domain h as the input for the prediction model. The model operates and outputs the predicted velocity for the following p seconds, continuously updated throughout cyclic prediction.

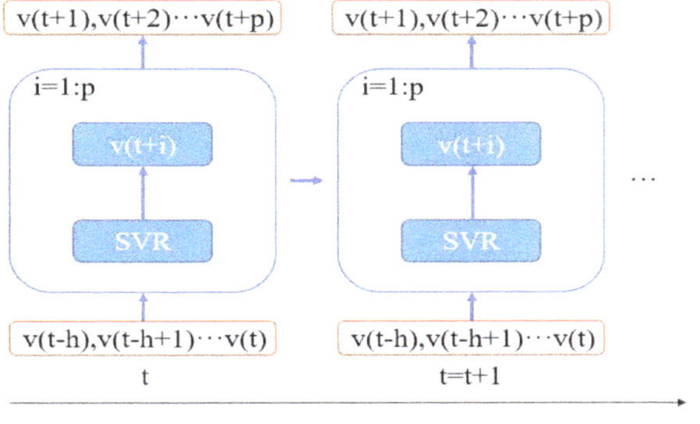

Figure 8. SVR speed prediction flowchart.

3.3. Multi-Signal Vehicle Speed Prediction Based on LSTM

Compared to a normal RNN, LSTM has a more extended short-term memory function. The key components of the LSTM are the memory unit and gate, and the input–output gate adjusts the content of the memory unit. Because of the gate structure, LSTM can avoid the "gradient disappearing" issue that plagues most RNN models. Information can be retained and gradients can be transferred over multiple time steps.

Figure 9 depicts the LSTM network configuration. The LSTM includes a unit state C compared to the RNN to save the long-term state. The input of an LSTM consists of three components: x_t, h_{t-1}, and C_{t-1}. Here, x_t represents the current input value, while h_{t-1} and C_{t-1} represent the output and state values from the previous time step, respectively. Two outputs are obtained through computation: the current output value h_t and the unit state C_t. Unlike traditional RNNs, LSTM uses three gates—input, output, and forget gates—to regulate the memory C. Retention of the input x_t in C_t is determined by the input gate, while the output gate determines how much of h_{t-1} and x_t is retained in h_t. The calculation formula for each parameter in the structure diagram is shown in Equation (17):

$$\begin{cases} f_t = \sigma(W_f[h_{t-1}, x_t] + b_f) \\ i_t = \sigma(W_i[h_{t-1}, x_t] + b_i) \\ g_t = \tanh(W_g[h_{t-1}, x_t] + b_g) \\ C_t = f_t \cdot C_{t-1} + i_t \cdot g_t \\ o_t = \tanh(W_o[h_{t-1}, x_t] + b_o) \\ h_t = o_t \cdot \tanh(C_t) \end{cases} \quad (17)$$

where f_t is the forget gate output value; i_t is the input gate output value; o_t is the output gate output value; σ represents the corresponding gate function; and W is the parameter of the gates.

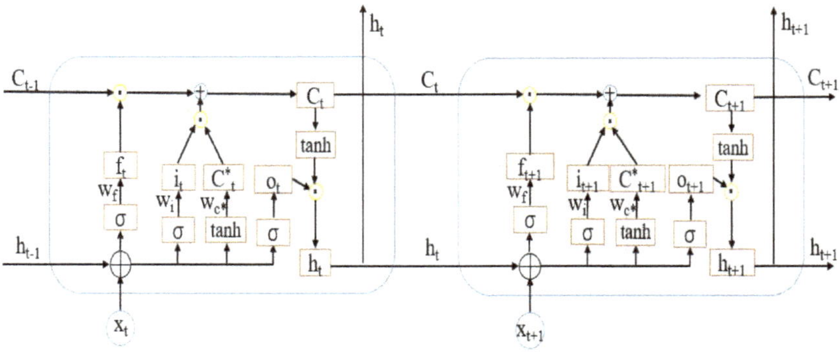

Figure 9. LSTM block diagram.

The process of predicting vehicle speed using LSTM is similar to the SVR method and is achieved through rolling prediction. As shown in Figure 10, N_h represents the historical temporal step size. By inputting the accelerator pedal opening, brake pedal opening, motor speed, air resistance, and historical vehicle speed information into the historical time domain, the neural network can calculate the change of speed within the subsequent p seconds. Then the rolling window is used for future speed prediction. During the training process, the network not only learns different features but also learns the noise in the training set which may cause the network to perform poorly on the test set due to overfitting. Therefore, this article adds a dropout layer. After adding the dropout layer, the neural network randomly deletes nodes in the hidden layer and all connections forward and backward with a probability p, thus forming a new network structure to ensure the model's generalization ability and prevent overfitting. This article uses a neural network with a dropout probability of 0.2, as shown in Figure 11. The green network nodes shown are the selected nodes, and the white nodes are abandoned nodes.

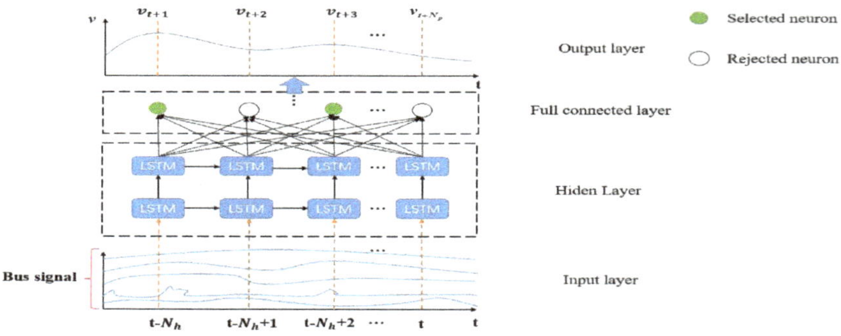

Figure 10. Multi-signal LSTM vehicle speed prediction schematic

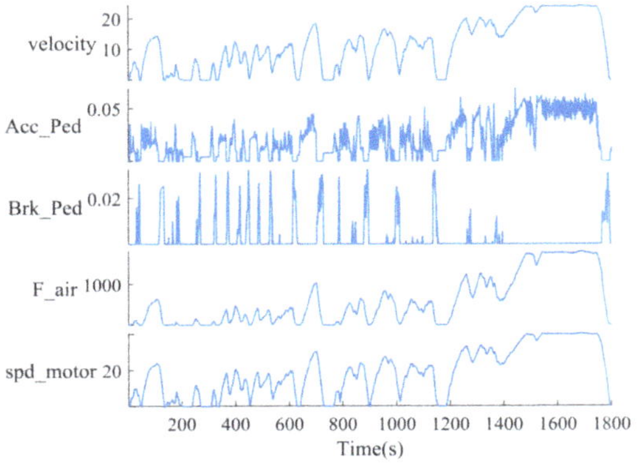

Figure 11. Multi-signal LSTM speed prediction input signal.

3.4. Vehicle Speed Prediction Results and Performance Comparison

We built a vehicle model and conducted a vehicle speed prediction simulation using Matlab2022b; the computer's CPU is Intel i7-9750H with a frequency of 2.60 GHz and 16 GB RAM. After the construction of two vehicle speed prediction models, both models were trained using the training set shown in Figure 6 and tested using the WTVC scenario as the test set. For the SVR model, only the speed with a historical time domain of 30 s was collected as prediction input. For the LSTM method, the historical vehicle speed, throttle pedal opening, brake pedal opening, and motor speed with a historical time domain of 30 s were collected for vehicle speed prediction inputs. Various vehicle signals collected from the whole vehicle model during the simulation of the training set are shown in Figure 11. When the historical time domain of both speed prediction methods was 30 s, and the prediction time domain was 5 s, the root mean square error (RMSE) of the multi-signal LSTM and SVR vehicle speed prediction methods were 3.0936 and 4.2482, respectively. The red part of the graph shows the predicted vehicle speed per second, and the black part shows the actual vehicle speed.The prediction results of the two methods are shown in Figure 12a. The performance of these two prediction methods under different prediction time domains while keeping the historical time domain at 30 s is shown in Table 3. It can be seen that the RMSE of the multi-signal LSTM speed prediction is smaller and more accurate, better reflecting the future trend of the vehicle speed. Therefore, we used the multi-signal LSTM to perform vehicle speed prediction and used the adaptive ECMS method as the energy management strategy based on the prediction. In this paper, the prediction time domain of 3, 5, and 7 s were performed to determine the most suitable prediction time domain size. The three vehicle speed prediction performances are shown in Figure 12b. We found that the larger the time domain of the prediction model, the relatively poorer the prediction accuracy. Additionally, considering the combination of energy management and vehicle speed prediction, it is not conducive to plan energy management strategies when the predicted future period is too short; therefore, 5 s was selected as the prediction time domain of the MPC framework for energy management so the MPC-ECMS could make optimal decisions.

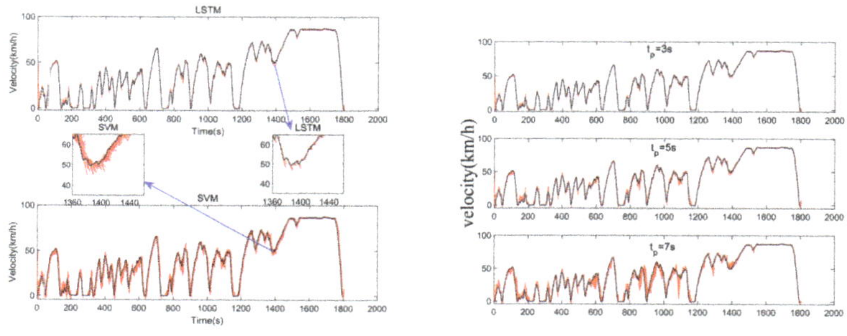

(**a**) (**b**)

Figure 12. Vehicle speed prediction effect of different prediction methods. (**a**) Comparison of multi-signal LSTM and SVM vehicle speed prediction; (**b**) comparison of multi-signal LSTM vehicle speed prediction in different prediction time domains.

Table 3. Performance comparison chart of the different vehicle speed prediction methods.

hp	**Multi-Signal LSTM**		**SVM**	
	***T_pre* (s)**	**RMSE**	***T_pre* (s)**	**RMSE**
3 s	0.0044	1.6085	0.00611	2.3735
5 s	0.0034	3.0936	0.009597	4.2482
7 s	0.0045	6.6171	0.0234	6.9404

hp is the predicted time domain size, T_pre is the prediction time.

Further research on the two methods of speed prediction yielded SVM and multi-signal LSTM speed prediction heat maps, as shown in Figure 13. The experimental results show that when the prediction time was 3 and 5 s, the speed point distribution predicted by the multi-signal LSTM method is more concentrated around the true value than the SVM method and has a better predictive performance. When the prediction time was 7 s, although there are a few scattered points in the LSTM method, most of them are still concentrated around the true value. The results in Table 3 show that when the prediction time was 7 s, the predictive performance of the two methods is relatively similar.

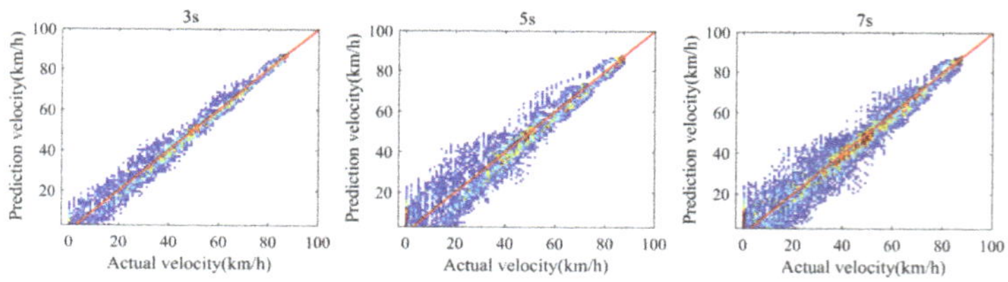

(**a**)

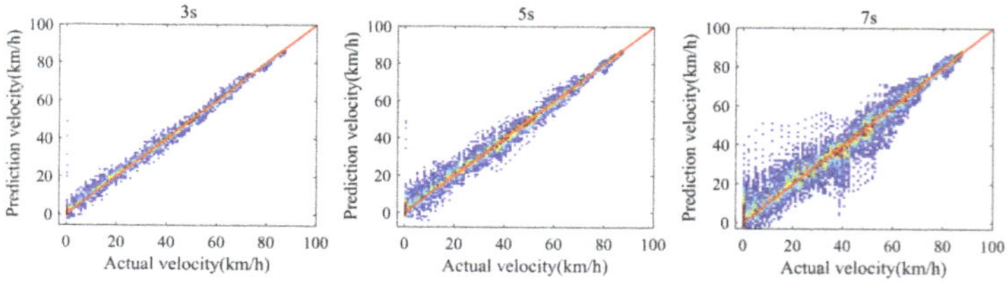

(**b**)

Figure 13. Vehicle speed prediction effect of the different prediction methods. (**a**) SVM speed forecast heat map comparison; (**b**) multi-signal LSTM speed prediction heat map comparison.

4. MPC-ECMS

The MPC-ECMS control strategy can use MPC to combine the speed prediction module and energy management module, achieving more accurate control. Compared with traditional ECMS control strategies, the MPC-ECMS energy management strategy can continuously adjust the equivalent factor by rolling optimization, ensuring that the SoC can always be maintained near the target value and avoiding the problem of the battery not operating optimally due to the SoC being too high or too low, caused by the inability of traditional ECMS equivalent factors to be changed. The MPC-ESMS energy management strategy utilizes both the flexibility of MPC and the real-time nature of ECMS to achieve real-time control by continuously adjusting the equivalence factors.

4.1. Energy Management Based on ECMS

The main idea of ECMS is the Pontryagin minimum principle (PMP). The PMP was proposed and proven by the former Soviet mathematician Pontryagin. Compared with the classical variational calculus method in solving the optimal control problem, the PMP has a wider application and fewer constraints, and its core conclusion is that in the time interval $[t_0, t_f]$, for any admissible control variable $u(t)$, there exists an optimal control $u^*(t)$ that minimizes the Hamiltonian function H, i.e.:

$$H[x^*, u^*, \lambda^*, t] \leq H[x^*, u, \lambda^*, t] \tag{18}$$

Range-extended vehicles mainly have two operating modes: energy consumption mode and SoC maintenance mode. This article proposes an energy management method under the energy maintenance mode, and we take the minimum fuel consumption under the electricity maintenance mode as the optimization objective. The SoC is regarded as the state variable, and the power generated by the range-extender as the control variable, which can transform this problem into a minimum principle problem. Its performance metric expression is as follows:

$$J(t_f) = \int_{t_0}^{t_f} \dot{m}_f(P_{RE})dt + \phi(SOC_{t_f} - SOC_{t\arg et}) \tag{19}$$

where $\dot{m}_f$ represents the instantaneous fuel consumption, kg; P_{RE} represents the range-extender output power, kW; $SOC(t_f)$ represents the SoC at the terminal time,%; SOC_target is the target SoC of the battery during battery sustaining mode,%; and ϕ is the penalty factor that enables the SoC to stay near the target value.

The equation of state and Hamiltonian function for the ECMS problem are shown in Equation (20):

$$\begin{cases} S\dot{O}C(t) = f[x, u, t] = -\frac{I_{bat}}{Q_{bat}} \\ H[SOC, P_{RE}, \lambda, t] = \dot{m}_f(P_{RE}) + \lambda\, S\dot{O}C \end{cases} \tag{20}$$

According to the PMP, the optimal output power solution formula for the range-extender can be obtained from Equation (21):

$$P_{RE}{}^*(t) = \arg\min\{H[x, u, \lambda, t]\}, P_{RE} \in U_{RE} \tag{21}$$

The necessary condition for solving the optimal output power sequence in the PMP is:

$$\begin{cases} S\dot{O}C(t) = f[SOC^*(t), P_{RE}{}^*(t)] \\ \dot{\lambda}^*(t) = -\lambda^*(t)[\frac{\partial f(SOC^*(t), P_{RE}{}^*(t))}{\partial SOC}] \\ SOC^*(t_0) = SOC_0 \\ SOC^*(t_F) = SOC_{t\arg et} \end{cases} \tag{22}$$

where SOC_0 represents the initial SoC, %; SOC_{target} represents the target SoC, %; t_0 represents the starting time of the operating condition, s; and t_f represents the simulation termination time, s. It is also known that in a range-extended electric vehicle operating in battery maintenance mode, the variation of the SoC is very small, and both the voltage and resistance are considered constant; thus, λ can be treated as a constant value.

$$\dot{\lambda}^*(t) = 0 \tag{23}$$

Therefore, the co-state variable can be approximated as a constant in the power maintenance mode. In the power sustaining mode, ECMS assumes that the output power of the range-extender and battery are all from fuel, and the battery is regarded as a buffer device for energy transfer. The equivalent fuel consumption can be calculated using the following equation:

$$\dot{m}_{eqv} = \dot{m}_f + \dot{m}_e = \dot{m}_f + s\frac{P_{bat}}{H_{LHV}} \tag{24}$$

where m_{eqv} is the equivalent fuel consumption, kg; m_f represents the fuel consumption, kg; P_{bat} represents the battery output power, kW; H_{LHV} is the heating value of th eengine fuel, kJ/kg; and s denotes the equivalent factor, representing the degree of conversion from fuel to electrical energy.

Based on the above derivation, the expression of the Hamiltonian function in the ECMS strategy can be obtained as follows:

$$H[SOC(t), P_{RE}(t), \lambda(t), t] == \dot{m}_f(t) - \frac{\lambda(t)H_{HLV}}{\eta_{bat}^{sign(P_{RE})}Q_{bat}U_{oc}}\frac{P_{bat}(t)}{H_{LHV}} \tag{25}$$

where U_{OC} is the voltage, V; and $\eta_{bat}^{sign(P_{RE})}$ varies depending on the working state of the battery. The equivalent factor expressions for batteries under different operating states can be obtained from Equations (24) and (25) as follows:

$$\begin{cases} s = -\frac{H_{LHV}}{\eta_{bat}^{sign(P_{RE})}Q_{bat}U_{OC}}\lambda(t) \\ s_{chg} = \eta_{dis}^2 s_{dis} \end{cases} \tag{26}$$

where s_{dis} is the discharge equivalent factor; s_chg is the charging equivalent factor; and η_{dis} is the discharge efficiency.

The range-extender has different working states when different equivalent factors are selected. This paper sets the reference SoC for power maintenance to 0.3. Figure 14 is the comparison diagram of SoC maintenance and vehicle fuel consumption when different equivalent factors are selected. Figure 14 shows that when the equivalent factor $s = 2.5$, the SoC change is mostly stable. As the value of s increases, the vehicle tends to use more fuel and continuously charge the battery.

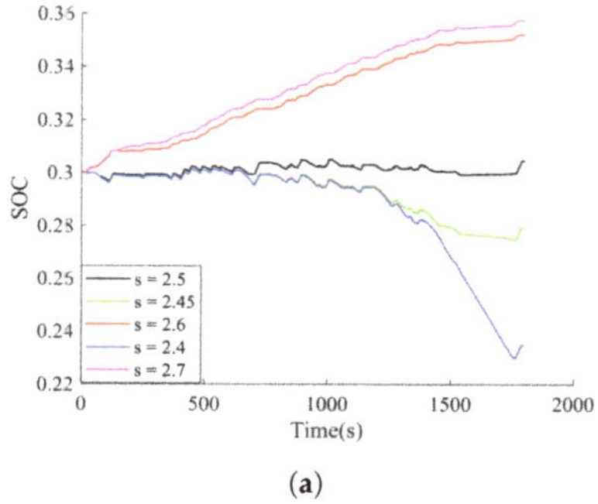

(**a**)

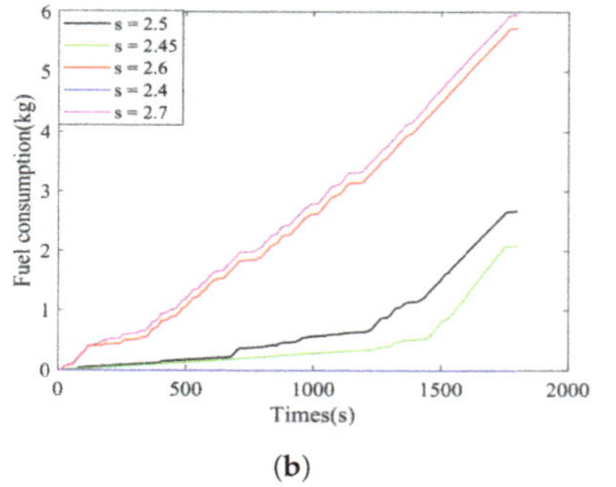

(**b**)

Figure 14. Simulation results of different equivalent factors. (**a**) Changes in the power battery SoC under different equivalent factors; (**b**) vehicle fuel consumption under different equivalent factors.

Conversely, when s decreases the car tends to use more electric power, resulting in a faster decrease in the SoC. In addition, in this paper, ECMS is based on the premise that the co-state variable $\lambda(t)$ remains constant. According to Equation (22), when the SoC changes too much, s is not constant, and the ECMS strategy cannot obtain the optimal solution. Therefore, when the SoC changes, an appropriate method is needed to make s to change adaptively.

4.2. MPC-ECMS Energy Management

According to the previous discussion, equivalent factors need to be adaptively adjusted. In the case of completing speed prediction, the combination of predictive speed and energy management can be achieved through the MPC framework, and the automatic adjustment of equivalent factors can be achieved through short-term planning. Combining any control method with the highly adaptable MPC control architecture can achieve real-time control. The fundamental premise of predictive control is to forecast the system's future output using the present model, the status of the system at hand, and upcoming control variables. Three elements comprise its management process: the rolling optimization, the feedback adjustment, and the forecast model. The MPC solution process is shown in Figure 15.

To achieve adaptive adjustment of the equivalent factors, the equivalent factors are first discretized with a suitable range at certain intervals. We use the multi-signal vehicle speed prediction model introduced in Section 3.3 to predict the speed in the next 5 s. The power generation range of the generator is from 0 to 285 kW, while the range of discretized equivalent factors is from 2.48 to 2.68. Then, the predicted velocity sequence is input into the control strategy by the prediction model. The control strategy calculates the required power sequence in 5 s based on the speed sequence. The best control sequence is calculated using PMP under different equivalent factors, and its fuel consumption and performance indicators are calculated accordingly. Since the performance indicator already considers the impact of the target SoC in the minimum principle, the calculated control variable enables the SoC satisfy the constraints. Finally, among all the corresponding relationships between the equivalent factors and the performance indicators, the equivalent factor that minimizes the performance indicator is selected as the designated equivalent factor s for the next moment. Meanwhile, the first control variable of the output sequence corresponding to the equivalent factor s is output as the output power of the range-extender at the next moment. The formula to solve the equivalent factor s is shown in Equation (27).

$$s = \arg\min(J) \tag{27}$$

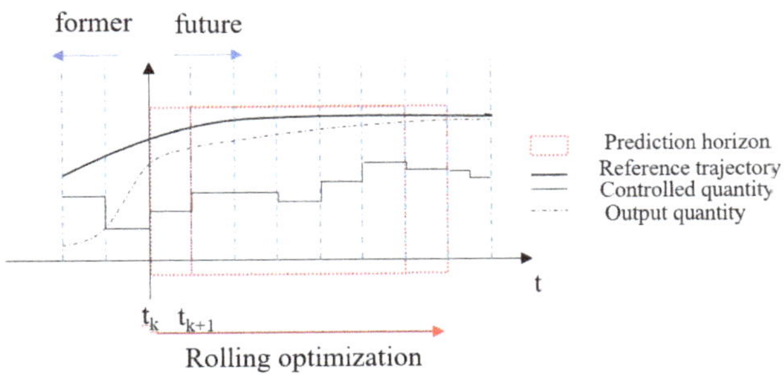

Figure 15. Principle of the MPC.

Although the control strategy can calculate the optimal control variable at each moment, there are a lot of fluctuations and inaccurate speed predictions in the cycling conditions, causing noise when calculating the required power of the whole vehicle. These noises cause an increase in fuel consumption. To eliminate the noise in the required power calculated by the control strategy without causing a large range of battery SoC changes, it is necessary to filter the required power of the range-extender calculated by the control strategy. This is because rapidly changing the engine power in a short time is not conducive to improving the engine's fuel economy. Experimental results show that this filtering method improves the fuel economy while having little effect on the SoC. The filtering method adopted here was first-order filtering, which is a simple control method with a good real-time performance and very beneficial for eliminating power noise. The discrete formula of the first-order filtering is shown as Equation (28).

$$Y(n) = aU(n) + (1 - a)Y(n - 1) \tag{28}$$

where Y represents the required power of the range-extender after filtering, n represents the time step, U is the power calculated by the control strategy,and a is a coefficient.

Here, we mainly focus on the multi-signal LSTM vehicle speed prediction MPC-ECMS (multi-signal-LSTM-MPC-ECMS). We set the initial SoC to 0.3 for the conducted simulation, and compared the control effect of the proposed strategy with the MPC-ECMS under SVM vehicle speed prediction (SVM-MPC-ECMS), dynamic programming (DP), ECMS ($s = 2.5$) and power-following control strategy. Each control strategy used the same coefficient for filtering, and the comparison before and after power noise filtering is shown in Figure 16.In the figure, the red line indicates the signal change after filtering and the blue line indicates the signal change before filtering. The final SoC and fuel consumption before and after filtering for each control strategy are shown in Table 4. When the entire working condition is in the pure electric drive mode, the final SoC is 0.2349. The experimental results under various control strategies show that the SoC is still within the allowable range after filtering. A greater reduction in fuel consumption is achieved relative to the reduction in SoC.

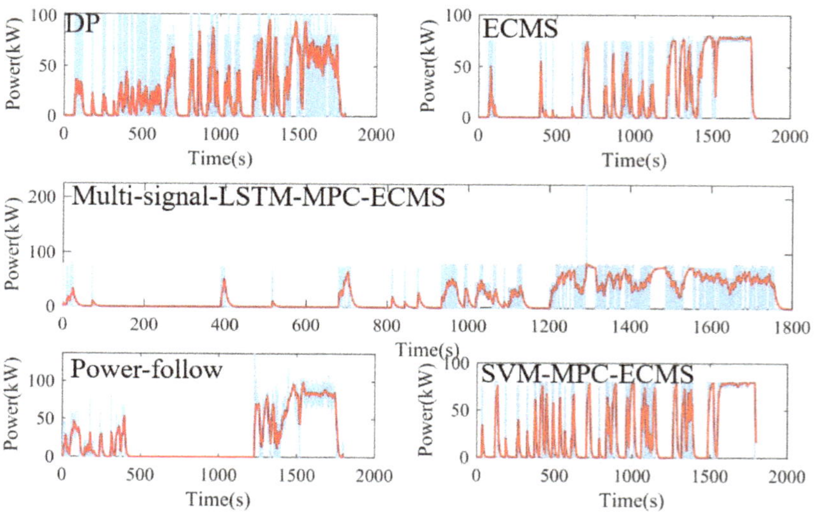

Figure 16. Comparison of the output power before and after filtering for each control strategy.

Table 4. Performance comparison before and after filtering under the different control strategies.

Control Strategy	SoC		Fuel Consumption (kg)	
	Pre-Filter	After-Filter	Pre-Filter	After-Filter
DP	0.2938	0.2902	2.9637	2.536
ECMS ($s = 2.5$)	0.308	0.3045	3.112	2.663
Multi-signal-LSTM-MPC-ECMS	0.3031	0.2976	3.02	2.627
Power-follow	0.306	0.3052	3.103	2.902
SVM-MPC-ECMS	0.3013	0.2985	3.505	3.115

As shown in Figure 17, the ordinate is $\Delta SOC/C_{fuel}$, ΔSOC is the difference between the final SoC under this control strategy and the final SoC under the pure electric drive mode, and C_{fuel} is the total fuel consumption under this control strategy. The results show that SoC variation under unit fuel consumption after filtering is significant, so filtering is helpful for fuel saving.

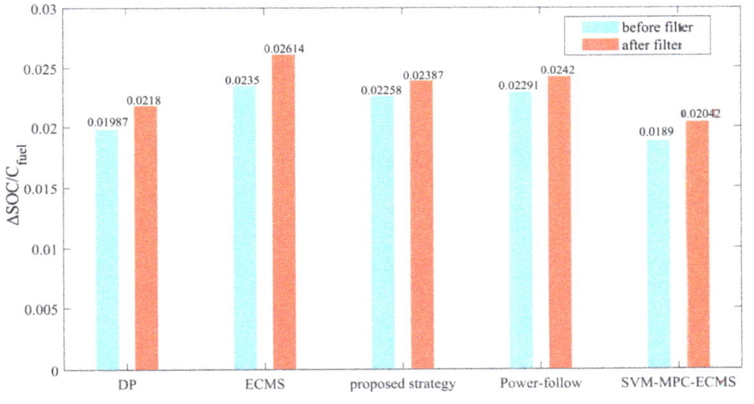

Figure 17. Performance comparison chart before and after output power filtering for each control strategy.

4.3. HIL Simulation Experiment

In order to validate the precision and real-time effectiveness of the control strategy, a HIL simulation experimental platform was built to simulate the multi-signal LSTM-MPC-ECMS control strategy. A schematic diagram of the HIL is shown in Figure 18. The HIL simulation experimental platform equipment mainly consists of an upper computer, a MicroAutoBox controller, and a SCALEXIO real-time simulation hardware system. The upper computer is responsible for monitoring the entire simulation process and saving the results, the controller is responsible for simulating the electronic control unit, making decisions according to the control strategy, and the real-time simulation hardware system is used to simulate the behaviour of the whole vehicle, providing a complete vehicle environment for HCU. SCALEXIO and AUTOBOX are connected to the upper computer through network cables, and the harness interface connects SCALEXIO to the expansion version. SCALEXIO is connected to the expansion board of AUTOBOX through the CAN high/low lines for signal transmission.

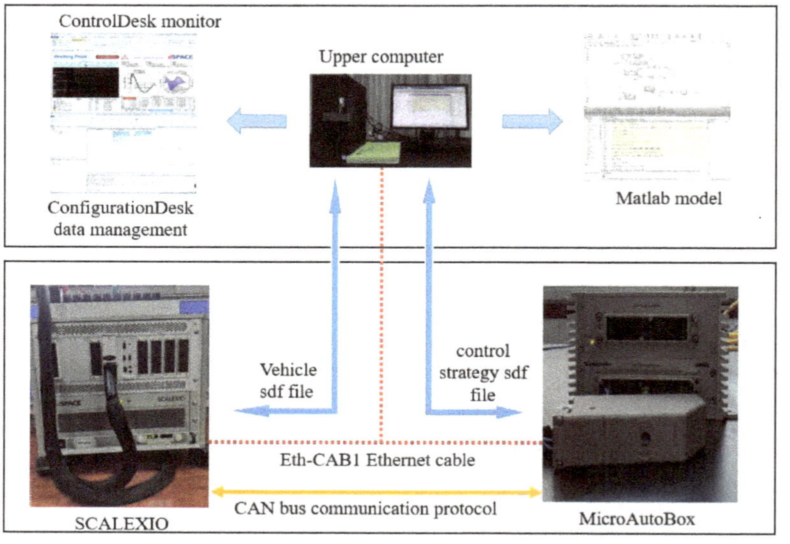

Figure 18. Schematic of the hardware-in-the-loop simulation.

5. Results and Discussion

After the HIL simulation, the results of the proposed strategy are compared with other control strategies using WTVC as the cycle condition; the initial SoC was 0.3. The change in the battery SoC and the fuel consumption under various control strategies are shown in Figure 19. We found that the other control strategies result in relatively a stable battery SoC, except for the DP and SVM-MPC-ECMS control strategies. Because the DP algorithm only considers the final SoC value, it does not pay attention to the changes in the SoC during the process. The vehicle speed prediction of the SVM-MPC-ECMS is worse than the proposed control strategy. Among all the control strategies, the DP is the least fuel consumptive because it is the theoretical optimal solution; however, the DP is difficult to implement in the actual driving process, making it commonly used as a theoretical reference for the economy of other control strategies. From Figure 19b, it can be seen that the proposed strategy is the closest to the DP algorithm, while MPC-ECMS under LSTM multi-signal vehicle speed prediction fuel consumption was reduced by 1.352% relative to the ECMS ($s = 2.5$) strategy, 9.476% relative to the power-following control strategy, and a decrease of 15.7% relative to the SVM-MPC-ECMS. Both multi-signal-LSTM-MPC-ECMS and SVM-MPC-ECMS are MPC-ECMS control strategies. The difference between them is their method of predicting vehicle speed. Figure 20 shows the economic changes of these two control strategies compared to the other control strategies. The numbers represent the ratio of fuel savings compared to the other control strategies. We found that the multi signal-LSTM-MPC-ECMS control strategy is more economical than the SVM-MPC-ECMS. The reason is that in the vehicle speed prediction part, the Multi-signal-LSTM prediction achieves higher accuracy, which shows that the prediction accuracy of the prediction part directly affects the control effect of the MPC-ECMS control strategy.

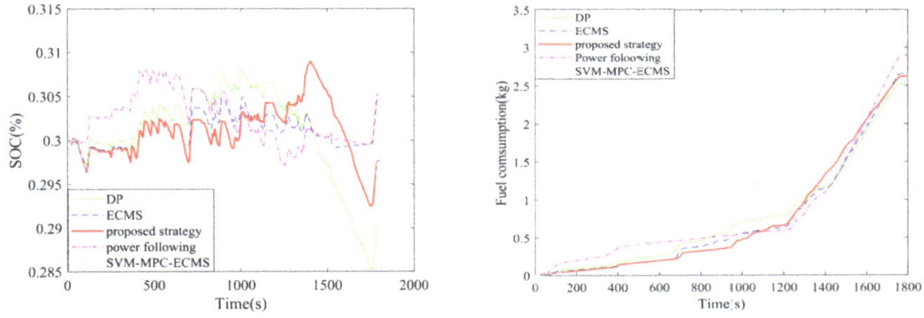

(**a**) (**b**)

Figure 19. Graph of changes under various control strategy SOCs. (**a**) Comparison of the changes in SoC for different control strategies; (**b**) the fuel consumption of the different control strategies.

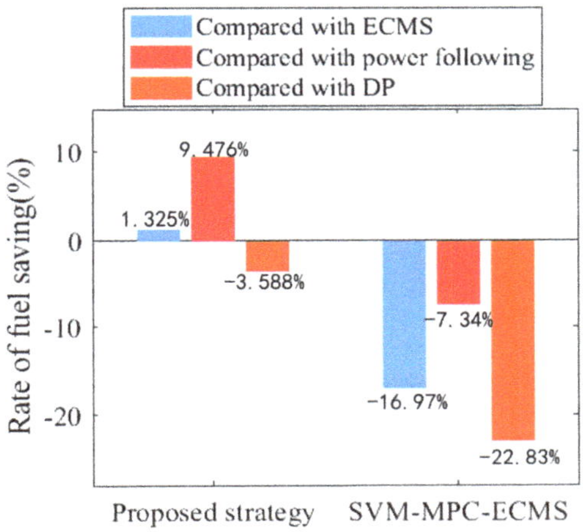

Figure 20. Comparison of different MPC-ECMS control strategies in terms of economic efficiency.

Compared with rule-based control strategies, such as power-following, MPC-ECMS control strategies can provide better control in the time domain. Compared with the ECMS, the MPC-ECMS can adapt to various changing working conditions, and this fully reflects the advantage of the MPC-ECMS control strategy to change the generator output power in real-time. The change in the MPC-ECMS equivalent factor under LSTM multi-signal vehicle speed prediction is shown in Figure 21. The diagram shows that the MPC-ECMS energy management strategy can adapt to different working conditions by constantly adjusting the equivalence factor.

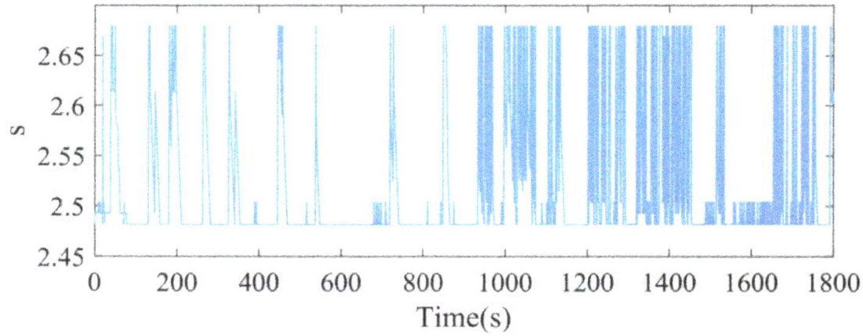

Figure 21. Proposed control strategy equivalent factor change chart.

Figure 22 shows the working points of the engine and generator under various control strategies; the dotted line in the figure is the optimal working curve of the range-extender. The engine and generator jointly influence the optimal working curve of a range-extender; therefore, the optimal working speed and torque should be calculated based on this curve. From the distribution diagram of the working points, we find that most of the working points of the MPC-ECMS control strategy are concentrated near the optimal curve, whether it is the MPC-ECMS method proposed in this article or the SVM-MPC-ECMS, in which the points that are not on the optimal curve are mainly the operating points in the speed transition stage. It can be seen that improving the prediction accuracy and extending the prediction time domain in order to optimize over a longer period of time can enable the range-extender to work nearer the optimal curve. We believe that improving the prediction accuracy while extending the prediction time domain will not only allow control strategies to have longer planning time and greater control accuracy, but also make full use of the high-efficiency range of the engine, reducing the rapid changes in speed and torque in a short time, and thereby improve the overall fuel efficiency. However, the optimal working curve of the range-extender is significantly influenced by the generator's efficiency, resulting in a difference between the optimal operating curves of the range-extender and the engine. This difference can be seen from the fuel consumption and working point distribution of engines with different control strategies.

The control strategy proposed in this paper calculates the demand power sequence in the future by predicting the vehicle speed in the future time domain, facilitating the controller to plan better. The experimental results show that the fuel consumption of the whole vehicle decreases significantly under this control strategy, and the engine works in the efficient zone most of the time. This control strategy is applicable to many types of vehicles In addition, the study of speed prediction can be applied to the vehicle following conditions by predicting the speed of the preceding vehicle to predict the operating condition of the main vehicle.

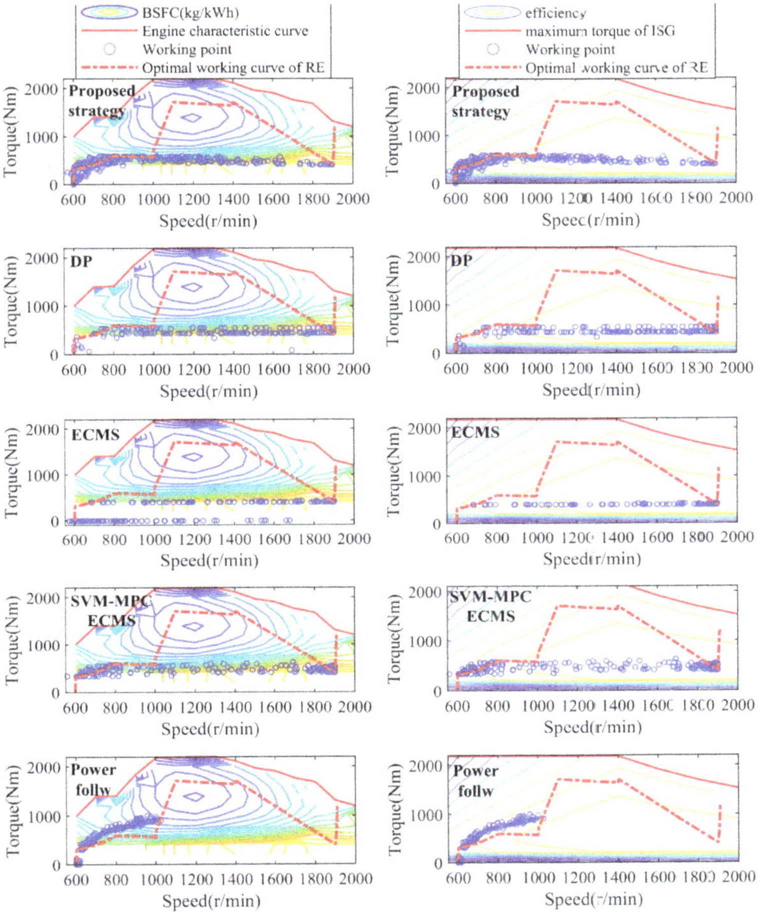

Figure 22. Distribution of the engine and generator operating points for the different control strategies.

6. Conclusions

In this paper, the multi-signal LSTM speed prediction model is combined with MPC-ECMS for energy management to improve the fuel economy.

Firstly, the proposed vehicle speed prediction method has a better prediction accuracy when compared to the SVM vehicle speed prediction method. The LSTM-based vehicle speed prediction model can train a large amount of training data, but the prediction speed will be slower when the SVM training data increases. SVM parameter adjustment will also have a great impact on the prediction effect.

Secondly, the control effect of the MPC-ECMS based on speed prediction is closely related to the accuracy of prediction; therefore, improving the prediction accuracy is beneficial to expand the prediction time domain, allowing the control strategy to make plans for longer periods of time. A prediction time domain that is too short is not only difficult to plan for in time, but it cannot take full advantage of the efficient area of the range-extender. Compared with other LSTM-based speed prediction models, the proposed multi-signal LSTM speed prediction model considers the influence of multiple other vehicle signals on the future speed, which is more conducive to improving the prediction accuracy.

Finally, although the prediction model used in this article can predict vehicle speed with high accuracy, actual driving environments still contain complex operating conditions, such as sharp changes in torque demand during climbing. Therefore, in future work we consider torque prediction, enabling a more accurate calculation of the demand power by variations in torque, thus resulting in greater adaptivity of the equivalence factor.

Author Contributions: Software, methodology ,validation, writing—original draft preparation, visualization, L.L.; writing—review and editing, supervision, project administration, funding acquisition, H.Z.; investigation, X.L.; data curation, C.S.; formal analysis , X.Z.; conceptualization, H.Y. All authors have read and agreed to the published version of the manuscript.

Funding: This work was supported by the Key Research and Development Plan of Shandong Province of China (No. 2018GGX105004) and Qingdao Civi Science and Technology Plan (No. 19-6-1-88-nsh).

Institutional Review Board Statement: Not applicable.

Informed Consent Statement: Not applicable.

Data Availability Statement: The data presented in this study are available on request from the corresponding author.

Acknowledgments: The authors thank the editors and the anonymous reviewers for their constructive feedback and valuable advice.

Conflicts of Interest: The authors declare no conflicts of interest.

Abbreviations

The following abbreviations are used in this manuscript:

LSTM	Long short-term memory neural networks
MPC	Model predictive control
ECMS	Equivalent fuel consumption minimum strategy
SVM	Support vector machine
SVR	Support vector regression
SoC	State of charge
DP	Dynamic programming
WTVC	World transient vehicle cycle
HIL	Hardware-in-the-loop

References

1. Zhao, H.; Mu, L.; Li, Y.; Qiu, J.; Sun, C.; Liu, X. Unregulated Emissions from Natural Gas Taxi Based on IVE Model. *Atmosphere* **2021**, *12*, 478. [CrossRef]
2. Liu, C.; Wang, Y.; Wang, L.; Chen, Z. Load-adaptive real-time energy management strategy for battery/ultracapacitor hybrid energy storage system using dynamic programming optimization. *J. Power Sources* **2019**, *438*, 227024. [CrossRef]
3. Zhang, N.; Ma, X.; Jin, L. Energy management for parallel HEV based on PMP algorithm. In Proceedings of the IEEE 2017 2nd International Conference on Robotics and Automation Engineering (ICRAE), Shanghai, China, 29–31 December 2017; pp. 177–182
4. Musardo, C.; Rizzoni, G.; Guezennec, Y.; Staccia, B. A-ECMS: An adaptive algorithm for hybrid electric vehicle energy management. *Eur. J. Control* **2005**, *11*, 509–524. [CrossRef]
5. Yuan, Z.; Teng, L.; Fengchun, S.; Peng, H. Comparative study of dynamic programming and Pontryagin's minimum principle on energy management for a parallel hybrid electric vehicle. *Energies* **2013**, *6*, 2305–2318. [CrossRef]
6. Zhang, H.; Fu, L.; Song, J.; Yang, Q. Power energy management and control strategy study for extended-range auxiliary power unit. *Energy Procedia* **2016**, *104*, 32–37. [CrossRef]
7. Chen, B.C.; Wu, Y.Y.; Tsai, H.C. Design and analysis of power management strategy for range extended electric vehicle using dynamic programming. *Appl. Energy* **2014**, *113*, 1764–1774. [CrossRef]
8. Pan, C.; Liang, Y.; Chen, L.; Chen, L. Optimal control for hybrid energy storage electric vehicle to achieve energy saving using dynamic programming approach. *Energies* **2019**, *12*, 588. [CrossRef]
9. Rezaei, A.; Burl, J.B.; Zhou, B. Estimation of the ECMS equivalent factor bounds for hybrid electric vehicles. *IEEE Trans. Control Syst. Technol.* **2017**, *26*, 2198–2205. [CrossRef]
10. Yu, X.; Lin, C.; Tian, Y.; Zhao, M.; Liu, H.; Xie, P.; Zhang, J. Real-time and hierarchical energy management-control framework for electric vehicles with dual-motor powertrain system. *Energy* **2023**, *272*, 127112. [CrossRef]
11. Shen, P.; Zhao, Z.; Zhan, X.; Li, J.; Guo, Q. Optimal energy management strategy for a plug-in hybrid electric commercial vehicle based on velocity prediction. *Energy* **2018**, *155*, 838–852. [CrossRef]

12. Wang, W.; Guo, X.; Yang, C.; Zhang, Y.; Zhao, Y.; Huang, D.; Xiang, C. A multi-objective optimization energy management strategy for power split HEV based on velocity prediction. *Energy* **2022**, *238*, 121714. [CrossRef]
13. Xing, J.; Chu, L.; Hou, Z.; Sun, W.; Zhang, Y. Energy Management Strategy Based on a Novel Speed Prediction Method. *Sensors* **2021**, *21*, 8273. [CrossRef]
14. Chen, R.; Yang, C.; Han, L.; Wang, W.; Ma, Y.; Xiang, C. Power reserve predictive control strategy for hybrid electric vehicle using recognition-based long short-term memory network. *J. Power Sources* **2022**, *520*, 230865. [CrossRef]
15. Han, J.; Shu, H.; Tang, X.; Lin, X.; Liu, C.; Hu, X. Predictive energy management for plug-in hybrid electric vehicles considering electric motor thermal dynamics. *Energy Convers. Manag.* **2022**, *251*, 115022. [CrossRef]
16. Lin, X.; Wu, J.; Wei, Y. An ensemble learning velocity prediction-based energy management strategy for a plug-in hybrid electric vehicle considering driving pattern adaptive reference SOC. *Energy* **2021**, *234*, 121308. [CrossRef]
17. Ritter, A.; Widmer, F.; Duhr, P.; Onder, C.H. Long-term stochastic model predictive control for the energy management of hybrid electric vehicles using Pontryagin's minimum principle and scenario-based optimization. *Appl. Energy* **2022**, *322*, 119192. [CrossRef]
18. Li, M.; He, H.; Feng, L.; Chen, Y.; Yan, M. Hierarchical predictive energy management of hybrid electric buses based on driver information. *J. Clean. Prod.* **2020**, *269*, 122374. [CrossRef]
19. Wei, C.; Chen, Y.; Li, X.; Lin, X. Integrating intelligent driving pattern recognition with adaptive energy management strategy for extender range electric logistics vehicle. *Energy* **2022**, *247*, 123478. [CrossRef]
20. Chen, Z.; Gu, H.; Shen, S.; Shen, J. Energy management strategy for power-split plug-in hybrid electric vehicle based on MPC and double Q-learning. *Energy* **2022**, *245*, 123182. [CrossRef]
21. Lin, X.; Zhang, J.; Su, L. A trip distance adaptive real-time optimal energy management strategy for a plug-in hybrid vehicle integrated driving condition prediction. *J. Energy Storage* **2022**, *52*, 105055. [CrossRef]
22. Zhao, Z.; Xun, J.; Wan, X.; Yu, R. Mpc based hybrid electric vehicles energy management strategy. *IFAC-PapersOnLine* **2021**, *54*, 370–375. [CrossRef]
23. Wang, Y.; Zhang, Y.; Zhang, C.; Zhou, J.; Hu, D.; Yi, F.; Fan, Z.; Zeng, T. Genetic algorithm-based fuzzy optimization of energy management strategy for fuel cell vehicles considering driving cycles recognition. *Energy* **2023**, *263*, 126112. [CrossRef]
24. Zhou, Y.; Ravey, A.; Péra, M.C. Multi-mode predictive energy management for fuel cell hybrid electric vehicles using Markov driving pattern recognizer. *Appl. Energy* **2020**, *258*, 114057. [CrossRef]
25. Türker, E.; Bulut, E.; Kahraman, A.; Çakıcı, M.; Öztürk, F. Estimation of Energy Management Strategy Using Neural-Network-Based Surrogate Model for Range Extended Vehicle. *Appl. Sci.* **2022**, *12*, 12935. [CrossRef]
26. Al-Saadi, Z.; Phan Van, D.; Moradi Amani, A.; Fayyazi, M.; Sadat Sajjadi, S.; Ba Pham, D.; Jazar, R.; Khayyam, H. Intelligent Driver Assistance and Energy Management Systems of Hybrid Electric Autonomous Vehicles. *Sustainability* **2022**, *14*, 9378. [CrossRef]
27. Wang, H.; Huang, Y.; Khajepour, A.; Song, Q. Model predictive control-based energy management strategy for a series hybrid electric tracked vehicle. *Appl. Energy* **2016**, *182*, 105–114. [CrossRef]
28. Mu, L.; Zhao, H.; Li, Y.; Liu, X.; Qiu, J.; Sun, C. Traffic flow statistics method based on deep learning and multi-feature fusion. *CMES Comput. Model. Eng. Sci.* **2021**, *129*, 465–483. [CrossRef]
29. Hong, J.; Wang, Z.; Qu, C.; Zhou, Y.; Shan, T.; Zhang, J.; Hou, Y. Investigation on overcharge-caused thermal runaway of lithium-ion batteries in real-world electric vehicles. *Appl. Energy* **2022**, *321*, 119229. [CrossRef]
30. Hong, J.; Zhang, H.; Xu, X. Thermal fault prognosis of lithium-ion batteries in real-world electric vehicles using self-attention mechanism networks. *Appl. Therm. Eng.* **2023**, *226*, 120304. [CrossRef]
31. Sun, X.; Fu, J.; Yang, H.; Xie, M.; Liu, J. An energy management strategy for plug-in hybrid electric vehicles based on deep learning and improved model predictive control. *Energy* **2023**, *269*, 126772. [CrossRef]

electronics

MDPI

Article

A Skin Cancer Classification Method Based on Discrete Wavelet Down-Sampling Feature Reconstruction

Qing-e Wu [1,*], Yao Yu [1] and Xinyang Zhang [2,3]

1 School of Electrical and Information Engineering, Zhengzhou University of Light Industry, Zhengzhou 450002, China; 332101030027@email.zzuli.edu.cn
2 School of Mechanical Engineering, University of Science and Technology Beijing, Beijing 100083, China
3 Shunde Innovation School, University of Science and Technology Beijing, Foshan 528399, China
* Correspondence: wqe2008@zzuli.edu.cn or wqe969699@163.com

Abstract: Aiming at the problems of feature information loss during down-sampling, insufficient characterization ability and low utilization of channel information in skin cancer diagnosis of melanoma, a skin pathological mirror classification method based on discrete wavelet down-sampling feature reconstruction is proposed in this paper. The wavelet down-sampling method is given first, and the multichannel attention mechanism is introduced to realize the pathological feature reconstruction of high-frequency and low-frequency components, which reduces the loss of pathological feature information due to down-sampling and effectively utilizes the channel information. A skin cancer classification model is given, using a combination of depth-separable convolution and 3 × 3 standard convolution and wavelet down-sampling as the input backbone of the model to ensure the perceptual field while reducing the number of parameters; the residual module of the model is optimized using wavelet down-sampling and Hard-Swish activation function to enhance the feature representation capability of the model. The network weight parameters are initialized on ImageNet using transfer learning and then debugged on the augmentation HAM10000 dataset. The experimental results show that the accuracy of the proposed method for skin cancer pathological mirror classification is significantly improved, reaching 95.84%. Compared with the existing skin cancer classification methods, the proposed method not only has higher classification accuracy but also accelerates the classification speed and enhances the noise immunity. The method proposed in this paper provides a new classification method for skin cancer classification and has some practical value.

Keywords: melanoma; down-sampling; wavelet transform; image classification; transfer learning; data augmentation

Citation: Wu, Q.-e.; Yu, Y.; Zhang, X. A Skin Cancer Classification Method Based on Discrete Wavelet Down-Sampling Feature Reconstruction. *Electronics* **2023**, *12*, 2103. https://doi.org/10.3390/electronics12092103

Academic Editor: Hyunjin Park

Received: 11 April 2023
Revised: 30 April 2023
Accepted: 1 May 2023
Published: 4 May 2023

1. Introduction

Skin cancer is one of the most widespread threats to human health worldwide, and one malignant skin tumor, also known as melanoma, is one of the swift-growing cancers in the world. This disease can be easily treated if detected early [1]. A report has shown that the five-year survival rate of localized malignant melanoma is 99% when diagnosed and treated early, whereas the survival rate of advanced melanoma is only 25%; therefore, whether the disease can be diagnosed early has a crucial impact on whether patients can improve their survival rate [2]. The traditional examination method is to first go through a doctor's visual inspection and then use dermoscopic imaging to aid the diagnosis [3]. However, benign and malignant skin cancers are extremely similar in characteristics, and the varying levels of professional doctors and the uncertainty of subjective decision-making have led to numerous skin cancer patients not being diagnosed early and treated timely. With the development of artificial intelligence in the medical field, deep learning has been widely used for the detection and classification of medical images over the past few years [4]. The application of artificial intelligence in medical image-assisted diagnosis to provide a second opinion to help professional doctors diagnose skin cancer has become an inevitable trend [5]. Deep convolutional neural networks (CNNs), an important research topic in deep learning, have been widely used in the medical field [6].

Codella et al. [7] used features from sparse coding features, low-level manual features and deep residual networks to adjust the output of the AlexNet network and use a support vector machine (SVM) to perform classification experiments on the ISIC database and finally obtained 93.1% accuracy, 92.8% specificity, and 94.9% sensitivity. Pomponiu et al. [8] used a migration learning-based AlexNet convolutional neural network and data augmentation to classify skin lesion data after feature extraction using the K nearest neighbor (KNN) algorithm and finally obtained 93.64% accuracy, 95.18% specificity, and 92.1% sensitivity. Esteva et al. [9] used a skin lesion classification method based on a pre-trained GoogleNet architecture and performed classification experiments on clinical images and finally obtained 72.1% accuracy. Khan et al. [10] improved a novel deep CNN based on VGG and AlexNet models to classify skin lesions, and classification tests that were performed on ISIC 2016 and ISIC 2017 datasets finally obtained 92.1% and 96.5% accuracy, respectively. Ahmed et al. [11] used three models based on deep CNN (InceptionResnetV2, Xception and NASNetLarge) ensembles to test the pre-trained models on the ISIC2019 dataset, and finally, the CNN ensembles obtained 93.70% accuracy and 0.931 AUC and exceeded the top-2 rankings on the ISIC2019 classification challenge leaderboard for that year. Abayomi-Alli et al. [12] improved a new data augmentation model for skin cancer classification, and a migration learning based pre-trained squeeze net architecture was used to classify melanoma skin cancer. Classification experiments were performed on the PH2 dataset and finally obtained 92.18% accuracy. Although convolutional neural networks can classify dermoscopy images well, the issue of partial feature information loss caused by down-sampling will be difficult to avoid as the depth and breadth of the network increase. As the performance characteristics of malignant and benign skin diseases are too similar, the problem of partial loss of feature information due to down-sampling may affect the accuracy of the classification of dermoscopy images. Pooling operation is widely used in convolutional neural networks as the main down-sampling method, which often ignores some useful feature information and causes poor generalization ability, which affects the performance of the model. To deal with this problem, Zeiler et al. [13] proposed a stochastic pooling strategy that replaced the traditional deterministic pooling operation with a stochastic process and was able to combine it with other arbitrary forms of regularization, abandoning the dropout-like approach of discarding information in favor of selecting from the network's existing information, obtaining state-of-the-art performance at the time. He et al. [14] designed a stride convolution as a learnable down-sampling method, which was widely used in the model but introduced more parameters and overhead at the same time. Zhao et al. [15] embedded random shifting in the down-sampling layer during the training process and dynamically adjusted receptive fields by shifting the kernel centers on the feature maps in different directions, effectively mitigating the loss of feature information during down-sampling. Jiang et al. [16] introduced parameters into the pooling layer and used parametric pooling to replace the traditional pooling process, which preserves the features that are desired to be preserved in the convolutional neural network with a small increase in network parameters. Although the above methods alleviate the problem of losing feature information due to pooling operations to a certain extent, they are still traditional pooling operations in nature, and no matter what pooling methods are used they will cause feature information loss and thus affect the performance of convolutional neural networks.

Aiming at the problems of feature information loss, low utilization of channel information and insufficient characterization capability in traditional down-sampling, this paper proposes a discrete wavelet-based down-sampling feature reconstruction method applied to skin cancer mirror classification. The method uses discrete wavelet transform as the down-sampling operation and adds a multichannel attention mechanism to effectively combine the high-frequency components and low-frequency components after wavelet decomposition and also effectively utilizes the channel information to enhance the characterization ability of the model; a skin cancer classification model is given, based on the ResNet50 network model, using a combination of 3 × 3 convolutional kernel and depth-separable convolution and wavelet down-sampling as the input backbone of the model, which reduces the number of parameters while maintaining the perceptual field of the network. Finally, the network weight parameters are initialized on ImageNet using transfer learning, and fine-tuned training is performed on the augmentation HAM10000 dataset to further improve the model classification performance.

The main contributions of the proposed method to the classification of melanoma skin cancer are as follows:

- In the current study, we developed a wavelet down-sampling feature reconstruction method to address the problem of traditional down-sampling feature information loss, introducing a multichannel attention mechanism to effectively combine low-frequency components with high-frequency components, fully utilizing channel information to reveal finer details.
- This paper developed a wavelet down-sampling feature reconstruction method-based convolutional neural network as a feature extractor to classify melanoma skin cancer using skin pathological mirror.
- A data augmentation and hair removal algorithm is used to pre-process the skin pathological mirror dataset HAM1000, and the pre-processing method is tested on the pre-processing HAM10000 dataset using transfer learning. The experimental results show that the proposed method has high accuracy for melanoma skin cancer classification.

2. Related Work

The wavelet transform is capable of focusing on arbitrary details of the signal by multiscale refinement through telescopic translation operations and is widely used in signal processing and pattern recognition [17–19]. In recent years, wavelets have been analyzed by many scholars in combination with convolutional neural networks as emerging down-sampling methods, such as Mallat et al. [20] by linking wavelet transform with average pooling cascade, which preserves the detail information of images to a great extent with shift-invariance features. Zhang et al. [21] introduced multi-level wavelets into convolutional neural networks, which greatly preserves the texture information of images by using the inverse transform of wavelets. Li et al. [22] used the low-frequency component of the discrete wavelet transform output to replace the traditional commonly used down-sampling method and obtained strong robustness and better network performance while confirming the effectiveness of using the low-frequency component of the two-dimensional discrete wavelet to replace the down-sampling layer output in the convolutional neural network. The low-frequency component of the wavelet transform output contains the main feature information of the image, and the high-frequency component contains detailed information such as the texture features of the image [23]. In the existing forms of combining wavelets and convolutional neural networks, mainly the high-frequency components are discarded, and only the low-frequency components are used, or the high and low-frequency components are used in equal proportions. However, both of them have certain disadvantages: if the high-frequency components are discarded, it may cause the loss of important texture information, leading to the problem of poor results in detail recovery and insufficient characterization ability, while the high- and low-frequency components contain very different amounts of information and have different effects on the loss function, so if they are used in equal proportions, it is not reasonable enough and there is also the problem of low utilization of channel information [24].

3. Wavelet Down-Sampling Reconstruction

The input skin mirror image (X) is feature decomposed using a 2D discrete wavelet transform to obtain a low-frequency component (X_{LL}) containing information on the main pathology feature and three high-frequency components containing information on pathological details: the horizontal component (X_{LH}), the vertical component (X_{HL}) and the diagonal component (X_{HH}). 2D discrete wavelet transform can be formulated as

$$[X_{LL}, X_{LH}, X_{HL}, X_{HH}] = 2D_{DWT(X)} \tag{1}$$

where $(X_{LL}, X_{LH}, X_{HL}, X_{HH}) \in \mathrm{c} \times \frac{h}{2} \times \frac{w}{2}$; c is the number of channels of the input skin mirror image; and h and w are the height and width, respectively. The new high-frequency components $X'_{LH}, X'_{HL}, X'_{HH}$, are obtained by summing the low-frequency components with each high-frequency component separately and can be formulated as

$$\begin{cases} X'_{LH} = \alpha 1 X_{LH} + X_{LL} \\ X'_{HL} = \alpha 2 X_{HL} + X_{LL} \\ X'_{HH} = \alpha 3 X_{HH} + X_{LL} \end{cases} \tag{2}$$

where αi is the absolute value of the high-frequency component to which it belongs, reflecting the severity of the pathology, $i = 1, 2, 3$.

Based on SKNet [25] multi-branch structure selective channel attention mechanism, each new high-frequency component in Equation (2) is integrated to obtain the texture feature F of pathology; then, global average pooling (GAP) operation is performed to obtain the global information s on each channel, and then the fully connected (FC) operation is performed on s to obtain the size of the weight z accounted for by each channel; the normalization function softmax is used to reaggregate the weights Q after normalizing z to obtain the new pathological texture features F' and finally obtain the integrated pathology comprehensive feature M by jumping connection X_{LL} to improve the characterization ability. The whole process is shown in Figure 1. The integrated pathology comprehensive feature M can be formulated as

$$Q = \mathrm{Softmax}(\sigma_{\mathrm{relu}}(BN(W \cdot GAP(F_{(i,j)})))) \tag{3}$$

$$F' = Q_1 \times X'_{LH} + Q_2 \times X'_{HL} + Q_3 \times X'_{HH} \tag{4}$$

$$M = \mathrm{concat}([F', X_{LL}]) \tag{5}$$

where $(i, j) \in (h, w)$; $W \in d \times c$; $d = \max(\frac{c}{r}, L)$; r is the scaling ratio; and L is the minimum value of d to control the dimensionality of the output. FC is composed of the batch normalized BN and σ_{relu} activation function; in turn, $M \in 2\mathrm{c} \times \frac{h}{2} \times \frac{w}{2}$.

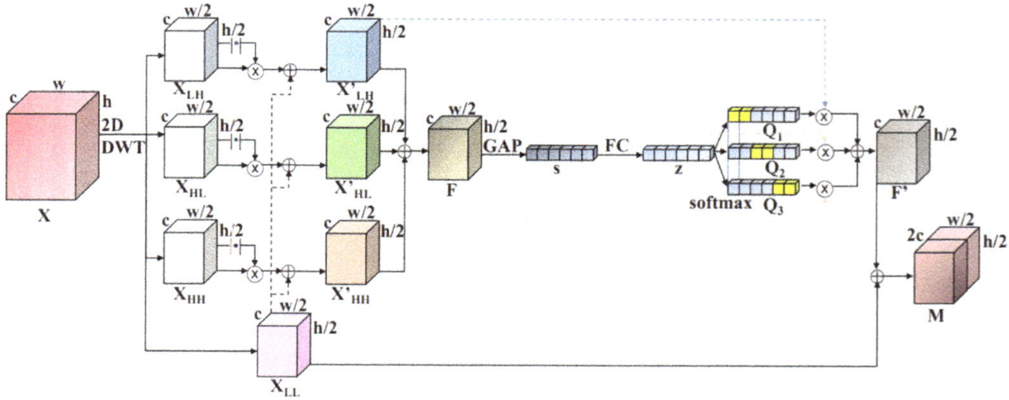

Figure 1. 2D discrete wavelet down-sampling reconstruction.

Finally, the pathology comprehensive feature M is reconstructed by using 1D discrete wavelet for feature decomposition to output the final pathology feature map. Firstly, the comprehensive feature M is downscaled using the reshape operation to obtain a feature map of the form $1 \times n$, and then the 1D discrete wavelet decomposition is performed to obtain the low-frequency component X_L and the high-frequency component X_H. Finally, the weights of the low-frequency component and the high-frequency component are calculated using the selective channel attention mechanism, and the final reconstructed feature *rec_X* is composed according to the weights; the whole process is shown in Figure 2. The final reconstructed feature *rec_X* can be formulated as

$$[X_L, X_H] = 1D_{DWT(M)} \tag{6}$$

$$q = \mathrm{Softmax}(\sigma_{\mathrm{relu}}(BN(W \cdot GAP(X_L \oplus X_H)))) \tag{7}$$

$$rec\_X = q_1 \times X_L + q_2 \times X_H \tag{8}$$

where q_1 and q_2 are the weights of the low-frequency component and the high-frequency component features within each channel, $q_1 + q_2 = 1$, $rec\_X \in c \times \frac{h}{2} \times \frac{w}{2}$.

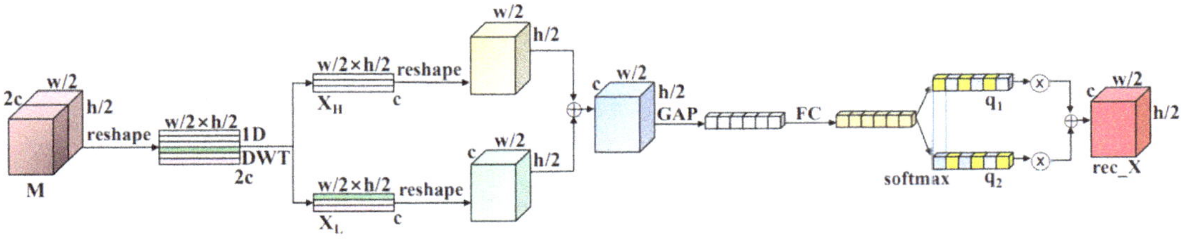

Figure 2. 1D discrete wavelet down-sampling reconstruction.

4. Skin Cancer Classification Model

In this paper, the skin cancer classification model is based on the ResNet50 network model. The input backbone of the ResNet50 network is mainly composed of a 7×7 convolutional kernel and a maximum pooling layer. In this paper, a combination of a 3×3 standard convolution and a depth-separable convolution [26] is used to replace the 7×7 convolutional kernel, and wavelet down-sampling is used to replace the maximum pooling operation, which reduces the computational effort while deepening the network depth during convolution. The depth-separable convolution and input backbone structures are shown in Figures 3 and 4.

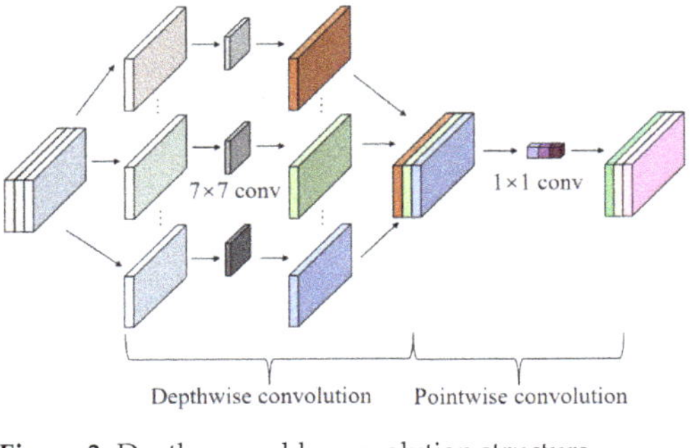

Figure 3. Depth-separable convolution structure.

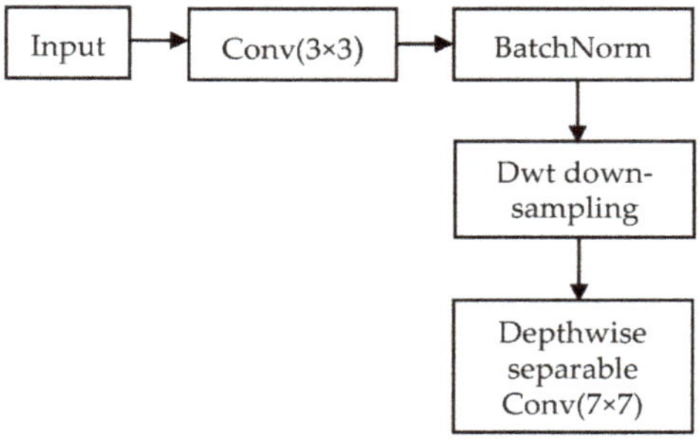

Figure 4. Input backbone structure.

As shown in Figures 3 and 4, the wavelet down-sampling output feature map is fed into the depth-separable convolution, and three convolution kernels of size $7 \times 7 \times 1$ are used. Each convolution kernel does convolution for one input channel only, and the output feature map of size $224 \times 224 \times 3$ is obtained. Then, using 64 point-by-point convolutions with $1 \times 1 \times 3$ convolution kernels, each convolution kernel will obtain a mapping of size $224 \times 224 \times 1$, resulting in an output feature map of size $224 \times 224 \times 64$. The improved input backbone not only maintains the original perceptual field but also reduces the number of parameters by 32.4% and deepens the depth of the network.

The Hard-Swish function [27] is used instead of the ReLU function in the backbone network to avoid the "dead ReLU" problem and the problem that the output is reduced to zero when the input is used. The Hard-Swish function is unbounded, smooth and non-monotonic and can be formulated as

$$\mathrm{Hard-Swish}(x) = \begin{cases} 2x, x \geq \frac{5}{2\lambda} \\ 0.4\beta x^2 + x, \frac{5}{-2\lambda} < x < \frac{5}{2\lambda} \\ 0, x \leq \frac{5}{-2\lambda} \end{cases} \tag{9}$$

In Equation (9), λ is a trainable parameter or a custom parameter, and the output is more likely to be set to 0 when x is smaller rather than directly equal to 0 as in the ReLU function, which not only increases the randomness but also increases the dependence on the input, effectively improving the robustness of the model. To avoid excessive activation and BN layers that degrade the model performance [28], the ReLU function in the residual block is replaced by the Hard-Swish function; the activation and BN layers after the first convolution are discarded; the activation function before the output of the residual block is moved to the input; and then the wavelet down-sampling method is used to improve the learning ability of the model. The improved residual block is shown in Figure 5.

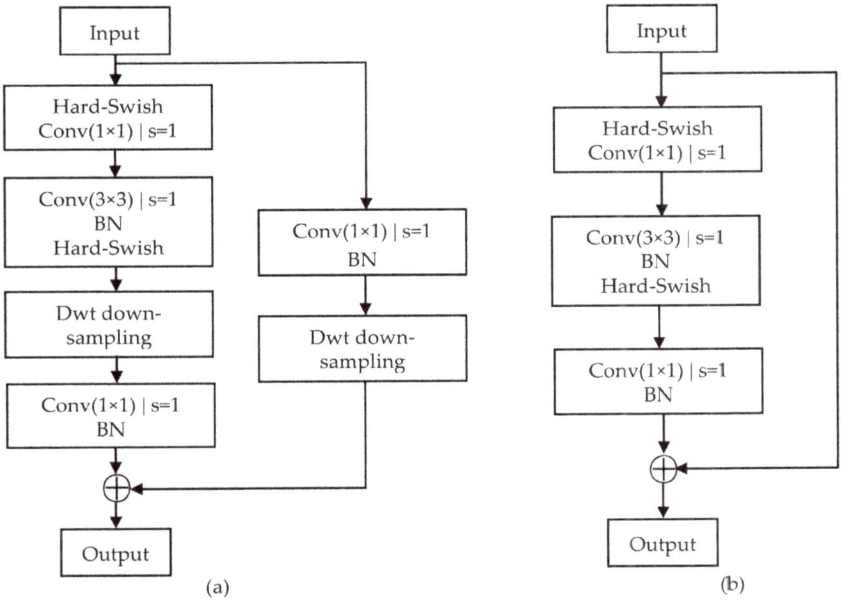

Figure 5. 1D discrete wavelet down-sampling reconstruction: (**a**) Conv Block; (**b**) Identity Block.

5. Experiments and Analysis of Results

5.1. Dataset and Pre-Processing

The publicly available skin lesion dataset HAM10000 [29] was used to perform experiments on skin cancer classification models. The dataset consists of seven skin diseases, namely benign keratosis (BKL), actinic keratosis and intra-epithelial carcinoma (AKIEC), dermatofibrosarcoma (DF), melanocytic nevus (NV), vascular lesion (VASC), melanoma (MEL), and basal cell carcinoma (BCC), with a total of 10,015 images, as shown in Figure 6.

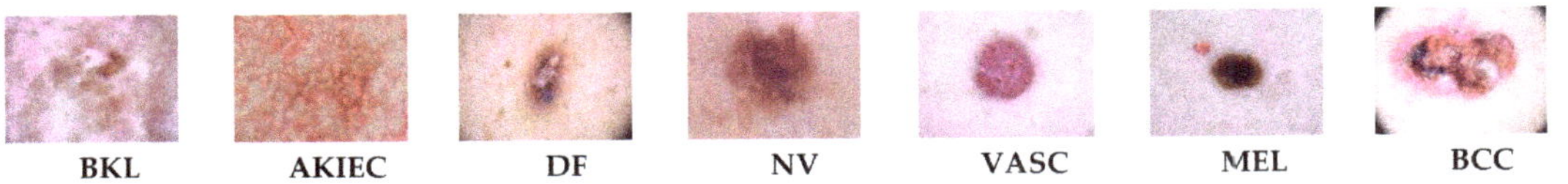

Figure 6. Example of HAM10000 dataset.

Aiming at the problem that hair occlusion in the skin mirror image can affect the diagnosis results, this paper gives a hair removal algorithm [30] to automatically remove the hair occlusion in the skin mirror image and restore the information of the hair occlusion area, the Algorithm 1 shows the hair removal algorithm and Figure 7 shows the comparison before and after hair removal in the skin mirror image.

Algorithm 1: hair removal algorithm

Input: skin mirror image (img)
Output: restoration skin mirror image (img1)

1. use morphological closed top-hat operator to enhance the hairs with varying strength in the skin mirror image img.
2. extract the effective measure of hairs based on the defined extension characteristic function to describe the extension state of the stripe-like connected region.
3. the image restoration technique using partial differential equations is used to restore the hair-obscured areas to obtain the skin mirror image img1 after hair removal.

Figure 7. Comparison of before and after hair masking restoration. (**a**) Before restoration. (**b**) After restoration.

Since the sample distribution in the HAM10000 dataset is extremely uneven, this paper uses data augmentation methods including rotation, cropping and random flipping [31] to ensure the data category balance and enhance the generalization ability of the model. The augmentation dataset consists of 16,002 samples, which are divided into two independent and non-intersecting datasets, the training set and the test set, with a division ratio of 8:2.

5.2. Experimental Environment and Evaluation Indicators

The experiments were run on a Windows 10 64-bit platform environment with a 12th generation i7-12700H, 32G RAM, NVIDIA GTX3080Ti discrete graphics card with 12G video memory and Pytorch as the deep learning framework.

AC, *F*1-Score was used as the evaluation indicators of the model and defined as follows:

$$AC = \frac{TP + TN}{TP + FP + TN + FN} \tag{10}$$

$$F1 = \frac{2 \times TP}{2 \times TP + FP + FN} \tag{11}$$

where *AC* is the classification accuracy; *F*1-Score is the summed average of accuracy and completeness; and the higher the *F*1-Score value is, the better the classification performance of individual categories is. *TP*, *TN*, *FP* and *FN* represent the number of samples with true positives, true negatives, false positives and false negatives, respectively.

5.3. Comparative Experiments

To verify the effectiveness of the proposed method and to also determine that the improved input backbone and wavelet down-sampling methods have improved the model performance, ablation experiments were conducted on the test set, and the experimental results are shown in Table 1.

Table 1. Ablation experiments of the proposed method.

Base Model	Input Backbone	Residual Module	Data Augmentation	AC/%	F1/%
ResNet50			√	90.48	90.69
	√		√	93.45	93.67
		√	√	94.51	94.64
	√	√		93.94	94.03
	√	√	√	95.28	95.39

As can be seen from Table 1, the model classification performance is improved by using the improved input backbone in this paper, and the classification accuracy is improved by 2.97%. However, the classification performance of the model decreases significantly when data augmentation is not used to balance the data categories. It is verified that the data augmentation method can effectively improve the classification performance of the network and the effectiveness of the method proposed in this paper.

The performance of the improved ResNet50 model in the paper is compared with five mainstream deep learning network models, AlexNet [32], VGG19 [33], MobileNet-V2 [34], DenseNet-121 [35] and EfficientNet-B0 [36], all of which are pre-trained on the ImageNet dataset. Pre-training is performed to initialize some of the weight parameters, and then experiments are conducted on the augmentation HAM10000 dataset, and the experimental results are shown in Table 2.

Table 2. Performance comparison of different classification models on the HAM1000 dataset.

Model	AC/%	F1/%
AlexNet	87.81	88.13
VGG19	90.34	90.61
MobileNet-V2	88.23	88.86
DenseNet-121	92.93	93.14
EfficientNet-B0	93.62	93.78
Proposed model	95.84	95.96

From Table 2, the classification accuracy and F1 values of the improved model in this paper on the HAM10000 dataset are better than those of other network models, which are 95.84% and 95.96%, respectively, among which AlexNet has the lowest accuracy and F1 values, which are 87.81% and 88.13%, respectively, and EfficientNet-B0 has the highest accuracy and F1 values except for the model in this paper, which are 93.62% and 93.78%, respectively.

For the HAM10000 dataset, many scholars have proposed research methods, and the research methods proposed in this paper were compared with the existing methods, and the comparison results are shown in Tables 3 and 4.

Table 3. Comparison with existing methods on the HAM1000 dataset (AC, F1 and Parameters).

References	AC/%	F1/%	Parameters/MB
ResNet50 + SA [37]	91.55	91.50	91.2
CNN ensembles [11]	93.09	-	-
IRv2-SA [37]	93.47	93.65	181.3
Loss balance + ensemble [38]	92.60	-	-
Low-cost augmentation + CNN [39]	95.79	86.14	42.0
FixCaps [40]	96.49	-	1.86
Our method	95.84	95.96	60.9

Table 4. Comparison with existing methods on the HAM1000 dataset (AUC).

References	AUC	Category						
		AKIEC	BCC	BKL	DF	MEL	NV	VASC
ResNet50 + SA [37]	0.980	0.981	0.996	0.964	0.971	0.973	0.979	0.999
CNN ensembles [11]	0.929	0.902	0.934	0.885	0.968	0.925	0.951	0.941
IRv2-SA [37]	0.985	0.981	0.998	0.982	0.973	0.974	0.984	1.000
Loss balance + ensemble [38]	0.941	0.919	0.947	0.908	0.980	0.931	0.960	0.942
Low-cost augmentation + CNN [39]	0.976	0.988	0.989	0.968	0.972	0.943	0.970	0.995
FixCaps [40]	-	-	-	-	-	-	-	-
Our method	0.982	0.983	0.992	0.975	0.976	0.971	0.983	0.992

In Table 3, our method was compared with six previously reported melanoma skin cancer classification methods, among which FixCaps method achieved the highest classification accuracy of 96.49% because its method discarded the down-sampling operation and used the dynamic routing algorithm, which not only directly avoided the problem of feature information loss caused by traditional down-sampling but also could read useful feature information to the maximum extent. In addition, the other five methods and our method all contain down-sampling operation, among which our method achieves the highest classification accuracy and F1 value of 95.84% and 95.96%, respectively. In terms of the number of model parameters, FixCaps has a minimum of 1.86 MB, and IRv2-SA has a maximum of 181.3 MB, while the model in our method is in the middle with 60.9 MB.

In Table 4, we compared the AUC scores of our method with previous melanoma skin cancer classification methods, and the AUC score is the golden indicator of classification model performance; the higher the AUC score is, the better the model performance is, and the AUC score is also the ranking indicator of melanoma skin cancer classification ranking. From Table 4, we can see that the highest AUC score of the IRv2-SA method is 0.985; the lowest AUC score of CNN ensembles method is 0.929; and the AUC score of our method is 0.982, which is in the second place.

Combined with Tables 3 and 4, the proposed method in this paper has excellent performance in classifying melanoma skin cancer, which exceeds most of the previously reported melanoma skin cancer methods. Figure 8 shows the confusion matrix at the best classification case. Figure 9 shows the ROC curve that is plotted with true positive rate against the false positive rate of each lesion category, individually, and Table 5 shows the corresponding classification metrics. As seen in Table 5, the highest classification accuracy of 99.16% was achieved for dermatofibrosarcoma (DF), and the lowest classification accuracy of 93.31% was achieved for benign keratosis (BKL).

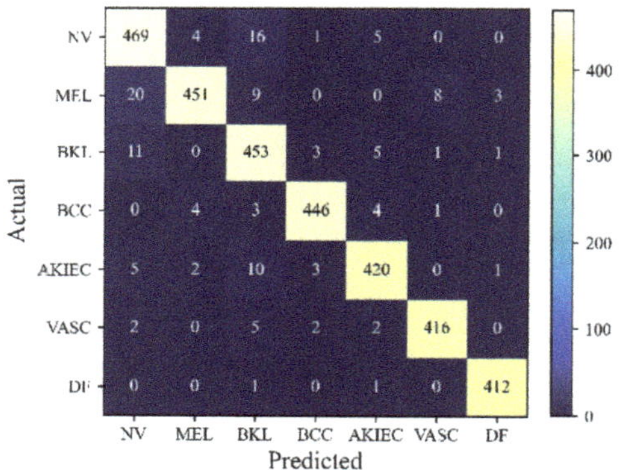

Figure 8. Best classification confusion matrix.

Table 5. Classification indicators.

Category	Precision/%	Recall/%	F1-Score/%
NV	92.50	94.75	93.61
MEL	97.83	91.85	94.75
BKL	91.15	95.57	93.31
BCC	98.02	97.38	97.70
AKIEC	96.11	95.24	95.67
VASC	97.65	97.42	97.54
DF	98.80	99.52	99.16

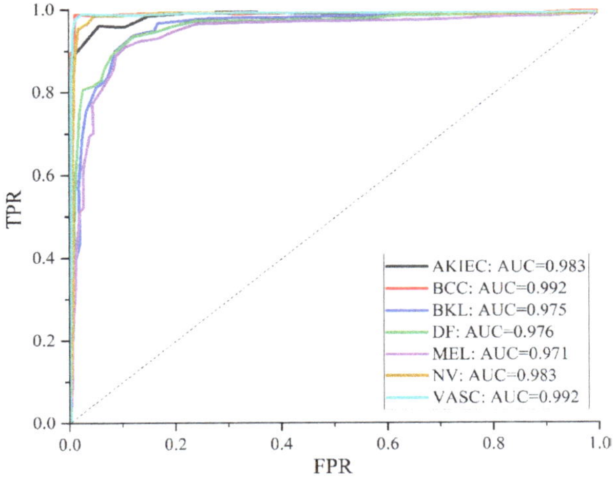

Figure 9. ROC curve of the best classification.

6. Conclusions

This paper introduces the implementation and application of wavelet analysis as down-sampling in convolutional neural networks to solve the problem of insufficient network characterization ability due to the loss of feature information by down-sampling in skin cancer classification and to solve the problem of low channel information utilization and inability to effectively combine low-frequency and high-frequency components by introducing the multichannel attention mechanism into wavelet analysis. Firstly, a combination of depth-separable convolution and 3×3 standard convolution and wavelet down-sampling is used as the input backbone of the ResNet50 network, and then the Hard-Swish function and wavelet down-sampling are used to improve the residual module of the ResNet50 network, which not only ensures the perceptual field of the original network but also reduces the number of parameters and the feature loss of down-sampling and improves the expression ability of the network. The experimental results show that the method achieves excellent classification results on the augmentation HAM10000 dataset, and the classification accuracy reaches 95.84% after combining with migration learning, which effectively improves the skin cancer diagnosis accuracy.

Author Contributions: Methodology, Q.-e.W., Y.Y., X.Z.; writing-original draft, Q.-e.W., Y.Y.; writing—review and editing, Q.-e.W., X.Z.; funding acquisition, Q.-e.W. All authors have read and agreed to published version of the manuscript.

Funding: Qing-e Wu for funding this work through the Key Science and Technology Program of Henan Province number (222102210084), Key Science and Technology Project of Henan Province University number (23A413007), respectively.

Data Availability Statement: Dataset used in the experiment is available on web publicly at https://www.kaggle.com/datasets/kmader/skin-cancer-mnist-ham10000 (accessed on 17 January 2023).

Conflicts of Interest: The authors declare no conflict of interest.

References

1. Siegel, R.L.; Miller, K.D.; Jemal, A. Cancer statistics, 2019. *CA A Cancer J. Clin.* **2019**, *69*, 7–34. [CrossRef] [PubMed]
2. Koh, H.K. Melanoma screening: Focusing the public health journey. *Arch. Dermatol.* **2007**, *143*, 101–103. [CrossRef] [PubMed]
3. Zunair, H.; Hamza, A.B. Melanoma detection using adversarial training and deep transfer learning. *Phys. Med. Biol.* **2020**, *65*, 135005. [CrossRef] [PubMed]
4. LeCun, Y.; Bengio, Y.; Hinton, G. Deep learning. *Nature* **2015**, *521*, 436–444. [CrossRef]

5. Qiu, C.H.; Huang, C.F.; Xia, S.R.; Kong, D.X. Application review of artificial intelligence in medical images aided diagnosis. *Space Med. Med. Eng.* **2021**, *34*, 407–414.
6. Chen, Y.D.; Zhang, Q.; Lan, L.; Peng, L.; Yin, J. A Review of Deep Convolutional Neural Networks in Medical Image Segmentation. *Chin. J. Health Inform. Manag.* **2021**, *18*, 278–284.
7. Codella, N.; Cai, J.; Abedini, M.; Garnavi, R.; Halpern, A.; Smith, J.R. Deep learning, sparse coding, and SVM for melanoma recognition in dermoscopy images. In *Machine Learning in Medical Imaging: 6th International Workshop, MLMI 2015, Proceedings of the MICCAI 2015, Munich, Germany, 5 October 2015;* Held in Conjunction with MICCAI 2015; Springer International Publishing: Cham, Switzerland, 2015; pp. 118–126.
8. Pomponiu, V.; Nejati, H.; Cheung, N.M. Deepmole: Deep neural networks for skin mole lesion classification. In Proceedings of the 2016 IEEE International Conference on Image Processing (ICIP), Phoenix, AZ, USA, 25–28 September 2016; pp. 2623–2627.
9. Esteva, A.; Kuprel, B.; Novoa, R.A.; Ko, J.; Swetter, S.M.; Blau, H.M.; Thrun, S. Dermatologist-level classification of skin cancer with deep neural networks. *Nature* **2017**, *542*, 115–118. [CrossRef]
10. Khan, M.A.; Sharif, M.I.; Raza, M.; Anjum, A.; Saba, T.; Shad, S.A. Skin lesion segmentation and classification: A unified framework of deep neural network features fusion and selection. *Expert Syst.* **2022**, *39*, e12497. [CrossRef]
11. Ahmed, S.A.A.; Yanikoğlu, B.; Göksu, Ö.; Aptoula, E. Skin lesion classification with deep CNN ensembles. In Proceedings of the 2020 28th Signal Processing and Communications Applications Conference (SIU), Gaziantep, Turkey, 5–7 October 2020; pp. 1–4.
12. Abayomi-Alli, O.O.; Damasevicius, R.; Misra, S.; Maskeliunas, R.; Abayomi-Alli, A. Malignant skin melanoma detection using image augmentation by oversamplingin nonlinear lower-dimensional embedding manifold. *Turk. J. Electr. Eng. Comput. Sci.* **2021**, *29*, 2600–2614. [CrossRef]
13. Zeiler, M.D.; Fergus, R. Stochastic pooling for regularization of deep convolutional neural networks. In Proceedings of the 1st International Conference on Learning Representations, ICLR 2013, Scottsdale, AZ, USA, 2–4 May 2013.
14. He, K.; Zhang, X.; Ren, S.; Sun, J. Deep residual learning for image recognition. In Proceedings of the IEEE Conference on Computer Vision and Pattern Recognition, Las Vegas, NV, USA, 27–30 June 2016; pp. 770–778.
15. Zhao, G.; Wang, J.; Zhang, Z. Random Shifting for CNN: A Solution to Reduce Information Loss in Down-Sampling Layers. In Proceedings of the International Joint Conference on Artificial Intelligence (IJCAI), Melbourne, Australia, 25 August 2017; pp. 3476–3482.
16. Jiang, Z.T.; Qin, J.Q.; Zhang, S.Q. Parameterized pooling convolution neural network for image classification. *Acta Electron. Sin.* **2020**, *48*, 1729.
17. Daubechies, I. Ten Lectures on Wavelets. *Comput. Phys.* **1992**, *6*, 697. [CrossRef]
18. Hong, J.; Wang, Z.; Qu, C.; Zhou, Y.; Shan, T.; Zhang, J.; Hou, Y. Investigation on overcharge-caused thermal runaway of lithium-ion batteries in real-world electric vehicles. *Appl. Energy* **2022**, *321*, 119229. [CrossRef]
19. Hong, J.; Zhang, H.; Xu, X. Thermal fault prognosis of lithium-ion batteries in real-world electric vehicles using self-attention mechanism networks. *Appl. Therm. Eng.* **2023**, *226*, 120304. [CrossRef]
20. Bruna, J.; Mallat, S. Invariant scattering convolution networks. *IEEE Trans. Pattern Anal. Mach. Intell.* **2013**, *35*, 1872–1886. [CrossRef]
21. Liu, P.; Zhang, H.; Zhang, K.; Lin, L.; Zuo, W. Multi-level wavelet-CNN for image restoration. In Proceedings of the IEEE Conference on Computer Vision and Pattern Recognition Workshops, Salt Lake City, UT, USA, 18–22 June 2018; pp. 773–782.
22. Li, Q.; Shen, L.; Guo, S.; Lai, Z. Wavelet integrated CNNs for noise-robust image classification. In Proceedings of the IEEE/CVF Conference on Computer Vision and Pattern Recognition, Seattle, WA, USA, 13–19 June 2020; pp. 7245–7254.
23. Xu, D. Application of wavelet transform-based image processing techniques. *J. Soochow Univ. (Nat. Sci.)* **2002**, *1*, 45–49.
24. Xu, K.; Qin, M.; Sun, F.; Wang, Y.; Chen, Y.K.; Ren, F. Learning in the frequency domain. In Proceedings of the IEEE/CVF Conference on Computer Vision and Pattern Recognition, Seattle, WA, USA, 13–19 June 2020; pp. 1740–1749.
25. Li, X.; Wang, W.; Hu, X.; Yang, J. Selective kernel networks. In Proceedings of the IEEE/CVF Conference on Computer Vision and Pattern Recognition, Long Beach, CA, USA, 15–20 June 2019; pp. 510–519.
26. Dang, L.; Pang, P.; Lee, J. Depth-wise separable convolution neural network with residual connection for hyperspectral image classification. *Remote Sens.* **2020**, *12*, 3408. [CrossRef]
27. Avenash, R.; Viswanath, P. Semantic Segmentation of Satellite Images using a Modified CNN with Hard-Swish Activation Function. In Proceedings of the nternational Joint Conference on Computer Vision, Imaging and Computer Graphics (VISIGRAPP) Prague, Czech Republic, 25–27 February 2019; pp. 413–420.
28. Bronskill, J.; Gordon, J.; Requeima, J.; Nowozin, S.; Turner, R. Tasknorm: Rethinking batch normalization for meta-learning. In Proceedings of the International Conference on Machine Learning, Virtual, 13–18 July 2020; pp. 1153–1164.
29. Tschandl, P.; Rosendahl, C.; Kittler, H. The HAM10000 dataset, a large collection of multi-source dermatoscopic images of common pigmented skin lesions. *Sci. Data* **2018**, *5*, 180161. [CrossRef]
30. Xie, F.; Qin, S.; Jiang, Z.; Meng, R. Unsupervised repair of hair-occluded information for skin melanoma image. *Chin. J. Sci Instrum.* **2009**, *30*, 699–705.
31. Youwen, G.; Benjun, Z.; Xiaofei, H. Research on image recognition of convolution neural network based on data augmentation *Comput. Technol. Dev.* **2018**, *28*, 62–65.
32. Krizhevsky, A.; Sutskever, I.; Hinton, G.E. Imagenet classification with deep convolutional neural networks. *Commun. ACM* **2017** *60*, 84–90. [CrossRef]

33. Bansal, M.; Kumar, M.; Sachdeva, M.; Mittal, A. Transfer learning for image classification using VGG19: Caltech-101 image data set. *J. Ambient. Intell. Humaniz. Comput.* **2021**, *14*, 3609–3620. [CrossRef] [PubMed]
34. Sandler, M.; Howard, A.; Zhu, M.; Zhmoginov, A.; Chen, L.C. Mobilenetv2: Inverted residuals and linear bottlenecks. In Proceedings of the IEEE Conference on Computer Vision and Pattern Recognition, Salt Lake City, UT, USA, 18–22 June 2018; pp. 4510–4520.
35. Huang, G.; Liu, Z.; Van Der Maaten, L.; Weinberger, K.Q. Densely connected convolutional networks. In Proceedings of the IEEE Conference on Computer Vision and Pattern Recognition, Honolulu, HI, USA, 21–26 July 2017; pp. 4700–4708.
36. Tan, M.; Le, Q. Efficientnet: Rethinking model scaling for convolutional neural networks. In Proceedings of the 36th International Conference on Machine Learning, Long Beach, CA, USA, 10–15 June 2019; Volume 97, pp. 6105–6114.
37. Gao, M. Soft Attention Improves Skin Cancer Classification Performance. In *Interpretability of Machine Intelligence in Medical Image Computing, and Topological Data Analysis and Its Applications for Medical Data: 4th International Workshop, Proceedings of the iMIMIC 2021, and 1st International Workshop, Strasbourg, France, 27 September 2021*; TDA4MedicalData 2021, Held in Conjunction with MICCAI 2021; Springer Nature: Berlin, Germany, 2021; Volume 12929.
38. Gessert, N.; Nielsen, M.; Shaikh, M.; Werner, R.; Schlaefer, A. Skin lesion classification using ensembles of multi-resolution EfficientNets with meta data. *MethodsX* **2020**, *2020*, 100864. [CrossRef] [PubMed]
39. Shen, S.; Xu, M.; Zhang, F.; Shao, P.; Liu, H.; Xu, L.; Zhang, C.; Liu, P.; Yao, P.; Xu, R.X. Erratum to "A Low-Cost High-Performance Data Augmentation for Deep Learning-Based Skin Lesion Classification". *BME Front.* **2023**, *4*, 0011. [CrossRef]
40. Lan, Z.; Cai, S.; He, X.; Wen, X. FixCaps: An improved capsules network for diagnosis of skin cancer. *IEEE Access* **2022**, *10*, 76261–76267. [CrossRef]

electronics

Article

Research on the Effects of Different Electrolyte Ratios on Heat Loss Control in Lithium-Ion Batteries

Xiaoming Xu [1,2,3], Xinyang Zhang [1,2,3] and Jichao Hong [1,2,3,*]

1 School of Mechanical Engineering, University of Science and Technology Beijing, Beijing 100083, China
2 Beijing Key Laboratory of Lightweight Metal Forming, Beijing 100083, China
3 Shunde Graduate School, University of Science and Technology Beijing, Foshan 528000, China
* Correspondence: hongjichao@ustb.edu.cn

Abstract: As the demand for high-performance battery technology increases, the new energy vehicle industry has an urgent need for safer and more efficient battery systems. A model combining five side reactions was developed to be applied to the studies related to this paper. In this paper, the thermal runaway triggering process of Li-ion batteries is simulated, and the relationship between the local heating of the cathode collector surface and the change of the high-temperature area distribution of the diaphragm layer is analyzed. The thermal runaway mechanism is further revealed. Based on the simulation results, the following conclusions can be drawn: phosphonitene compounds can delay the decomposition of the solid electrolyte interphase membrane and reduce the energy yield of battery-side reactions. Compared with the phosphonitene compound, the optimized structure of adding phosphonitene has little effect on the thermal stability of the battery.

Keywords: lithium-ion battery; electro-hydraulic ratio; heat loss; thermodynamic simulation; thermal stability

Citation: Xu, X.; Zhang, X.; Hong, J. Research on the Effects of Different Electrolyte Ratios on Heat Loss Control in Lithium-Ion Batteries. *Electronics* **2023**, *12*, 1876. https://doi.org/10.3390/electronics12081876

Academic Editor: Fabio Corti

Received: 26 February 2023
Revised: 1 April 2023
Accepted: 14 April 2023
Published: 16 April 2023

1. Introduction

Currently, most of the new energy vehicle battery systems use refilled lead-acid batteries. The technology for these two types of batteries is relatively mature [1–3], and the search for higher-performance batteries is imminent. Due to the increasing demand for higher-performance batteries, lithium-ion batteries are challenging these new battery systems. Compared to conventional lead-zinc batteries, lithium-ion batteries have a better energy density [4–6]. Given their superior performance, lithium-ion batteries have attracted much attention recently as a power source for electric vehicles and electronic load devices [7–10]. While chemical properties such as battery capacity and high-magnification performance have improved due to the thermal runaway prognosis of battery systems using the modified multi-scale entropy in real-world electric vehicles, the thorny issue of lithium-ion battery safety has not yet been addressed [11–14].

The safety of lithium-ion batteries has been a serious impediment to their widespread use in electric and hybrid vehicles [15–17]. Because the electrolyte in high-energy batteries is a flammable organic solvent, lithium-ion batteries can suffer thermal runaway under various electromechanical-electrical-thermal abuse conditions, such as overcharging, external shocks, and thermal shock. Thermal runaways in lithium-ion batteries can cause high temperatures, smoke, explosions, and fires [18,19]. In recent years, many efforts have been made to improve the safety of battery materials. In order to obtain a more stable and safer high-energy lithium-ion battery, a more stable and high-performance electrolyte needs to be applied to the battery system. Systematic approaches and technological breakthroughs based on electrolyte research are essential in studying lithium-ion batteries, and better electrolytes facilitate stable improvements in high-energy lithium-ion battery systems [20].

The addition of flame retardants to the electrolyte is one of the most effective ways to improve the safety of lithium-ion batteries today. It has elucidated the mechanism of action of trimethyl phosphate (TMP) by studying the effect of TMP on the thermal stability of electrolytes. The results showed that TMP, when subjected to thermal decomposition and vaporization, produces free radicals of both phosphorus, which can combine with hydrogen radicals and reduce the content of hydrogen radicals in the reaction system, effectively inhibiting the combustion process of electrolytes [21]. A large number of new organic films have been used as flame retardants for lithium-ion batteries and have been verified to have better flame retardancy, providing a safer electrolyte for lithium-ion batteries [22,23]. From the above studies, it can be seen that conducting research on composite flame retardants with characteristics such as a low melting point, a high flash point, a low viscosity, stable electrochemical properties, and efficient flame retardant properties is one of the critical development directions for flame retardant additives for lithium-ion batteries.

Therefore, in this paper, a thermal model combining five side reactions is established in COMSOL to simulate the thermal runaway process of lithium-ion battery cells, and the effect of local heating of lithium-ion batteries under different heat dissipation conditions on the change in the distribution of the high-temperature region of the diaphragm is mainly analyzed to reveal further the effect of varying electrolyte ratios on the thermal runaway of lithium-ion batteries.

2. Model Design and Construction

2.1. Numerical Model

Consider a cylindrical lithium-ion cell of radius R, radial thermal conductivity $\boldsymbol{k_r}$, heat capacity C_p, and mass density r. The cell experiences a temperature-dependent internal heat generation rate $Q(T)$ throughout its volume and is being cooled at the outside surface with a convective heat transfer h due to a mechanism such as a coolant flow. The interest is in determining the parameter space within which the cell will not undergo thermal runaway, i.e., the cell temperature does not become unbounded. In this case, the governing energy equation for the temperature rise $T(r,t)$ in the cell is given by:

$$Q(T) = Q_{SEI}(t) + Q_{anode}(t) + Q_{PVDF}(t) + Q_{cathode}(t) + Q_e$$

where $Q(T)$ is the total heat yield of the reaction. Q_{SEI} is the heat generated by the decomposition of the SEI membrane, and $Q_{anode}/Q_{cathode}$ is the heat generated by the reaction between the anode/cathode and the electrolyte, respectively. Q_{PVDF} is the heat generated by the reaction of the binder. Q_e is the heat generated by the reaction of the electrolyte.

$$k_r\left(\frac{\partial^2 T}{\partial r^2} + \frac{1}{r}\frac{\partial T}{\partial r}\right) + Q(T) = \rho \cdot C_P \frac{\partial T}{\partial t} \tag{1}$$

$$\frac{\partial T}{\partial r} = 0 \text{ at } r = 0 \tag{2}$$

$$-k_r\left(\frac{\partial T}{\partial r}\right) = h(T - T_0) \text{ at } r = R \tag{3}$$

Equations (1)–(3) can be used to determine whether a set of preconditions can prevent thermal runaway by ensuring that T is a bounded solution at all times. A Taylor series expansion with $T = 1/4\ T_0$ is first performed for $Q(T)$, where the terms of second-order and higher are ignored.

$$k_r\left(\frac{\partial^2 T}{\partial r^2} + \frac{1}{r}\frac{\partial T}{\partial r}\right) + Q(T_0) + \beta(T - T_0) = \rho C_p \frac{\partial T}{\partial t} \tag{4}$$

where $\beta = \frac{dQ}{dT}$ is the slope of $Q(T)$.

To solve Equation (4), we note that the heat generation term can be linearly divided into two components, $(Q(T_0) - bT_0)$ and bT. The first component is a constant, which is known from heat conduction theory and will lead to a stable temperature field. However, the second heat generation component bT increases with temperature and may lead to an unbounded temperature. $T_2(r,t)$ represents the temperature increase caused by the second heat-generating component, using the variable separation technique in the following equation:

$$T_2(r,t) = \sum_{n=1}^{\infty} C_n J_0(\frac{\mu_n r}{R}) \cdot \exp(\frac{k_r}{\rho \cdot C_P}(\frac{\beta}{k_r} - \frac{\mu_n^2}{R^2}) \cdot t) \tag{5}$$

where J_0 is the Bessel function of the first kind of order 0, C_n is constant coefficients, and $\boldsymbol{\mu_n}$ are the non-dimensional eigenvalues given by the roots of the equation.

$$B_i \cdot J_0(x) - xJ_1(x) = 0 \tag{6}$$

where $\boldsymbol{B_i} = \frac{\boldsymbol{hR}}{\boldsymbol{K_r}}$ is the Biot number. Note that $\boldsymbol{C_n}$ is obtained using orthogonality and the initial condition of the temperature field. The temperature solution in Equation (5) may be either bounded or unbounded depending on the sign of the term within the exponential function in Equation (6).

The negative electrode is protected from direct reaction with solvent by an ionically conducting film called SEI. Solid electrolyte films are formed during the initial cycle of a lithium-ion battery. SEI films that are too thick or so thin that they are non-existent are not suitable for lithium battery applications. The presence of a reasonable SEI film protects the cathode active material from reacting with the electrolyte. Regarding the cycle life and safety of Li-ion batteries, when the internal temperature of the battery reaches about 130 °C, the SEI film decomposes, resulting in a completely exposed negative electrode and a large amount of exothermic decomposition of the electrolyte on the electrode surface, leading to a rapid increase in the internal temperature of the battery. This is the first exothermic side reaction inside the Li-ion battery and the starting point of a series of thermal runaway problems. The heat generation equation is shown in Equation (7):

$$R_{sei} = A_{sei} \exp[-\frac{E_{a,sei}}{RT}] c_{sei}^{m_{sei}} \tag{7}$$

where c_{SEI} is the dimensionless number of lithium-containing meta-stable products in the SEI.

At elevated temperatures (>120 °C), an exothermic reaction between intercalated lithium and electrolyte can occur. The heat generation equation is shown in Equation (8):

$$R_{ne} = A_{ne} \exp[-\frac{E_{a,ne}}{RT}] c_{ne}^{m_{ne}} \tag{8}$$

where c_{ne} is the dimensionless number of the lithium-containing meta-stable products in the reaction.

The cathode material reacts with the electrolyte under oxidizing conditions. It is mainly various types of lithium compounds, which always react with the electrolyte in trace amounts under different environmental conditions and with different intensities of reaction. The cathode material reacts with the electrolyte to produce insoluble products, making the reaction irreversible. The cathode material involved in the reaction loses its original structure, and the lithium power battery loses its corresponding power and permanent capacity. The reaction is called the positive-solvent reaction. The heat generation equation is shown in Equation (9):

$$R_{pe} = A_{pe} \exp[-\frac{E_{a,pe}}{RT}] c_{pe}^{m_{pe}} \tag{9}$$

where c_{pe} is the dimensionless number of the lithium-containing meta-stable products in the reaction.

Due to the exothermic side reactions of the electrolyte at the negative electrode, the internal temperature of the cell is increasing, which in turn leads to further thermal decomposition of LiPF6 and the solvent within the electrolyte. The electrolyte can decompose exothermically at elevated temperatures (>200 °C), as expressed in Equation (10):

$$R_e = A_e \exp[-\frac{E_{a,e}}{RT}]c_e^{m_e} \tag{10}$$

where c_e is the dimensionless number of the lithium-containing meta-stable products in the SEI.

When the temperature is about 513.15 K, the binder starts to react. The reaction can be expressed in Equation (11).

$$R_{pvdf} = A_{pvdf} \exp[-\frac{E_{a,pvdf}}{RT}]c_{pvdf}^{m_{pvdf}} \tag{11}$$

The five mentioned reactions are SEI-decomposition, negative-solvent, positive-solvent, electrolyte decomposition, and the binder reaction. These are the five reactions in the model that have the most significant effect on the thermal phenomena in the reaction. In this simulation, the thermal runaway phenomenon will be analyzed under the above five reactions and the heat transfer of the lithium battery itself. The specific simulation parameters for the five layers are shown in Table 1. The specific simulation parameters for the five layers are shown in Table 2 [24].

Table 1. Heating situations.

Label	Value	Description
Asei	1.667×10^{15} [1/s]	SEI-decomposition frequency factor
Ane	2.5×10^{13} [1/s]	Negative-solvent frequency factor
Ape1	1.75×10^{9} [1/s]	Positive-solvent 1 frequency factor
Ape2	1.077×10^{12} [1/s]	Positive-solvent 2 frequency factor
Ae	5.14×10^{25} [1/s]	Electrolyte decomposition frequency factor
Apvdf	1.917×10^{25} [1/s]	Binder frequency factor
Esei	1.3508×10^{5} [J/mol]	SEI-decomposition activation energy
Ene	1.3508×10^{5} [J/mol]	Negative-solvent activation energy
Epe1	1.1495×10^{5} [J/mol]	Positive-solvent 1 activation energy
Epe2	1.5888×10^{5} [J/mol]	Positive-solvent 2 activation energy
Ee	2.74×10^{5} [J/mol]	Electrolyte decomposition activation energy
Epvdf	2.86×10^{5} [J/mol]	Binder activation energy
Hsei	257 [J/g]	SEI-decomposition heat release
Hne	1714 [J/g]	Negative-solvent heat release
Hpe1	277 [J/g]	Positive-solvent 1 heat release
Hpe2	284 [J/g]	Positive-solvent 2 heat release
He	155 [J/g]	Electrolyte decomposition heat release
Hpvdf	1500 [J/g]	Binder decomposition heat release
Wc	6.104×10^{5} [g/m^3]	Specific carbon content
Wp	1.221×10^{6} [g/m^3]	Specific positive active content
We	4.069×10^{5} [g/m^3]	Specific electrolyte content

Table 2. Size of a three-dimensional model.

Parameters	d_1	d_2	d_3	d_4	d_5	d_6	W	H
Unit/mm	1.01	5.89	4.04	4.54	0.52	16	94	168

The SEI membrane is insoluble in organic solvents and effectively prevents the co-embedding of solvent molecules, thus greatly improving the cycling performance and service life of the electrode. The crystallinity of the material has a great influence on the formation of the SEI film during the first charging process. The electrolyte conductive agent inside the battery also has an important influence on the SEI film. If the conductive agent in the electrode is not uniformly dispersed, it will lead to an uneven SEI film, and different conductivity will be generated in each part.

2.2. Three-Dimensional Model

In the previous section, a heat transfer model was developed based on the object of study in this paper, which is a mathematical model created after taking into account the factors ignored by this paper. In this section, the three-dimensional physical model that was input into COMSOL is presented. The simulations involved in this study were then performed after inputting the mathematical model into the physical model. The parameters of the numerical model are set to describe the thermochemical reactions, such as heat production and heat transfer in the reaction. Based on the parameter settings and the numerical model described above, a 3D model that can simulate the 3D shape of the object and visualize the simulation results is created. Both use the same Cartesian coordinate system. The 3D model is an external feature of the numerical model simulation results.

The mathematical model established in the previous section can describe the variation of heat production and heat dissipation in the simplified lithium battery.

Actually, the internal structure of a lithium-ion battery has many layers, and the thickness of each layer is skinny. The computation will be extensive if it is modeled as the actual battery. As a result, a 3D model has been established in COMSOL Multiphysics that has five layers (the cathode current collector layer, the positive material layer, the electrolyte and separator layer, the negative material layer, and the anode current collector layer) and two terminals (the positive terminal and the negative terminal) according to the different side reactions of different parts of the lithium-ion battery in this paper. Figure 1 shows the simplified model geometry, the five temperature measuring points on the separator layer, and the computational mesh, and the total number is 375,793. The specific size of each layer and the battery cell are listed in Table 2.

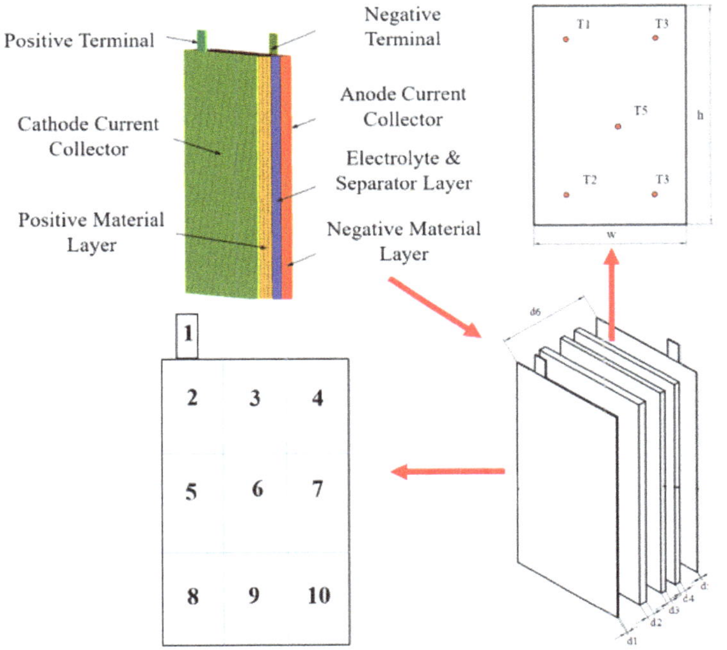

Figure 1. Schematic of model geometry and a computational mesh.

2.3. Boundary Condition

The heating location is the cathode current collector, which is divided into ten zones (shown in Figure 1 and numbered 1 to 10), and the heating is listed in Table 3. The boundary conditions for some of the simulations are shown in Table 4. In these tests, the test locations are 1, 2, and 3. The environment is assumed to be air with an ambient temperature of 293.15 K.

Table 3. Heating situations.

Test Number	Heating Temperature (K)	Density (kg/m^3)	Heat Capacity (J/(kg·K))	Heat Conductivity Coefficient (w/(m·K))
Test 1	423.15	1008.98	1978.16	0.344
Test 2	473.15	1008.98	1978.16	0.344
Test 3	523.15	1008.98	1978.16	0.344
Test 4	423.15	978.34	2467.34	0.344
Test 5	473.15	978.34	2467.34	0.344
Test 6	523.15	978.34	2467.34	0.344
Test 7	423.15	1006.45	2345.65	0.546
Test 8	473.15	1006.45	2345.65	0.546
Test 9	523.15	1006.45	2345.65	0.546

Table 4. Part of the simulation boundary conditions.

Label	Expression/Value
Lcell	94 [mm]
Hcell	168 [mm]
Tcell	16 [mm]
Wtab	10 [mm]
kT_pos	1.74 [W/(m × K)]
kT_neg	1.04 [W/(m × K)]
kT_sep	0.344 [W/(m × K)]
rho_pos	2362.36 [kg/m^3]
rho_neg	1347.33 [kg/m^3]
rho_sep	1008.98 [kg/m^3]
Cp_pos	1142.29 [J/(kg × K)]
Cp_neg	1437.4 [J/(kg × K)]
Cp_sep	1978.16 [J/(kg × K)]
Asei	1.667×10^{15} [s^{-1}]
Ane	2.5×10^{13} [s^{-1}]
Ape1	1.75×10^{9} [s^{-1}]
Ape2	1.077×10^{12} [s^{-1}]
Ae	5.14×10^{25} [s^{-1}]
Apvdf	1.917×10^{25} [s^{-1}]
Esei	1.3508×10^{5} [J/mol]
Ene	1.3508×10^{5} [J/mol]
Epe1	1.1495×10^{5} [J/mol]
Epe2	1.5888×10^{5} [J/mol]
Ee	2.74×10^{5} [J/mol]
Epvdf	2.86×10^{5} [J/mol]
Hsei	257 [J/g]
Hne	1714 [J/g]
Hpe1	277 [J/g]
Hpe2	284 [J/g]
He	155 [J/g]
Hpvdf	1500 [J/g]
Wc	6.104×10^{5} [g/m^3]

Table 4. *Cont.*

Label	Expression/Value
Wp	1.221×10^6 [g/m^3]
We	4.069×10^5 [g/m^3]
Wpvdf	8.14×10^4 [g/m^3]
csei0	0.15
cne0	0.75
a0	0.04
ce0	1
cpvdf0	1
Tsei	343.15 [K]
Tne	393.15 [K]
Tpe1	443.15 [K]
Tpe2	493.15 [K]
Te	523.15 [K]
Tpvdf	513.15 [K]
Tsp	403.15 [K]

The object of the modeling study is a cobalt-manganese lithium battery with a rated voltage of 3.4 V. The cathode material is a mixture of nickel, cobalt, and manganese graphite anode material. The electrolyte is LiPF6.

3. Thermal Runaway Simulation

Tests 1, 2, and 3 are thermal runaway simulations for a battery without any flame retardant. Tests 4, 5, and 6 are thermal runaway simulation results for a battery with the addition of a phosphonitene compound. Tests 7, 8, and 9 had phosphorus and nitrogen added [25–29].

3.1. Locally Heating Tests 1, 2, and 3

The subplots (a, b, and c) of Figure 2 show the heat generation curves for each side reaction for tests 1, 2, and 3. It can be seen that SEI membrane decomposition and negative electrolyte reactions were the only side reactions that occurred in both trials. The other three curves did not change significantly during the heating process. However, in test 3, the SEI membrane completely decomposed, and the negative electrolyte reaction reached a heat generation of 1500 W/m^3 [30,31].

As shown in Figure 2, the heat transfer process is affected by the different temperatures of the heating band while the reaction is taking place. When the heating band temperature increases, the SEI membrane temperature rises faster and decomposes more quickly, so the heat generation increases faster. The same law is used after replacing the electrolyte solute, so it is not repeated in the following two parts of the explanation. The changes in the content of SEI membranes in the model were then analyzed. SEI membrane content (SEI_{MC}) was analyzed by COMSOL simulation. To further investigate the thermal behavior of the battery cell, SEI film curves are obtained and shown in Figure 3. It can be seen from Figure 3 that the SEI film starts to change at about 250 s under the heating condition of 423.15 K. The SEI film starts to change around 200 s under the heating condition of 473.15 K, and the SEI film starts to change around 180 s under the heating condition of 523.15 K. The curve of the SEI film in 423.15 K decreases linearly from 250 s, while in 473.15 K and 573.15 K, the curve of the SEI film decreases from 200 s to 180 s, respectively. Combining the three lines, it can be seen that as the temperature increases, the time point at which the SEI film begins to decrease becomes more and more advanced, and the decreasing curve of the SEI film becomes more and more tortuous [32,33].

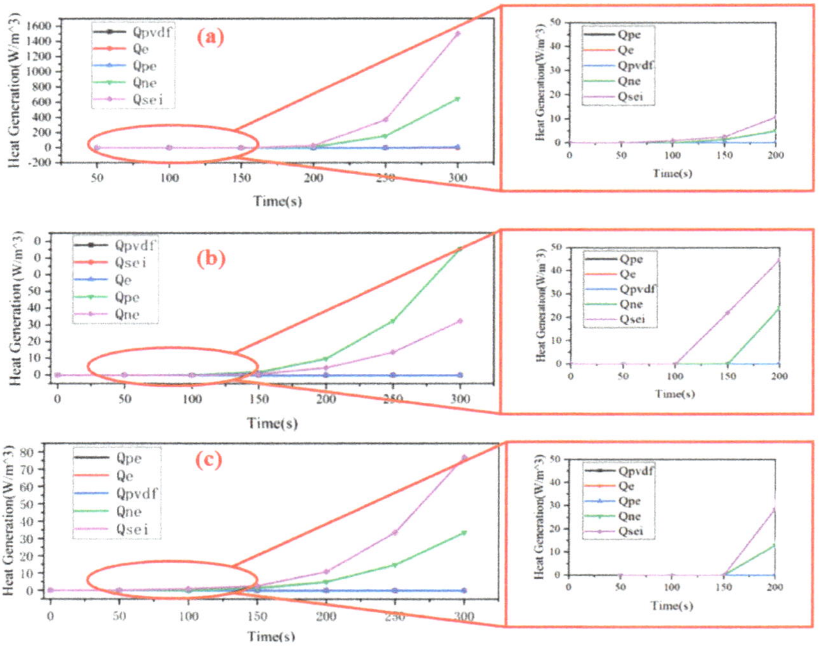

Figure 2. Variation of each component of the different side reactions: (**a**) 423.15 K; (**b**) 473.15 K; and (**c**) 523.15 K.

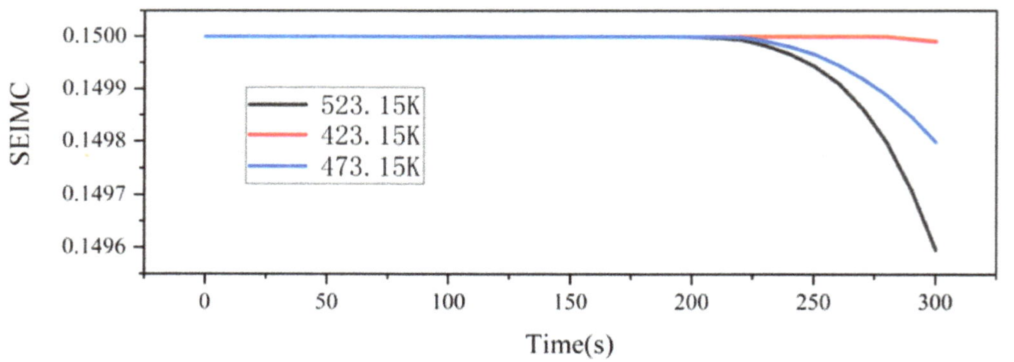

Figure 3. Conformational curves for changes in cell composition of SEI films at different temperatures in tests 1, 2, and 3.

The temperature distribution of each part of the battery heated at 423.15 K, 473.15 K and 523.15 K, respectively, is shown in Figure 4. It can be seen from Figure 4 that after heating for 300 s, the battery has thermal runaway under various working conditions, and the temperature of each part of the battery reaches above 673.15 K. The temperature of the battery near the heating zone rises most sharply. In Figure 4a, the highest temperature is in the positive electrode of the battery, and the temperature has reached 693.15 K or more. In Figure 4b, the highest temperature is still the positive battery, and the temperature reached 1023.15 K or more. In Figure 4c, the area with the highest battery temperature is at the negative pole of the battery, and the temperature reaches 753.15 K. It can be seen from

Figure 4a–c that the temperature change of each part of the battery does not increase with the increase in the heating temperature.

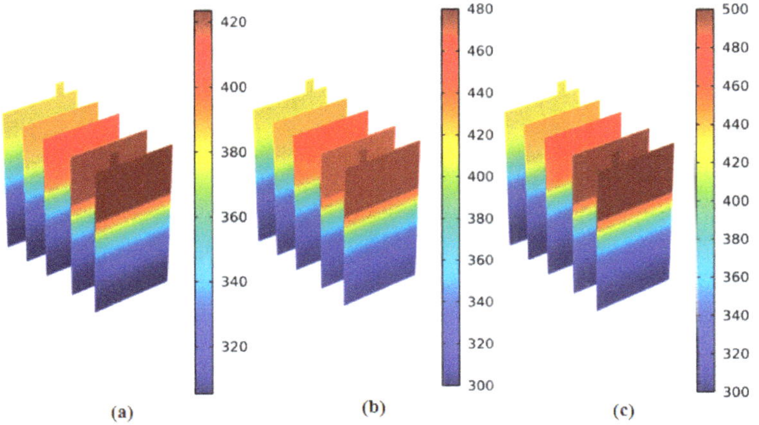

Figure 4. Temperature distribution curves for each part of the cell in tests 4, 5, and 6, for 300 s (**a**) 423.15 K; (**b**) 473.15 K; and (**c**) 523.15 K.

3.2. Locally Heating Tests 4, 5, and 6

The heat generation curves of each side reaction of tests 4, 5, and 6 are shown in Figure 5a–c. It can be seen that the SEI film decomposition and the negative electrolyte reaction are the only two side reactions that occur in the two tests. The other three curves did not change significantly during the heating process. However, the SEI film decomposes completely, and the heat generation of the negative electrolyte reaction reaches 1300 W/m^3 in test 6, but the two side reactions are just in the beginning state in test 4.

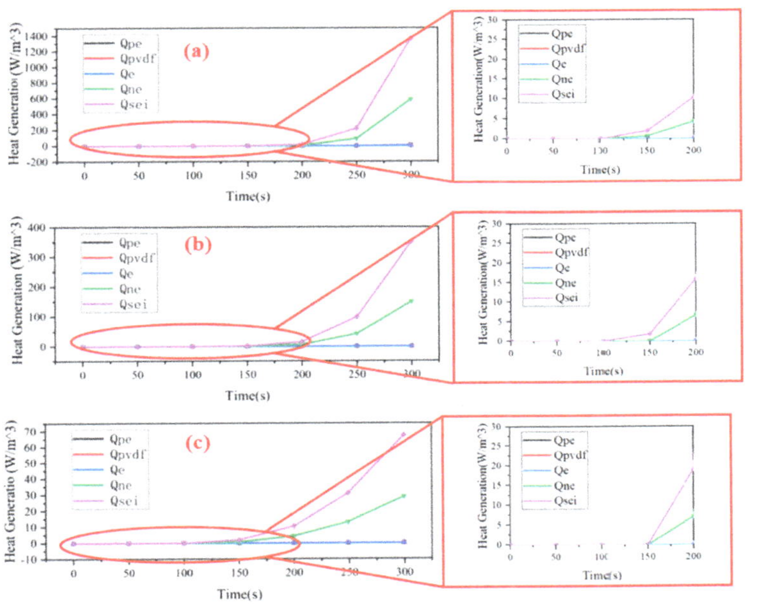

Figure 5. Heat production curves for each component in tests 4, 5, and 6: (**a**) 423.15 K; (**b**) 473.15 K; and (**c**) 523.15 K.

To further investigate the thermal behavior of the battery cell, SEI film curves are obtained and shown in Figure 6. It can be seen from Figure 6 that the SEI film starts to change after about 500 s under the heating condition of 423.15 K. The SEI film starts to change around 400 s under the heating condition of 473.15 K, and the SEI film starts to change around 300 s under the heating condition of 523.15 K. The curve of the SEI film in 423.15 K decreases linearly from 500 s, while in 473.15 K and 573.15 K, the curve of the SEI film decreases from 400 s to 300 s, respectively.

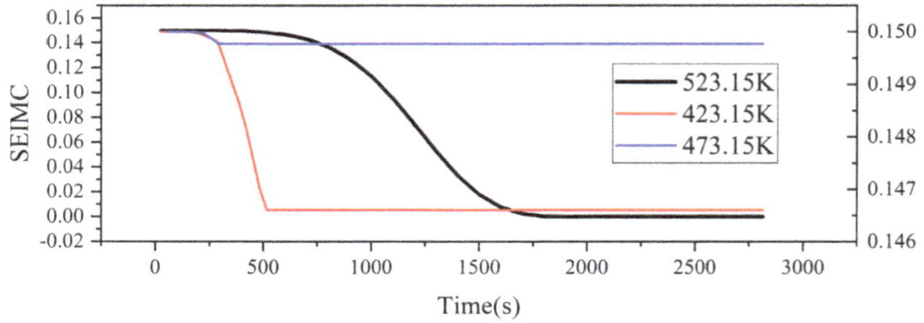

Figure 6. Conformational curves for changes in cell composition of SEI films at different temperatures in tests 4, 5, and 6.

The temperature distribution of each part of the battery heated at 423.15 K, 473.15 K, and 523.15 K, respectively, is shown in Figure 7. It can be seen from Figure 7 that after heating for 300 s, the battery has thermal runaway under various working conditions. The temperature of the battery near the heating zone rises most sharply. In (a), the highest temperature is in the negative electrode of the battery, and the temperature has reached 693.15 K or more. In (b), the highest temperature is still the negative battery, and the temperature reached 773.15 K or more. In (c), the area with the highest battery temperature is at the negative pole of the battery, and the temperature reaches 793.15 K. It can be seen from (a), (b), and (c) that the temperature change of each part of the battery does increase with the increase in the heating temperature.

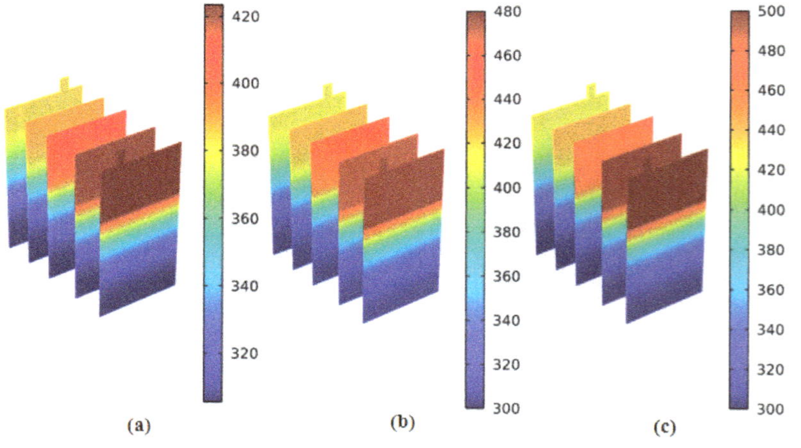

Figure 7. Temperature profiles of each part of the cell after heating for 300 s: (**a**) 423.15 K; (**b**) 473.15 K and (**c**) 523.15 K.

3.3. Locally Heating Tests 7, 8, and 9

The heat generation curves for each side reaction of tests 7, 8, and 9 are shown in the subplots (a, b, and c) of Figure 8. It can be seen that SEI membrane decomposition and negative electrolyte reactions are the only two side reactions in both tests. The other three curves did not change significantly during the heating changes. However, the heat generation for the complete decomposition of the SEI membrane and the negative electrolyte reaction reached 850 W/m^3 in test 9, but these two side reactions were only at the beginning of test 7.

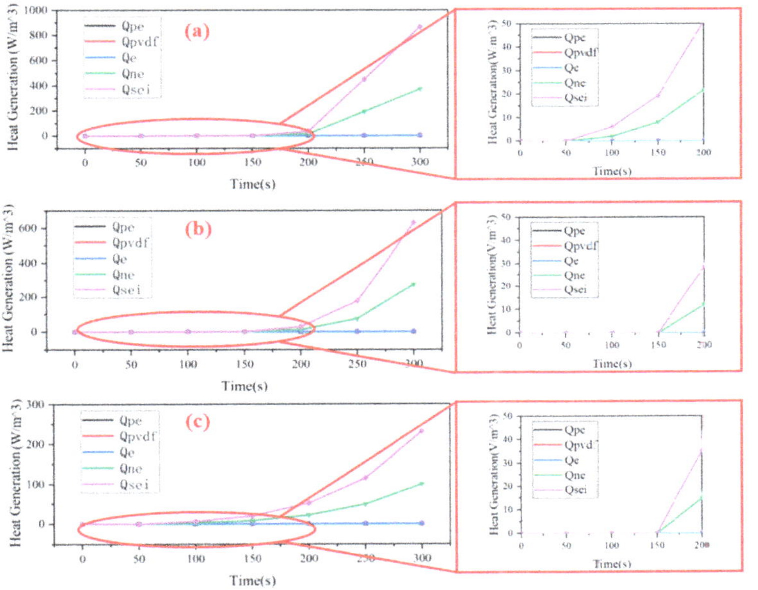

Figure 8. Heat production curves for the side reactions of each component: (**a**) 423.15 K; (**b**) 473.15 K; and (**c**) 523.15 K.

To further investigate the thermal behavior of the battery cell, SEI film curves are obtained and shown in Figure 9. It can be seen from Figure 9 that the SEI film starts to change at about 400 s under the heating condition of 423.15 K. The SEI film starts to change around 300 s under the heating condition of 473.15 K, and the SEI film starts to change around 200 s under the heating condition of 523.15 K. The curve of the SEI film in (a) decreases linearly from 400 s, while in Figure 8b,c, the curve of the SEI film decreases from 300 s to 200 s, respectively.

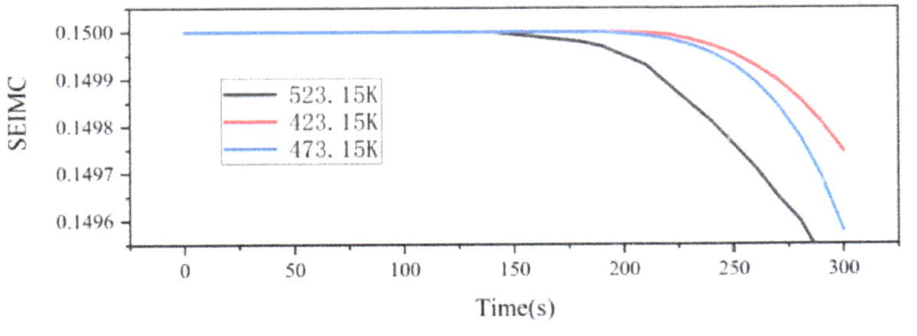

Figure 9. Conformational curves for changes in cell composition of the SEI films at different temperatures in tests 7, 8, and 9.

The temperature distribution of each part of the battery heated at 423.15 K, 473.15 K, and 523.15 K, respectively, is shown in Figure 10. It can be seen from Figure 10 that after heating for 300 s, the battery has thermal runaway under various working conditions. The temperature of the battery near the heating zone rises most sharply. In (a), the highest temperature is in the positive electrode of the battery, and the temperature has reached 693.15 K or more. In (b), the highest temperature is still the positive battery, and the temperature reached 793.15 K or more. In (c), the area with the highest battery temperature is at the negative pole of the battery, and the temperature reaches 773.15 K. It can be seen from (a), (b), and (c) that the temperature change of each part of the battery does increase with the increase in the heating temperature.

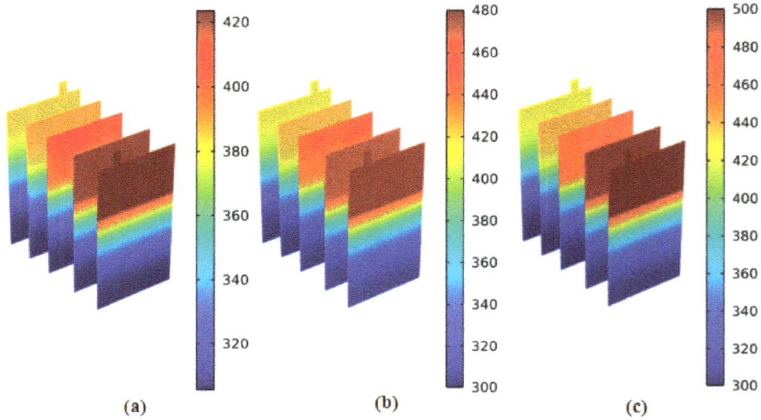

Figure 10. Temperature distribution of each part of the battery for 300 s: (**a**) 423.15 K; (**b**) 473.15 K; and (**c**) 523.15 K.

3.4. Discussion

From the test results, the rate of change of the components of the battery and the heat production decreased after the addition of the flame retardant, mainly because the addition of the phosphonitene compound can effectively improve the thermal stability of the battery, the SEI film is not easily decomposed, and the heat production of the side reaction during heating is significantly reduced [32–34].

However, it can be seen from the heat loss control map that there is no significant difference between the two groups of test results when the phosphonate compound and the optimized structure are added, which indicates that the optimized design will not affect the thermal stability of the battery.

Above all, five thermal models concerning the side reactions of lithium-ion batteries were established, and numerical simulations of thermal runaway were carried out. The triggering condition for thermal runaway in this experiment is local heating, which can effectively simulate the heat exposure of the battery during a real thermal runaway.

4. Conclusions

In this paper, a thermal model containing five lateral reactions is developed to simulate the thermal runaway triggering process of lithium-ion batteries. The effect of local heating of the cathode collector surface on the change in the distribution of the high-temperature region of the diaphragm layer is analyzed by this model, and the thermal runaway mechanism is further revealed. From the simulation results, the following conclusions can be drawn: (1) Phosphonitene compounds can delay the decomposition of the cell SEI membrane; and (2) phosphonitene compounds can reduce the energy yield of the cell-side reaction.

In this paper, the changes in the distribution of the high-temperature region of the spacer when the lithium-ion battery is locally heated under different heat dissipation conditions are investigated. In the subsequent analysis, the thermal equilibrium capability of lithium-ion itself will be further investigated. Meanwhile, a separate mechanistic study of the reactions occurring on individual layers will be one of the subsequent research directions of this paper.

Author Contributions: Conceptualization, X.X. and X.Z.; methodology, X.Z.; software, X.Z.; validation, X.X., X.Z. and J.H.; resources, X.Z.; data curation, X.Z.; writing—original draft preparation, X.X.; writing—review and editing, J.H.; visualization, X.Z.; supervision, X.Z.; project administration, J.H.; funding acquisition, J.H. All authors have read and agreed to the published version of the manuscript.

Funding: The project is supported partly by the National Natural Science Foundation of China (No. 52107220), the Postdoctoral Research Fund Project of China (No. 2021M690353), the Foundation of State Key Laboratory of Automotive Simulation and Control (No. 20210211), the Scientific and Technological Innovation Foundation of Foshan (No. BK21BE012), the Postdoctoral Research Foundation of the Shunde Innovation School of the University of Science and Technology Beijing (No. 2021BH007), the Fundamental Research Funds for the Central Universities (No. FRF-BD-20-08A, NO. FRF-IDRY-21-013), and the Opening Foundation of the Key Laboratory of Advanced Manufacture Technology for Automobile Parts, Ministry of Education (No. 2022 KLMT03).

Data Availability Statement: Data sharing not applicable. No new data were created or analyzed in this study. Data sharing is not applicable to this article.

Conflicts of Interest: The authors declare no conflict of interest.

References

1. Michele, B. Damage Investigation on the Upper Skin Composite Panels of Eurofighter Aircraft. *Key Eng. Mater.* **2019**, *812*, 9–16.
2. Choi, W. A study on state of charge and state of health estimation in consideration of lithium-ion battery aging. *Sustainability* **2020**, *12*, 10451. [CrossRef]
3. Doshi, N. Modeling of Thermal Dynamics in Chevrolet Volt Gen II Hybrid Electric Vehicle for Integrated Powertrain and HVAC Optimal Operation Through Connectivity. Master's Thesis, Michigan Technological University, Houghton, MI, USA, 2019.
4. Feng, X.; Ren, D.; Zhang, S.; He, X.; Wang, L.; Ouyang, M. Influence of aging paths on the thermal runaway features of lithium-ion batteries in accelerating rate calorimetry tests. *Int. J. Electrochem. Sci.* **2019**, *14*, 44–58. [CrossRef] [PubMed]
5. Feng, Y.; Dong, Z. Optimal energy management with balanced fuel economy and battery life for large hybrid electric mining truck. *J. Power Sources* **2020**, *454*, 227948. [CrossRef]
6. Gandoman, F.H.; Jaguemont, J.; Goutam, S.; Gopalakrishnan, R.; Firouz, Y.; Kalogiannis, T.; Omar, N.; Van Mierlo, J. Concept of reliability and safety assessment of lithium-ion batteries in electric vehicles: Basics, progress, and challenges. *Appl. Energy* **2019**, *251*, 113343. [CrossRef]
7. Hong, J.; Wang, Z.; Qu, C.; Zhou, Y.; Shan, T.; Zhang, J.; Hou, Y. Investigation on Overcharge-caused Thermal Runaway of Lithium-Ion Batteries in Real-world Electric Vehicles. *Appl. Energy* **2022**, *321*, 119229. [CrossRef]
8. Hong, J.; Wang, Z.; Chen, W.; Wang, L.; Lin, P.; Qu, C. Online accurate state of health estimation for battery systems on real-world electric vehicles with variable driving conditions considered. *J. Clean. Prod.* **2021**, *294*, 125814. [CrossRef]
9. Hong, J.; Wang, Z.; Qu, C.; Ma, F.; Xu, X.; Yang, J.; Zhang, J.; Zhou, Y.; Shan, T.; Ho, Y. Fault Prognosis and Isolation of Lithium-ion Batteries in Electric Vehicles Considering Real-Scenario Thermal Runaway Risks. *IEEE J. Emerg. Sel. Top. Power Electron.* **2023**, *11*, 88–99. [CrossRef]
10. Liu, Y.; Elias, Y.; Meng, J.; Aurbach, D.; Zou, R.; Xia, D.; Pang, Q. Electrolyte solutions design for lithium-sulfur batteries. *Joule* **2021**, *5*, 2323–2364. [CrossRef]
11. Zou, F.; Manthiram, A. A review of the design of advanced binders for high-performance batteries. *Adv. Energy Mater.* **2020**, *10*, 2002508. [CrossRef]
12. Shahjalal, M.; Roy, P.K.; Shams, T.; Fly, A.; Chowdhury, J.I.; Ahmed, R.; Liu, K. A review on second-life of Li-ion batteries: Prospects, challenges, and issues. *Energy* **2022**, *241*, 122881. [CrossRef]
13. Li, W.; Xiong, S.; Zhou, X.; Shi, W.; Wang, C.; Lin, X.; Cheng, J. Design of Cylindrical Thermal Dummy Cell for Development of Lithium-Ion Battery Thermal Management System. *Energies* **2021**, *14*, 1357. [CrossRef]
14. Tran, M.K.; Panchal, S.; Khang, T.D.; Panchal, K.; Fraser, R.; Fowler, M. Concept review of a cloud-based smart battery management system for lithium-ion batteries: Feasibility, logistics, and functionality. *Batteries* **2022**, *8*, 19. [CrossRef] [PubMed]
15. Li, W.; Liao, K.; He, Q.; Xia, Y. Performance-aware cost-effective resource provisioning for future grid IoT-cloud system. *J. Energy Eng.* **2019**, *145*, 04019016. [CrossRef]
16. Ma, F.; Fu, C.; Yang, J.; Yang, Q. Control strategy for adaptive active energy harvesting in sediment microbial fuel cells. *J. Energy Eng.* **2020**, *146*, 04019034. [CrossRef]

17. Madani, S.S.; Schaltz, E.; Kær, S.K. Study of temperature impacts on a lithium-ion battery thermal behaviour by employing isothermal calorimeter. *ECS Trans.* **2018**, *87*, 295. [CrossRef]
18. Rogall, M.; Barai, A.; Brucoli, M.; Luk, P.; Bhagat, R.; Greenwood, D. DREMUS: A Data-Restricted Multi-Physics Simulation Model for Lithium-Ion Battery Storage. *J. Energy Storage* **2020**, *32*, 102051. [CrossRef]
19. Shen, Y.W.; Liang, L.Q.; Cui, M.; Shen, F.; Zhang, B.; Cui, T. Advanced control of DFIG to enhance the transient voltage support capability. *J. Energy Eng.* **2018**, *144*, 04018009. [CrossRef]
20. Lin, W.; Zhu, M.; Fan, Y.; Wang, H.; Tao, G.; Ding, M.; Liu, N.; Yang, H.; Wu, J.; Fang, J.; et al. Low temperature lithium-ion batteries electrolytes: Rational design, advancements, and future perspectives. *J. Alloy. Compd.* **2022**, *905*, 164163. [CrossRef]
21. Lu, D.; Lei, X.; Weng, S.; Li, R.; Li, J.; Lv, L.; Zhang, H.; Huang, Y.; Zhang, J.; Zhang, S.; et al. A self-purifying electrolyte enables high energy Li ion batteries. *Energy Environ. Sci.* **2022**, *15*, 3331–3342. [CrossRef]
22. Kim, I.H.; Liu, Y.H.; Pallister, S.; Pol, W.; Roberts, S.; Lee, E. Fault-tolerant resource estimate for quantum chemical simulations: Case study on Li-ion battery electrolyte molecules. *Phys. Rev. Res.* **2022**, *4*, 023019. [CrossRef]
23. Yang, W.; Zhou, F.; Zhou, H.; Wang, Q.; Kong, J. Thermal performance of cylindrical lithium-ion battery thermal management system integrated with mini-channel liquid cooling and air cooling. *Appl. Therm. Eng.* **2020**, *175*, 115331. [CrossRef]
24. Xu, X.; Fu, J.; Jiang, H.; He, R. Research on the heat dissipation performance of lithium-ion cell with different operating conditions. *Int. J. Energy Res.* **2017**, *41*, 1642–1654.
25. Shim, J.; Kostecki, R.; Richardson, T.; Song, X.; Striebel, K. Electrochemical analysis for cycle performance and capacity fading of a lithium-ion battery cycled at elevated temperature. *J. Power Sources* **2002**, *112*, 222–230. [CrossRef]
26. Xiong, R.; Yu, Q.; Shen, W.; Lin, C.; Sun, F. A Sensor Fault Diagnosis Method for a Lithium-Ion Battery Pack in Electric Vehicles. *IEEE Trans. Power Electron.* **2019**, *34*, 9709–9718. [CrossRef]
27. Yamanaka, T.; Takagishi, Y.; Tozuka, Y.; Yamaue, T. Modeling lithium ion battery nail penetration tests and quantitative evaluation of the degree of combustion risk. *J. Power Sources* **2019**, *416*, 132–140. [CrossRef]
28. Singh, G.; Kaur, M.; Singh, B. Detection of epileptic seizure EEG signal using multiscale entropies and complete ensemble empirical mode decomposition. *Wirel. Pers. Commun.* **2021**, *116*, 845–864. [CrossRef]
29. Fang, F.; Yu, S.; Xin, X. Data-Driven-Based Stochastic Robust Optimization for a Virtual Power Plant With Multiple Uncertainties. *IEEE Trans. Power Syst.* **2021**, *37*, 456–466. [CrossRef]
30. Cho, S.; Gao, Z.; Torgeir, M. Model-based fault detection, fault isolation and fault-tolerant control of a blade pitch system in floating wind turbines. *Renew. Energy* **2018**, *120*, 306–321. [CrossRef]
31. Sun, Z.; Wang, Z.; Liu, P.; Zhang, Z.; Chen, Y.; Qu, C. Overview of Fault Diagnosis in New Energy Vehicle Power Battery System. *J. Mech. Eng.* **2021**, *57*, 87–104.
32. Chen, R.; Nolan, A.M.; Lu, J.; Wang, J.; Yu, X.; Mo, Y.; Chen, L.; Huang, X.; Li, H. The thermal stability of lithium solid electrolytes with metallic lithium. *Joule* **2020**, *4*, 812–821. [CrossRef]
33. Vedhanarayanan, B.; Ji, X.; Lakshmi, K.S.; Lin, T.-W. Engineering solid-electrolyte interface from aqueous deep-eutectic solvent to enhance the capacity and lifetime of self-assembled heterostructures of 1T-MoS2/graphene. *Chem. Eng. J.* **2022**, *427*, 130966. [CrossRef]
34. Zhang, Y.; Kitchaev, D.A.; Yang, J.; Chen, T.; Dacek, S.T.; Sarmiento-Pérez, R.A.; Marques, M.A.L.; Peng, H.; Ceder, G.; Perdew, J.P.; et al. Efficient first-principles prediction of solid stability: Towards chemical accuracy. *Npj Comput. Mater.* **2018**, *4*, 9. [CrossRef]

electronics

MDPI

Article

An Integrated Vibration Elimination System with Mechanical-Electrical-Magnetic Coupling Effects for In-Wheel-Motor-Driven Electric Vehicles

Ze Zhao [1,*], Liang Gu [1], Jianyang Wu [2], Xinyang Zhang [3,4] and Haixu Yang [3,4]

1 School of Mechanical Engineering, Beijing Institute of Technology, Beijing 100081, China
2 Beijing Institute of Space Launch Technology, Beijing 100076, China
3 School of Mechanical Engineering, University of Science and Technology Beijing, Beijing 100083, China
4 Shunde Innovation School, University of Science and Technology Beijing, Foshan 528399, China
* Correspondence: zhaoze1990@bit.edu.cn

Abstract: This study aims to improve the vehicle vertical dynamics performance in the sprung and unsprung state for in-wheel-motor-driven electric vehicles (IWMD EVs) while considering the unbalanced electric magnetic force effects. An integrated vibration elimination system (IVES) is developed, containing a dynamic vibration-absorbing structure between the IWM and the suspension. It also includes an active suspension system based on a delay-dependent $H\infty$ controller. Further, a novel frequency-compatible tire (FCT) model is constructed to improve IVES accuracy. The mechanical-electrical-magnetic coupling effects of IWMD EVs are theoretically analyzed. A virtual prototype for the IVES is created by combining the CATIA, ADAMS, and MatLab/Simulink, resulting in a high-fidelity multi-body model, validating the IVES accuracy and practicability. Simulations for the IVES considered three different suspension structure types and time delay considerations were performed. Analyses in frequency and time domains for the simulation results have shown that the root mean square of sprung mass acceleration and the eccentricity are significantly reduced via the IVES, indicating an improvement in ride comfort and IWM vibration suppression.

Keywords: in-wheel-motor; unbalanced electric magnetic force; vertical-longitudinal dynamics; road-tire-rotor force; multi-optimization method; virtual prototype

Citation: Zhao, Z.; Gu, L.; Wu, J.; Zhang, X.; Yang, H. An Integrated Vibration Elimination System with Mechanical-Electrical-Magnetic Coupling Effects for In-Wheel-Motor-Driven Electric Vehicles. *Electronics* **2023**, *12*, 1117. https://doi.org/10.3390/electronics12051117

Academic Editor: Gianpaolo Vitale

Received: 28 January 2023
Revised: 17 February 2023
Accepted: 23 February 2023
Published: 24 February 2023

1. Introduction

Automotive electrification is rapidly expanding worldwide to tackle the challenges such as greenhouse gas emissions and fossil oil depletion. In-wheel-motor-drive electric vehicles (IWMD EVs), in which the drive motor is integrated directly into the wheel, have several benefits, including compactness, controllability, and efficiency. Recently, IWMD EVs started to attract researchers, and are an important research direction for future electric vehicles [1,2]. However, the use of wheel motors increases the unsprung mass, negatively affecting the vehicle dynamic performance, including ride comfort and road holding performance [3,4]. On the other hand, increasing expectations for improved noise-, vibration-, and harshness- (NVH) performance make it more important to deeply investigate correlated characteristics [5,6]. Additionally, IWM vibration might cause motor bearing wear and magnet gap eccentricity [7]. Further, eccentricity can cause an unbalanced electric magnetic force (UEMF) which further distorts the air gap distribution, exacerbating the in-wheel-motor (IWM) vibration, creating a vicious cycle [8] This is known as a typical mechanical-electrical-magnetic coupling system where UEMF has a key role [9,10].

Numerous studies were carried out to mitigate the adverse coupling effect on vehicle dynamics, and can generally be grouped into two categories. The first category includes designing the IWM as a dynamic vibration absorbing structure (DVAS), where the IWM is isolated from the axle by spring and damper elements [11]. The DVAS absorbs the vibration energy transmitted to the IWM by damper element [12]. Further, DVAS is generally divided into "chassis DVAS" and "tire DVAS", which flexibly connects the IWM to the sprung and unsprung masses, respectively [11]. Previous studies have shown that DVAS can partially suppress the IWM vibrations if the parameters of the additional spring dampers are properly selected [13,14]. Furthermore, an active DVAS control method is proposed in [15] through the installation of a controllable linear motor between the IWM stator and the wheel axle, effectively reducing the wheel vibration. Most existing DVAS installed in the IWM are passive vibration absorbers with fixed parameters, which cannot be adjusted to complex and variable road excitations. On the other hand, the active DVAS can achieve better performance but is seldom applied due to its cost and space constraints. In addition, when improving IWM vibrations, DVAS has difficulty optimizing sprung mass vibrations in the 4–8 Hz frequency range of the passenger-sensitive [16].

The second category considers the IWM suspension as a complete system, which utilizes conventional active suspension system (ASS) control algorithms to reduce the negative impacts of the increase in unsprung mass. Examples of such algorithms are explicit model predictive control method [17], fuzzy logic control [18], ceiling damping control [19], optimal sliding mode [20], and $H\infty$ control [21]. Active suspension systems inevitably have time delay, which is one of the most important factors affecting the stability of the system. Among the above control strategies, $H\infty$ control is widely used since it provides increased system robustness for time delay, and time constant perturbation [22]. Sun et al. [23] designed an adaptive robust $H\infty$ controller for electro-hydraulic actuator active suspension, robust to the uncertainty of actuator parameters and its nonlinearity. In [24], the robust $H\infty$ controllers were designed for ER suspension systems with parameter uncertainties. The dynamic bandwidth of the current-variable damper under the fast and slow response action was determined. Li et al. [25] proposed using a robust $H\infty$ control method to improve vehicle performance under load variations. The required vehicle suspension properties such as ride comfort, car handling, and suspension deflection are transformed into a continuous time system with input delay and sector bounded uncertainty. Shao et al. [26] studied the $H\infty$ control design method for an ASS of IWMD EVs while considering the actuator failure, time delay, and disturbances. Jing et al. [27] designed an ASS controller for IWMD EVs, aiming to isolate forces transmitted to the motor bearings and improve ride comfort. However, the ASS is a complex system with multiple parameters, and the mechanical-electrical-magnetic coupling effects will further aggravate this condition. Moreover, due to the time delay and power limitation of the actuator, the adjustable frequency range of the ASS is concentrated in the low and mid frequencies [28]

The DVAS and ASS significantly reduce the vibration of unsprung and sprung vehicle masses, respectively. However, few studies combined the ASS control strategies and DVAS approaches to mitigate the mechanical-electrical-magnetic coupling effects. Liu et al. [29] proposed a two-stage optimal control method to improve vehicle dynamics performance through ASS and DVAS. A linear quadratic regulator controller based on the particle swarm optimization and finite frequency $H\infty$ controller are designed for DVAS and ASS, respectively. However, the UEMF of IWM and the time delay of the ASS were not considered. Wang et al. [30] explored the $H\infty$ control strategy for an ASS system in the IWMD EVs, which reduced the IWM vibration and improved the ride comfort. However the IWM UEMF was not considered. Liu et al. [31] used the particle swarm optimization algorithm to optimize the DVAS parameters, while the alterable-domain-based fuzzy control method was used to control the DVAS actuator force. However, the proposed integration structure was limited by its complexity and was not effectively validated. Since DVAS and ASS have different roles in the IWMD EVs, their combination strategies should be studied to improve motor and suspension performances.

Generally, simplified tire models including only the spring characteristics are used in vehicle vertical control; however, the increase in unsprung mass and the high-frequency excitation generated by the motor directly affect tire working conditions [32]. In particular the increase in the unsprung mass causes an increasingly non-linear relationship between the contact force and the road roughness [33]. Typical tire models include physical and empirical tire models. The brush and ring models are two main forms of physical models; the magic formula tire model is attributed as an empirical tire model. However, brush and magic formula tire models are used for modeling the tangential tire force characteristics, not applicable for calculating vertical tire force [34,35]. The most commonly used models in vertical tire dynamics calculation is ring model, which is typically represented by the rigid ring model (RRM) [36] and the flexible ring model (FRM) [37]. For the former, the residual stiffness is introduced between the contact patch and the rigid ring to represent the static tire stiffness in the vertical directions. However, the tire belt deformation was not considered and the applicable frequency range during the analysis is usually low [36]. Further, the FRM uses a large number of segments interconnected by springs and dampers. The FRM bandwidth is thus up to 150 Hz, corresponding to the first flexible belt bending modes; however, it could rarely respond to low-frequency characteristics [37]. As to the IWMD EVs, there are complex maneuvers including high-frequency (50–100 Hz [38]) vibration of the IWM system, as well as low frequency excitations (under 20 Hz) from the road unevenness. Both groups exert serious influences on the vertical dynamics of the vehicle. Consequently, to improve the accuracy and adaptability, it is necessary to integrate the RRM and FRM.

As described above, the combination of DVAS and ASS should be comprehensively investigated to improve the vehicle vertical performance; the UEMF created by mechanical-electrical-magnetic coupling effects should also be considered. Additionally, an accurate road-tire-IWM model is necessary to simulate the real external excitation of the IWM suspension system. Finally, the scientific contributions of this study are:

1. A novel IVES was developed, containing a practical DVAS equipped between the IWM and the suspension, as well as a robust ASS based on the delay-dependent $H\infty$ controller. The UEMF was considered in this system and the mechanical-electrical-magnetic coupling effects of IWMD EVs were observed.
2. A frequency-compatible tire (FCT) model integrating the RRM and FRM was developed to ensure adaptability to different frequency ranges. It also improved the accuracy of vertical tire forces, which were further inputted as an external excitation to the IVES (instead of the road roughness), further improving the accuracy.
3. A novel virtual prototype was designed, combining CATIA, ADAMS, and MATLAB/Simulink environment to establish a high-fidelity multi-body model for the IVES. Particularly, the IVES structure was developed, aiming to maximize its integration ability and minimize the impact on the original chassis structure.

The remainder of the paper is organized as follows: the IVES mathematical model combined with the UEMF model, vertical vibration model, and driving model were introduced in Section 2. The proposed delay-dependent $H\infty$ controller was outlined in Section 3. In Section 4, performed simulations and validations were described for the IVES in a virtual prototype. Finally, study conclusions are given in Section 5.

2. The Mathematical Model of SIWMS

The IVES mathematical model was developed, containing the UEMF model in driving conditions, an FCT model, and a DVAS-ASS integrated model. The investigated UEMF and FCT model tire force are imported into the DVAS-ASS integrated model as internal and external excitation, respectively.

2.1. The UEMF Model in Driving Conditions

For a switched reluctance motor (SRM), the electromagnetic force can be decomposed into tangential and radial forces according to the electrical angle [39]. The material of stator and rotor is ferromagnetic, the radial force of electromagnetic force is much larger than the tangential force [40,41], and the eccentricity of the rotor and stator of IWM will cause a large deviation of radial force, which acts on the stator of the IWM to produce a violent vibration. From the above analysis, it is evident that the UEM is produced by the coupled action of electromagnetic and mechanical fields and represents the resultant global magnetic force acting on the rotor and stator due to an asymmetric magnetic field distribution in the air gap, The UEMF generation process is shown in Figure 1. In the figure, F_t is the tangential force, F_r is the radial force, F_e is the electromagnetic force.

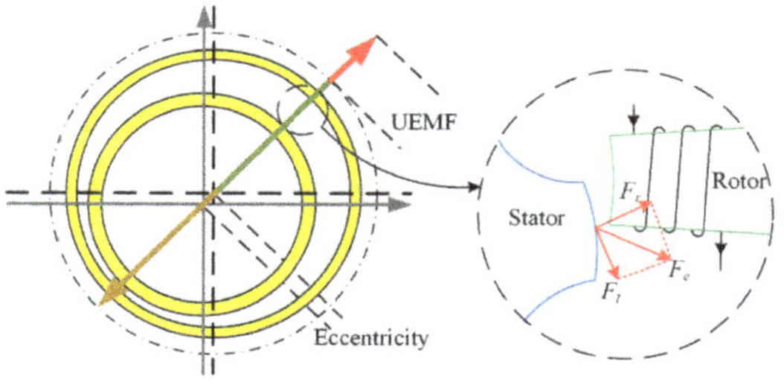

Figure 1. UEMF generated in a SRM.

In this paper, a 5-kW exterior rotor switched reluctance motor (SRM) prototype with 8/6-four phases was used [42]. The magnetic co-energy $W(i, \theta)$ is determined according to the current i and the phase inductance $L(\theta, i)$, where θ is the rotor angle. The first three terms of $L(\theta, i)$ Fourier expansion are given by:

$$L(\theta,i) = L_0(i) + L_1(i)\cos(N_r\theta + \pi) + L_2(i)\cos(2N_r\theta + 2\pi) \tag{1}$$

where L_0, L_1, and L_2 are calculated as:

$$\begin{cases} L_0(i) = \frac{1}{2}\left[\frac{1}{2}(L_a(i) + L_u) + L_m(i)\right] \\ L_1(i) = \frac{1}{2}(L_a(i) - L_u) \\ L_2(i) = \frac{1}{2}\left[\frac{1}{2}(L_a(i) + L_u) - L_m(i)\right] \end{cases} \tag{2}$$

where L_a, L_u, and L_m are inductances at fully-aligned (θ = 30°), unaligned (θ = 0°), and intermediate positions, respectively. These parameters can be fitted with polynomials based on either the finite element analysis or the experiment. Considering the relationship between the flux and the inductance, the k-th phase flux linkage can be derived as:

$$\begin{aligned} \psi(\theta,i_k) &= \textstyle\int_0^{i_k} L(i_k,\theta)\mathrm{d}i_k \\ &= \tfrac{1}{2}\left[\cos^2(N_r\theta) - \cos(N_r\theta)\right]\sum_{n=0}^{N} c_n i^n + \sin^2(N_r\theta)\sum_{n=0}^{N} \mathrm{d}_n i^n \\ &+ \tfrac{1}{2}L_u i_k\left[\cos^2(N_r\theta) + \cos(N_r\theta)\right] \end{aligned} \tag{3}$$

where $c_n = a_{n-1}/n$ and $d_n = b_{n-1}/n$. According to Faraday's law, the phase voltage is:

$$U_k = R_k i_k + \frac{\mathrm{d}\psi_k}{\mathrm{d}t} = R_k i_k + L_k(\theta,i_k)\frac{\mathrm{d}i_k}{\mathrm{d}t} + \omega_t\frac{\partial\psi_k}{\partial\theta} \tag{4}$$

where $_t$ is the angular rotor velocity. The phase current can be written as:

$$i_k = \int \frac{U_k - R_k i_k - \omega_t\frac{\partial\psi_k}{\partial\theta}}{L_k(\theta,i_k)}\mathrm{d}t. \tag{5}$$

For constant phase current i, relationships between the magnetic co-energy $W(i, \theta)$ torque T, and radial force F_r are defined as:

$$T = \left.\frac{\partial W(\theta,i)}{\partial\theta}\right|_{i=\text{ const}}, F_r = \left.\frac{\partial W(\theta,i)}{\partial l_g}\right|_{i=\text{ const}} \tag{6}$$

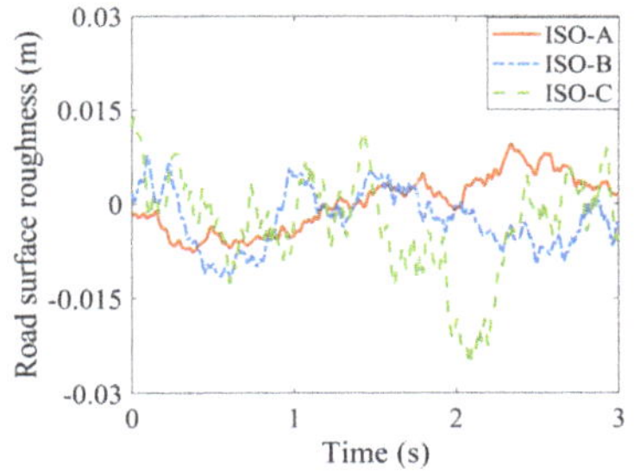

Figure 4. The road displacement spectrums.

When a tire is loaded on the road, a large deformation having a limited contact length occurs near the contact patch, which could be equated to a contact point [47]. Moreover, the vertical residual stiffness of RRM was introduced to determine the overall vertical tire stiffness [36]. The residual stiffness k_{rs} is equal to:

$$k_{rs} = z_r - z_t + q_{V1}{\omega_t}^2 \tag{12}$$

where z_t is the vertical tire displacement, ω_t is the angular tire velocity, q_{V1} is the vertical stiffness correlation coefficient of the tire.

The vertical contact point force F_{tzc} is directly related to the k_{rs}. The tire deformation within the contact patch can be equated to the contact point deformation using a deformation stiffness of k_{tr}, as shown in Figure 5. After neglecting the higher-order terms, a third-order polynomial was used to describe the vertical force due to the residual tire deflection [36]:

$$F_{tzc} = q_{Fzr3}k_{rs}^3 + q_{Fzr2}k_{rs}^2 + q_{Fzr1}k_{rs} \tag{13}$$

where q_{Fzr*} are polynomial coefficients expressed as:

$$\begin{cases} q_{Fzr1} = k_t\frac{q_{Fz1}(1+q_{V2}|\omega_t|)}{k_t-q_{Fz1}(1+q_{V2}|\omega_t|)} \\ q_{Fzr2} = k_t\frac{k_t(k_t\cdot q_{Fz2}+q_{Fzr1}\cdot q_{Fz2})(1+q_{V2}|\omega_t|)}{(k_t-q_{Fz1}(1+q_{V2}|\omega_t|))^2} \\ q_{Fzr3} = 2k_t\frac{q_{Fzr2}\cdot q_{Fz2}(1+q_{V2}|\omega_t|)}{(k_t-q_{Fz1}(1+q_{V2}|\omega_t|))^2} \end{cases} \tag{14}$$

where q_{V2}, q_{Fz1}, and q_{Fz2} are vertical stiffness correlation coefficients of tire and k_t is its sidewall stiffness.

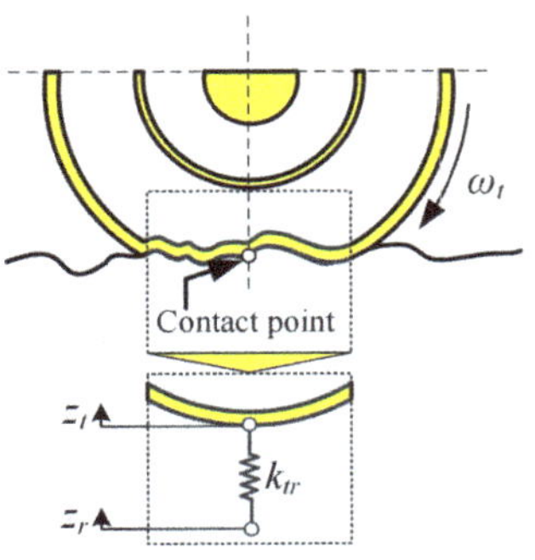

Figure 5. The equivalent process for deformation.

The RRM only considers the contact point deformation; the tire force generated by the tire belt in the contact patch is not calculated—the FRM was used to calculate it. The model includes a finite number of independent radial springs and damping elements evenly distributed in the lower semicircle, as shown in Figure 6 [48]. The total number of the discrete radial elements is denoted as N_{tr} [49].

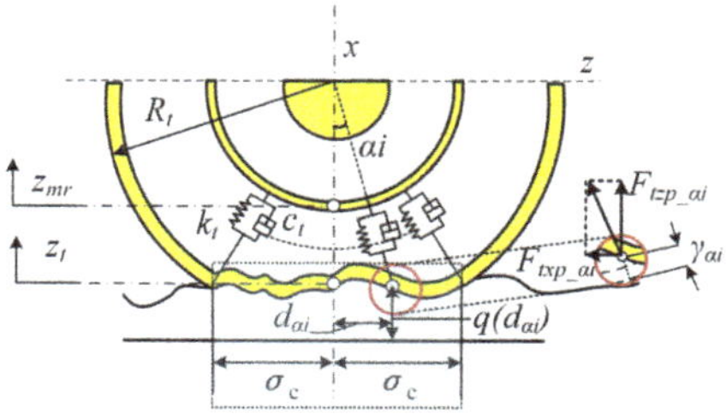

Figure 6. The illustrative representation of the FRM.

In Figure 6, σ_c is equal to the half of the contact patch, z_{mr} is the vertical coordinate of the rotor center, z_t is the vertical contact point displacement, c_t denotes the tire damping, and αi represents the angle between the arbitrarily chosen element and the vertical, ranging from 1 to N_{tr}.

For the positive x-axis, the subscript i indicates the element sequence number ranging from $-\arcsin(\sigma_c / R_t)$ to $+\arcsin(\sigma_c / R_t)$. Further, $\gamma_{\alpha i}$, $d_{\alpha i}$, and $q(d_{\alpha i})$ are the radial deformation, driving distance, and the element road elevation, respectively. As a result of tire deformation, the vertical component $F_{tzp\_\alpha i}$ of the radial spring and damping element forces are:

$$F_{tzp\_\alpha i} = \begin{cases} (\gamma_{\alpha i} k_{trd} + \dot{\gamma}_{\alpha i} c_{trd}) \cos \alpha i & \gamma_{\alpha i} > 0 \\ \dot{\gamma}_{\alpha i} c_{trd} \cos \alpha i & \gamma_{\alpha i} \leq 0 \end{cases} \quad (15)$$

The deformation and deformation rate of a certain radial element can be approximately calculated using:

$$\begin{cases} \alpha i = -\arcsin(\sigma_c / R_t) + 2i \arcsin(\sigma_c / R_t) / N_{tr} \\ d_{\alpha i} = x_t + \sqrt{(R_t{}^2 - \sigma_c{}^2)} \cdot \tan \alpha i \\ (R_t - \gamma_{\alpha i}) \cos \alpha i + q(dis_{\alpha i}) = R_t + z_{mr} + z_t \end{cases} \quad (16)$$

where x_t is the longitudinal displacement of the tire. The vertical tire force component F_{tzp} caused by tire deformation at the rotor center can be obtained by summing up the force components of each spring and damping element:

$$F_{tzp} = \sum_{i=1}^{N_{tr}} F_{txp\_\alpha i}. \quad (17)$$

In general, the vertical tire force calculated through the FTC model under road roughness excitation is:

$$F_{tz} = F_{tzc} + F_{tzp} \quad (18)$$

Finally, F_{tz} is further employed as an external excitation to the IVES (instead of the road roughness).

2.3. DVAS-ASS Integrated Model

The quarter models of three types of suspension structures were created in this section Since the sprung mass distribution coefficients in most modern vehicles are designed to be close to 1. Furthermore, in the suspension structure of the IWMD EVs configured with ASS, the four wheels are independent of each other in vertical motion. Therefore, the quarter model is representative when studying the vertical motion of the vehicle [50]. The sprung and unsprung mass are included, as well as the suspension and tires.

Figure 7a shows a passive-suspension system in which the IWM stator and rotor are rigidly connected to the axle and hub, respectively. Therefore, the IWM becomes the

The dynamic coupling relationships were investigated assuming the ISO-B road at a speed of 40 m/s. The drive lasted for 3 s and the Fast Fourier Transform (FFT) was used to derive spectral features. For observation, 1–2 s data points were selected for further statistical analysis. The responses of sprung mass acceleration (SMA) and eccentricity (ECC) with and without the UEMF under the pavement excitation are shown in Figure 9.

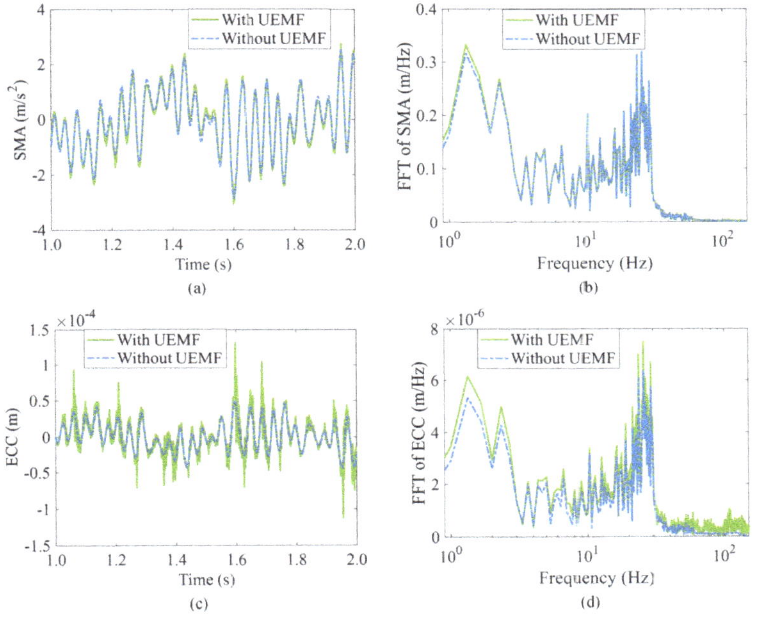

Figure 9. Coupling effects between the SMA, the ECC, and the UEMF: (**a**) the SMA time response; (**b**) the SMA frequency response; (**c**) the ECC time response; (**d**) the ECC frequency response.

The time-response results have shown that the SMA and ECC are larger when considering the UEMF effects. The same trend can be deduced from the frequency response—the effect of UEMF on the ECC at high frequencies is particularly pronounced.

The eccentricity between the stator and rotor was further exacerbated by the UEMF generated by the multi-field coupling effect. Such eccentricity, in turn, furthers the UEMF, which is complementary to it. This vicious cycle exacerbates motor vibration, shortening the IWM service life and reducing vehicle comfort. Therefore, it is worthwhile to investigate the issue of how to restrict the multi-field coupling effects.

3.2. Active Suspension Control

An $H\infty$ control scheme with strong robustness was used to design the active suspension controllers. The controllers were needed to ensure excellent control effect of the suspension system when given complicated electromagnetic force excitation and unmodeled dynamic perturbation. According to Equation (21), the following set of state variables was selected:

$$\begin{aligned}&\mathbf{x}(t) = \\ &[\dot{z}_s(t) \quad \dot{z}_{mr}(t) \quad \dot{z}_{ms}(t) \quad \dot{z}_{ax}(t) \quad \dot{z}_t(t)\ldots \\ &z_{ax}(t) - z_s(t) \quad z_{ms}(t) - z_{mr}(t) \quad z_{ax}(t) - z_{ms}(t) \quad z_{ax}(t) - z_t(t)^T]\end{aligned} \tag{22}$$

The dynamic equations of the ASS with the DVAS system can be written using the following state-space form:

$$\dot{\mathbf{x}}(t) = \mathbf{A}\mathbf{x}(t) + \mathbf{B}_w\mathbf{w}(t) + \mathbf{B}_u\mathbf{u}(t) \tag{23}$$

where:

$$\mathbf{A} = \begin{bmatrix} -c_s m_s^{-1} & 0 & 0 & c_s m_s^{-1} & 0 & k_s m_s^{-1} & 0 & 0 & 0 \\ 0 & 0 & 0 & 0 & 0 & 0 & k_b m_{mr}{}^{-1} & 0 & 0 \\ 0 & 0 & -c_d m_{ms}{}^{-1} & c_d m_{ms}{}^{-1} & 0 & 0 & -k_b m_{ms}{}^{-1} & k_d m_{ms}{}^{-1} & 0 \\ c_s m_{ax}^{-1} & 0 & c_d m_{ax}{}^{-1} & -(c_d + c_s + c_t) m_{ax}^{-1} & c_t m_{ax}^{-1} & -k_s m_{ax}^{-1} & 0 & -k_d m_{ax}^{-1} & -k_t m_{ax}^{-1} \\ 0 & 0 & 0 & c_t m_t^{-1} & c_t m_t^{-1} & 0 & 0 & 0 & k_t m_t^{-1} \\ -1 & 0 & 0 & 1 & 0 & 0 & 0 & 0 & 0 \\ 0 & -1 & 1 & 0 & 0 & 0 & 0 & 0 & 0 \\ 0 & 0 & -1 & 1 & 0 & 0 & 0 & 0 & 0 \\ 0 & 0 & 0 & 1 & -1 & 0 & 0 & 0 & 0 \end{bmatrix}$$

$$\mathbf{B}_w = \begin{bmatrix} 0 & 0 & 0 & 0 & -m_t^{-1} & 0 & 0 & 0 & 0 \\ 0 & -m_{mr}^{-1} & m_{ms}^{-1} & 0 & 0 & 0 & 0 & 0 & 0 \end{bmatrix}^T$$
$$\mathbf{B}_u = \begin{bmatrix} -m_s^{-1} & 0 & 0 & m_{ax}^{-1} & 0 & 0 & 0 & 0 & 0 \end{bmatrix}^T$$
$$\mathbf{u}(t) = \begin{bmatrix} u(t) \end{bmatrix} \mathbf{w} = \begin{bmatrix} F_{tz} & F_u \end{bmatrix}^T$$

Generally, the suspension performance is evaluated based on key factors such as ride comfort, suspension working space, and road-holding ability. Additionally, as for the IWMD EVs, the IWM UEMF is an essential performance factor. Thus, four system responses are investigated: SMA: $\ddot{z}_s$, rattle space (RS): $(z_s - z_{ax})$, tire deformation (TD): $(z_{ax} - z_t)$, and ECC: $(z_{ms} - z_{mr})$. Among these, the first three are used for evaluating the suspension performance, whereas the last one reflects the IWM vibration level.

Focusing on ride comfort improvement and the IWM working environment, the SMA and eccentricity should be minimized, while the other two factors are strict constraints that must be satisfied. Therefore, minimizing the disturbance transfer function norm (F_u and F_{tz}) to the control output ($\ddot{z}_s$ and $(z_{ms} - z_{mr})$) is our main goal. The constraints as follows:

$$\begin{aligned} |z_{ax} - z_s| &\leq z_{1\,\max} \\ |z_{ax} - z_t| &\leq z_{2\,\max} \\ |u(t)| &\leq u_{\max} \end{aligned} \tag{24}$$

where z_{1max} and z_{2max} are the maximum suspension and tire deflection, respectively. u_{max} is the maximum possible actuator output force.

Based on the above-presented conditions, the performance control output and constrained control output are defined next:

$$\mathbf{z}_1 = \begin{bmatrix} \ddot{z}_s & z_{ms} - z_{mr} \end{bmatrix}^T \tag{25}$$

$$\mathbf{z}_2 = \begin{bmatrix} \frac{z_{ax}(t) - z_s(t)}{z_{1\,\max}} & \frac{z_{ax}(t) - z_t(t)}{z_{2\,\max}} & \frac{u(t)}{u_{\max}} \end{bmatrix}^T \tag{26}$$

The state-space equations of the ASS and DVAS for the IWMD EVs can be described using:

$$\begin{aligned} \dot{\mathbf{x}}(t) &= \mathbf{A}\mathbf{x}(t) + \mathbf{B}_w\mathbf{w}(t) + \mathbf{B}_u\mathbf{u}(t) \\ \mathbf{z}_1 &= \mathbf{C}_1\mathbf{x}(t) + \mathbf{D}_1\mathbf{u}(t) \\ \mathbf{z}_2 &= \mathbf{C}_2\mathbf{x}(t) + \mathbf{D}_2\mathbf{u}(t) \end{aligned} \tag{27}$$

where:

$$\mathbf{C}_1 = \begin{bmatrix} -c_s m_s^{-1} & 0 & 0 & c_s m_s^{-1} & 0 & k_s m_s^{-1} & 0 & 0 & 0 \\ 0 & 0 & 0 & 0 & 0 & 0 & 1 & 0 & 0 \end{bmatrix}$$
$$\mathbf{D}_1 = \begin{bmatrix} -m_s^{-1} & 0 \end{bmatrix}^T$$
$$\mathbf{C}_2 = \begin{bmatrix} 0 & 0 & 0 & 0 & 0 & z_{1\max}^{-1} & 0 & 0 & 0 \\ 0 & 0 & 0 & 0 & 0 & 0 & 0 & 0 & z_{2\max}^{-1} \\ 0 & 0 & 0 & 0 & 0 & 0 & 0 & 0 & 0 \end{bmatrix}$$
$$\mathbf{D}_2 = \begin{bmatrix} 0 & 0 & u_{\max}^{-1} \end{bmatrix}^T$$

The ASS inevitably has operating time delays, with the majority being the response time delay of the actuator. Hence, time delay should not be neglected when designing the active suspension system.

Assuming that the time-varying delay $\tau(t)$ in a closed-loop controlled suspension system satisfies the condition:

$$0 < \tau(t) < \tau_{\max} \tag{28}$$

where τ_{max} represents the upper bound of the delay time. Considering the delay time, the output feedback controller can be described as:

$$\mathbf{u}(t) = \mathbf{u}(t-\tau) = \mathbf{K}_t\mathbf{x}(t-\tau) \tag{29}$$

where K_t is the output controller gain. Next, the ASS that considers the time delay in the control system can be expressed via:

$$\begin{aligned} \dot{\mathbf{x}}(t) &= \mathbf{A}\mathbf{x}(t) + \mathbf{B}_w\mathbf{w}(t) + \mathbf{B}_u\mathbf{K}_t\mathbf{x}(t-\tau) \\ \mathbf{z}_1 &= \mathbf{C}_1\mathbf{x}(t) + \mathbf{D}_1\mathbf{K}_t\mathbf{x}(t-\tau) \\ \mathbf{z}_2 &= \mathbf{C}_2\mathbf{x}(t) + \mathbf{D}_2\mathbf{K}_t\mathbf{x}(t-\tau) \end{aligned} \tag{30}$$

In this paper, the aim is to design a state feedback controller $\mathbf{u}(t) = \mathbf{K}_t\mathbf{x}(t-\tau)$ that will meet the following conditions:

(1) Without external perturbations, the closed-loop system shown in Equation (30) is asymptotically stable.
(2) The performance $\|\mathbf{z_1}(t)\|_\infty \leq \gamma w(t)_\infty$ is minimized subject to Equation (30), where γ is the bounded $H\infty$ parametrization.
(3) The time-domain constraint $|\mathbf{z_2}(t)| \leq 1$ must be satisfied.

Considering the time delay, the $H\infty$ controller was designed through the following steps:

Given positive scalars $\gamma > 0$, $\eta > 0$, and $\rho > 0$, for any time delay t satisfying $0 < \tau(t) < \tau_{\max}$, the system shown in Equation (26) with the controller from Equation (29) is asymptotically stable with $w(t) = 0$. In that case, it also satisfies the performance described in Equations (25) and (26) for $w(t) \in [0, \infty)$, given that symmetric matrices $\mathbf{P} > 0$, $\mathbf{Q} > 0$, and $\mathbf{R} > 0$ exist, and a general matrix $\mathbf{K}_t$ satisfying the following linear matrix equations (LMIs) is:

$$\begin{bmatrix} \mathbf{\Omega} & \mathbf{P}\mathbf{B}_u\mathbf{K}_t & 0 & \mathbf{P}\mathbf{B}_w & \sqrt{\tau_{\max}}\mathbf{A}^T\mathbf{P} & \mathbf{C}_1^T \\ * & -\tau_{\max}{}^{-1}\mathbf{Q} & 0 & 0 & \sqrt{\tau_{\max}}(\mathbf{B}_u\mathbf{K}_t)^T\mathbf{P} & (\mathbf{D}_1\mathbf{K}_t)^T \\ * & * & -\mathbf{R} & 0 & 0 & 0 \\ * & * & * & -\gamma^2\mathbf{I} & \sqrt{\tau_{\max}}\mathbf{B}_w{}^T\mathbf{P} & 0 \\ * & * & * & * & -2\eta\mathbf{P} + \eta^2\mathbf{Q} & 0 \\ * & * & * & * & * & -\mathbf{I} \end{bmatrix} < 0 \tag{31}$$

$$\begin{bmatrix} -\mathbf{P} & (\mathbf{C}_2 + \mathbf{D}_2\mathbf{K}_t)^T \\ * & -\frac{1}{\rho}\mathbf{I} \end{bmatrix} < 0 \tag{32}$$

where: $\mathbf{\Omega} = sys[\mathbf{P}(\mathbf{A} + \mathbf{B}_w\mathbf{K}_t)] + \mathbf{R}$

For comparison, the $H\infty$ controller without considering time delay is obtained as follows:

For a given scalar $\rho_c > 0$, if $\gamma_c > 0$ and a positive definite symmetric matrix $\mathbf{X}$ exists, making the following inequalities feasible under LMIs:

$$\begin{bmatrix} (\mathbf{AX}+\mathbf{B}_2\mathbf{W})^T+(\mathbf{AX}+\mathbf{B}_2\mathbf{W}) & \mathbf{B}_1 & (\mathbf{C}_1\mathbf{X}+\mathbf{D}_1\mathbf{W})^T \\ \mathbf{B}_1{}^T\mathbf{P} & -\gamma_c\mathbf{I} & \mathbf{0} \\ \mathbf{C}_1\mathbf{X}+\mathbf{D}_1\mathbf{W} & 0 & -\gamma_c\mathbf{I} \end{bmatrix} < 0 \quad (33)$$

$$\begin{bmatrix} -\mathbf{X} & \mathbf{C}_2\mathbf{X}+\mathbf{D}_2\mathbf{W} \\ * & -\frac{1}{\rho_c}\mathbf{I} \end{bmatrix} < 0 \quad (34)$$

The state feedback gain $\mathbf{K}_t$ can be given by $\mathbf{K}_t = \mathbf{WX}^{-1}$

Proof. Lemma 1 [53]: For any matrix $\mathbf{X} > 0$, matrices (or scalars) $\mathbf{M}$ and $\mathbf{N}$ with compatible dimensions, the following inequality holds:

$$-2\mathbf{M}^T\mathbf{N} \leq \mathbf{M}^T\mathbf{X}^{-1}\mathbf{M}+\mathbf{N}^T\mathbf{XN} \quad (35)$$

Since $x(t)-x(t-\tau)-\int_{t-\tau(t)}^{t}\dot{x}(\alpha)d\alpha = 0$, by adding it to Equation (30):

$$\dot{\mathbf{x}}(t) = (\mathbf{A}+\mathbf{B_2K})\mathbf{x}(t)+\mathbf{B_1w}(t)-\mathbf{B_2K}_t\int_{t-\tau(t)}^{t}\dot{\mathbf{x}}(t)\mathrm{d}t \quad (36)$$

Selecting the Lyapunov function shown below to analyze the system stability we obtain:

$$\mathbf{V}(t) = \mathbf{V}_1(t)+\mathbf{V}_2(t)+\mathbf{V}_3(t) \quad (37)$$

where:

$$\mathbf{V}_1(t) = \mathbf{x}^T(t)\mathbf{Px}(t)\mathbf{V}_2(t) = \int_{t-\tau_{\max}}^{t}\mathbf{x}^T(s)\mathbf{Rx}(s)d_s\mathbf{V}_3(t) = \int_{-\tau_{\max}}^{0}\int_{t+\theta}^{t}\mathbf{x}^T(s)\mathbf{Qx}(s)d_sd_\theta \quad (38)$$

with $\mathbf{P}$, $\mathbf{Q}$, and $\mathbf{R}$ being symmetric positive definite matrices. The derivative of $\mathbf{V}_1(t)$ is shown next:

$$\begin{aligned} \dot{\mathbf{V}}_1(t) &= \mathbf{x}^T(t)\left\{\mathbf{P}[\mathbf{A}+\mathbf{B_2K}_t]+[\mathbf{A}+\mathbf{B_2K}_t]^T\mathbf{P}\right\}\mathbf{x}(t) \\ &\quad +\mathbf{x}^T(t)\mathbf{PB_1w}(t)+\mathbf{w}^T(t)\mathbf{B}_1^T\mathbf{Px}(t)-2\mathbf{x}^T(t)\mathbf{PB_2K}_t\int_{t-\tau_{\max}}^{t}\dot{\mathbf{x}}(t)\mathrm{d}t \end{aligned} \quad (39)$$

Using **Lemma 1**, that gets:

$$\begin{aligned} \dot{\mathbf{V}}_1(t) &\leq \mathbf{x}^T(t)\left\{\mathbf{P}[\mathbf{A}+\mathbf{B_2K}_t]+[\mathbf{A}+\mathbf{B_2K}_t]^T\mathbf{P}\right\}\mathbf{x}(t) \\ &\quad +\mathbf{x}^T(t)\mathbf{PB_1w}(t)+\mathbf{w}^T(t)\mathbf{B}_1^T\mathbf{Px}(t)+ \\ &\quad \tau_{\max}\mathbf{x}^T(t)\mathbf{PB_2K_tQ}^{-1}(\mathbf{PB_2K}_t)^T\mathbf{x}(t)+\int_{t-\tau_{\max}}^{t}\dot{\mathbf{x}}(t)^T\mathbf{Q}\dot{\mathbf{x}}(t)\mathrm{d}t \end{aligned} \quad (40)$$

Similarly, derivatives of $\mathbf{V}_2(t)$ and $\mathbf{V}_3(t)$ are expressed as:

$$\begin{aligned} \dot{\mathbf{V}}_2(t) &= \mathbf{x}^T(t)\mathbf{Rx}(t)-\mathbf{x}^T(t-\tau_{\max})\mathbf{Rx}(t-\tau_{\max}) \\ \dot{\mathbf{V}}_3(t) &= \tau\mathbf{x}^T(t)\mathbf{Q}\dot{\mathbf{x}}(t)-\int_{t-\tau_{\max}}^{t}\dot{\mathbf{x}}(s)^T\mathbf{Q}\dot{\mathbf{x}}(s)\mathrm{d}s \end{aligned} \quad (41)$$

Further, the $\mathbf{V}(t)$ derivative is shown as follows:

$$\begin{aligned} &\dot{\mathbf{V}}(t) = \dot{\mathbf{V}}_1(t)+\dot{\mathbf{V}}_2(t)+\dot{\mathbf{V}}_3(t) \\ &\leq \mathbf{x}^T(t)\left\{\mathbf{P}[\mathbf{A}+\mathbf{B_2K}_t]+[\mathbf{A}+\mathbf{B_2K}_t]^T\mathbf{P}+\tau_{\max}(t)\mathbf{PB_2K_tQ}^{-1}(\mathbf{PB_2K})^T+\mathbf{R}\right\}\mathbf{x}(t) \\ &+\mathbf{x}^T(t)\mathbf{PBP}_1\mathbf{w}(t)+\mathbf{w}^T(t)\mathbf{B}_1^T\mathbf{Px}(t)-\mathbf{x}^T(t-\tau_{\max})\mathbf{Rx}(t-\tau_{\max})+\tau_{\max}\dot{\mathbf{x}}(t)^T\mathbf{Q}\dot{\mathbf{x}}(t) \end{aligned} \quad (42)$$

Defining $\xi^T(t) = \left[\mathbf{x}^T(t), \mathbf{x}^T(t-\tau), \mathbf{x}^T(t-\tau_{\max}), \mathbf{w}^T(t)\right]$, thus

$$\dot{\mathbf{V}}(t) \leq \xi^T(t)\Xi\xi(t) + \tau_{\max}\dot{\mathbf{x}}(t)^T\mathbf{Q}\dot{\mathbf{x}}(t) \tag{43}$$

where:

$$\Xi = \begin{bmatrix} \mathbf{\Theta} & 0 & 0 & \mathbf{PB}_1 \\ * & 0 & 0 & 0 \\ * & * & -\mathbf{R} & 0 \\ * & * & * & 0 \end{bmatrix}$$

$\mathbf{\Theta} = sys[\mathbf{P}(\mathbf{A} + \mathbf{B_2K}_t)] + \mathbf{R} + \tau_{\max}\mathbf{PB_2K}_t\mathbf{Q}^{-1}(\mathbf{PB_2K}_t)^T$

The system $H\infty$ performance can be described as:

$$\mathbf{z_1}^T(t)\mathbf{z_1}(t) \leq \gamma^2\mathbf{w}^T(t)\mathbf{w}(t) \tag{44}$$

By adding to Equation (43), it is possible to obtain:

$$\dot{\mathbf{V}}(t) + \mathbf{z}_1^T(t)\mathbf{z}_1(t) - \gamma^2\mathbf{w}^T(t)\mathbf{w}(t)\}\xi^T(t)\mathbf{\Pi}_1\xi(t) \tag{45}$$

where:

$$\mathbf{\Pi}_1 = \begin{bmatrix} \Omega & \mathbf{PB_2K}_t & 0 & \mathbf{PB}_1 & \sqrt{\tau_{\max}}^T\mathbf{A}^T\mathbf{P} & \mathbf{C}_1^T \\ * & -\tau_{\max}{}^{-1}\mathbf{Q} & 0 & 0 & \sqrt{\tau_{\max}}(\mathbf{B_2K}_t)^T\mathbf{P} & (\mathbf{D}_1\mathbf{K}_t)^T \\ * & * & -\mathbf{R} & 0 & 0 & 0 \\ * & * & * & -\gamma^2\mathbf{I} & \sqrt{\tau_{\max}}\mathbf{B}_1{}^T\mathbf{P} & 0 \\ * & * & * & * & \mathbf{PR}^{-1}\mathbf{P} & 0 \\ * & * & * & * & * & -\mathbf{I} \end{bmatrix}$$

For any positive scalar η, the following expression is true:

$$\mathbf{PR}^{-1}\mathbf{P} \leq -2\eta\mathbf{P} + \eta^2\mathbf{R} \tag{46}$$

By replacing the item $\mathbf{PR}^{-1}\mathbf{P}$ in $\prod_1$ with $-2\eta\mathbf{P} + \eta^2\mathbf{R}$, we get $\prod_2$. Moreover, it is true that $\prod_1 \leq \prod_2$, where:

$$\prod_2 = \begin{bmatrix} \Omega & \mathbf{PB_2K}_t & 0 & \mathbf{PB}_1 & \sqrt{\tau_{\max}}\mathbf{A}^T\mathbf{P} & \mathbf{C}_1^T \\ * & -\tau_{\max}{}^{-1}\mathbf{Q} & 0 & 0 & \sqrt{\tau_{\max}}(\mathbf{B_2K}_t)^T\mathbf{P} & (\mathbf{D}_1\mathbf{K}_t)^T \\ * & * & -\mathbf{R} & 0 & 0 & 0 \\ * & * & * & -\gamma^2\mathbf{I} & \sqrt{\tau_{\max}}\mathbf{B}_1{}^T\mathbf{P} & 0 \\ * & * & * & * & -2\eta\mathbf{P} + \eta^2\mathbf{Q} & 0 \\ * & * & * & * & * & -\mathbf{I} \end{bmatrix}$$

If the matrix inequality $\prod_2 < 0$, that $\mathbf{z_1}^T(t)\mathbf{z_1}(t) \leq \gamma^2\mathbf{w}^T(t)\mathbf{w}(t)$.

From inequality Equation (45), it can be derived that $\dot{\mathbf{V}}(t) + \mathbf{z}_1^T(t)\mathbf{z}_1(t) - \gamma^2\mathbf{w}^T(t)\mathbf{w}(t) \leq 0$. Integrating both sides of the inequality above from zero to any $t > 0$, the following is obtained:

$$\mathbf{V}(t) - \mathbf{V}(0) + \int_0^t \|\mathbf{z}_1(t)\|_2^2 dt - \gamma^2 \int_0^t \|\mathbf{w}(t)\|_2^2 dt \leq 0 \tag{47}$$

Considering the existence of $\int_0^t \mathbf{w}(t)_2^2 dt \leq w_{\max}$, the above inequality can be further rewritten:

$$\mathbf{V}(t) + \int_0^t \|\mathbf{z}_1(t)\|_2^2 \mathrm{d}t \leq \gamma^2 w_{\max} + \mathbf{V}(0) \tag{48}$$

Definition $\gamma^2 w_{\max} + \mathbf{V}(0) = \rho$, where w_{max} is upper perturbation energy bound, can be introduced with a value equal to $w_{\max} = [\rho - \mathbf{V}(0)]/\gamma^2$, Therefore, it can be introduced under $t = 0$ and within a given perturbation suppression regime γ. In combination with $\int_0^t \|\mathbf{z}_1(t)\|_2^2 dt > 0$, synthetically, the following inequalities hold:

$$\begin{aligned} &\rho\mathbf{x}^T[\mathbf{C}_2 + \mathbf{D}_2\mathbf{K}_t]^T[\mathbf{C}_2 + \mathbf{D}_2\mathbf{K}_t]\mathbf{x} \\ &= \rho[\mathbf{C}_2\mathbf{x} + \mathbf{D}_2\mathbf{u}]^T[\mathbf{C}_2\mathbf{x} + \mathbf{D}_2\mathbf{u}] < x^T\mathbf{P}x < \rho \end{aligned} \tag{49}$$

This causes matrix inequality $-\mathbf{P} + \rho[\mathbf{C}_2 + \mathbf{D}_2\mathbf{K}_t]^T[\mathbf{C}_2 + \mathbf{D}_2\mathbf{K}_t] < 0$. Using the Schur complementary theorem [54], the equation is converted into:

$$\begin{bmatrix} -\mathbf{P} & (\mathbf{C}_2 + \mathbf{D}_2\mathbf{K}_t)^T \\ * & -\frac{1}{\rho}\mathbf{I} \end{bmatrix} < 0 \tag{50}$$

For a case without a time delay controller, the proof can be found in [55].

The proof is completed. □

4. Simulation and Analyses

In this section, the influence of the ASS and DVAS on the IWMD EVs performance, as well as the effectiveness of the delay-dependent $H\infty$ active suspension controller is illustrated.

4.1. Performance Comparison of Different Structures

Four system responses, the SMA, the RS, the TD, and the ECC were investigated in Section 3. For the maximum time-delay τ_{max}= 40 ms, Equations (31) and (32) are LMIs of variables **P**, **Q**, and **R**. Thus, the program can be solved via the solver-mincx within the LMI toolbox. Through LMI algorithms, the control gain matrix K_t of the ASS controller (considering the time delay) is obtained with the minimum guaranteed closed-loop $H\infty$ performance index γ_t = 8.82. Such output implies that for any time-delay satisfying $0 \leq \tau \leq 40$ ms, the controller can stabilize the system with the $H\infty$ performance.

$$K_t = [-18480.29 \;\; -1392.86 \;\; -587.89 \;\; -1687.74 \;\; -261.17 \;\; 39039.25 \;\; 12710.53 \;\; 6087.22 \;\; -88442.36]$$

Similarly, a conventional $H\infty$ controller without considering the control time delay can be derived. The gain matrix $\mathbf{K}_c$ of the Controller with the minimum guaranteed closed-loop $H\infty$ performance index γ_c = 6.28 is obtained as:

$$K_c = [-19081.62 \; -1443.61 \;\; -609.36 \;\; -1746.70 \;\; -276.04 \;\; 40139.99 \;\; 13055.35 \;\; 6380.52 \; -92866.90]$$

After the method proposed in this study to solve the multi-objective control problem of the ASS, its effectiveness is illustrated through four different cases:

1. Case 1—equipped with the IWM and passive suspension;
2. Case 2—equipped with the DVAS in IWM and passive suspension;
3. Case 3—equipped with the DVAS in IWM and the ASS using the control gain matrix K_c;
4. Case 4—equipped with the DVAS in IWM and the ASS using the control gain matrix K_t.

Figure 10 shows the controller forces applied on the ASS in cases 3 and 4. Both controller forces are less than umax. Further, the controller force in Case 4 is smaller than that in Case 3.

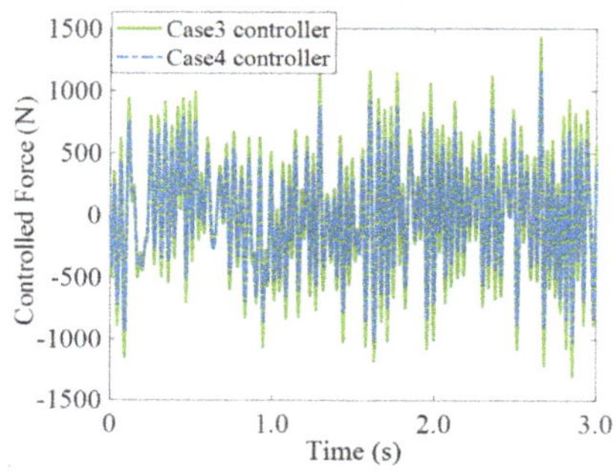

Figure 10. Controller forces in Case 3 and Case 4.

The response comparisons of the two-output time and frequency domains of the SMA and ECC are depicted in Figure 11.

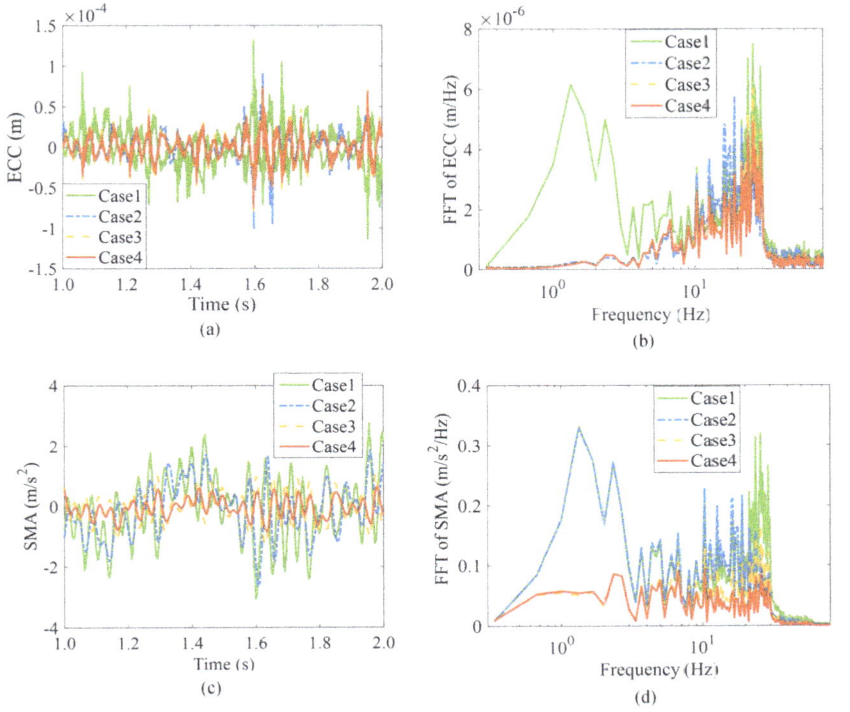

Figure 11. Responses of four cases under the road excitation: (**a**) SMA time response; (**b**) SMA frequency response; (**c**) ECC time response; (**d**) ECC frequency response.

The detailed simulation results are available in Figure 11. The responses of the SMA in time and frequency domains are shown in Figure 11a,b, respectively. As shown in Figure 11a, the SMA is significantly smaller in cases 3 and 4 compared to cases 1 and 2, mostly due to the introduction of the ASS. Figure 11b shows that Case 2 can reduce the amplitude of high-frequency compared to Case 1; however, there is no optimization effect at the human-sensitive frequency of 4–8 Hz. Cases 3 and 4 can significantly reduce the amplitude in both high and low frequencies (especially compared to Case 1), especially in the 4–8 Hz range.

The ECC responses in time and frequency domains are shown in Figure 11c,d, respectively. As shown in Figure 11c, in contrast with the SMA optimization effect, all three cases had reduced amplitude in the wide frequency domain concerning Case 1. The proposed control methods of the active suspension are effective for the IWM vibration. The SMA is smaller in cases 3 and 4 compared to in Case 2 in the frequencies between 3 Hz and 12 Hz, indicating that the proposed control method for the ASS can further reduce the impulse force acting on the IWM.

Simulation results for all four cases are provided in Table 2.

Table 2. Root Mean Square (RMS) optimization results.

RMS	SMA (m/s^2)-Decrement (%)	ECC (10^{-5} m)-Decrement (%)	RS (10^{-3} m)-Decrement (%)	TD (10^{-4} m)-Decrement (%)
Case 1	0.9941-	2.365-	4.229-	4.328-
Case 2	0.8001-↓19.5%	1.673-↓29.2%	4.221-↓0.2%	4.325-↓0.07%
Case 3	0.4148-↓58.2%	1.625-↓31.2%	4.223-↓0.1%	4.322-↓0.1%
Case 4	0.3217-↓67.6%	1.444-↓38.9%	4.222-↓0.2%	4.318-↓0.2%

As can be seen from Table 2, both the DVAS and the ASS can reduce the four-parameter indices, while having different roles in improving comfort and handling. The DVAS has an important role in reducing eccentricity, while the ASS has a decisive effect on spring acceleration. For the RS and TD, the optimization effect is not significant compared to the conventional suspension since they are not optimized in a targeted way. At the same time, the H-infinity algorithm, which considers the time delay, displayed a performance improvement compared to the algorithm not considering it.

Table 3 compares control methods in terms of simulation results considering the presence of time delay in the active suspension actuator. It also indicates that, with the increase in time delay, both Case 3 evaluation indexes deteriorated to different degrees; the SMA deterioration is obvious. In contrast, Case 4 adapted easier and the degree of deterioration was not significant—it was within acceptable limits.

Table 3. The RMS optimization results for both cases.

Case	Time Delay	SMA (m/s^2) -Deteriorate	ECC (10^{-5} m) -Deteriorate
Case 3	$\tau = 0$ ms	0.4848	1.625
	$\tau = 15$ ms	0.6279-↑29.5%	1.667-↑2.5%
controller	$\tau = 30$ ms	0.7634-↑57.5%	1.694-↑4.4%
Case 4	$\tau = 0$ ms	0.3217	1.444
	$\tau = 15$ ms	0.3467-↑7.7%	1.449-↑0.3%
controller	$\tau = 30$ ms	0.3545-↑10.1%	1.447-↑0.2%

4.2. Virtual Prototype Validation for the IVES

A virtual prototype (VP) enabled the proof testing before assembling the hardware, reducing both the manufacturing cost and time. The design possibilities of the IVES can be explored through the VP, and the study of tradeoffs between IVES component sizes becomes feasible. In this part, an VP, combining CATIA, ADAMS, and MatLab/Simulink environment was constructed to establish a high-fidelity multi-body model for the IVES.

The build process used to create the VP model is shown in Figure 12. Firstly, a complete vehicle model was parametrically modeled in CATIA, using the vehicle model parameters obtained from an actual IWMD EV. Then, the IVES model was imported into the ADAMS and the constraints of each component were established. The collision model of each component was established, the corresponding material was defined, and the loads and drives were added. Finally, the driving torque and controller force of the ASS was taken from the IWM system modules and ASS controller modules available in MATLAB/Simulink, respectively. For the UEMF, the vertical force F_u was applied to the IWM rotor and stator surfaces.

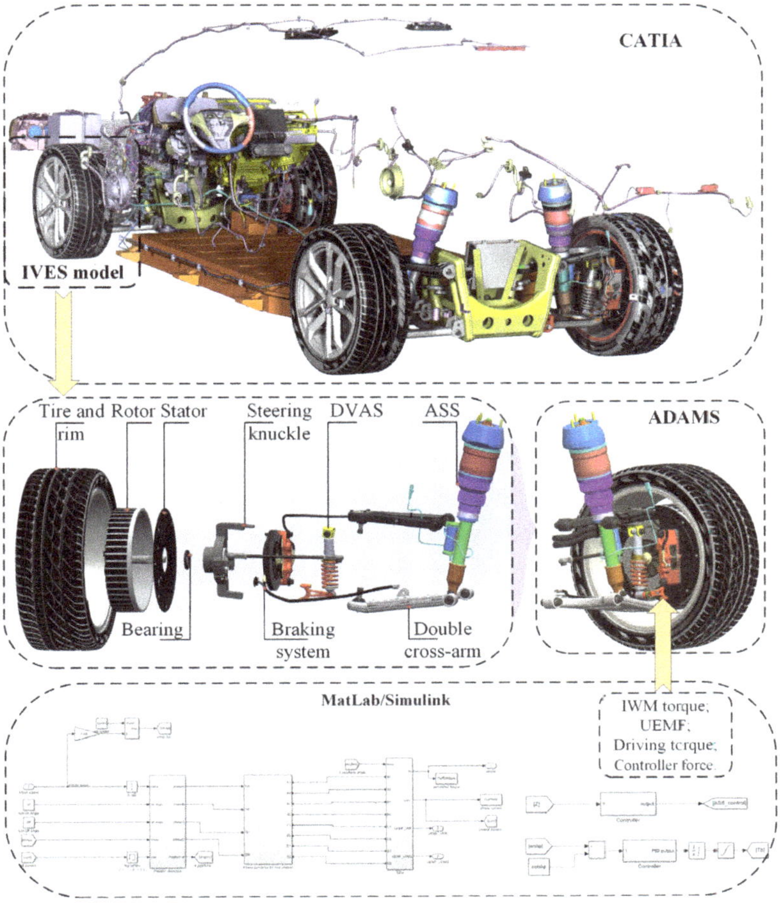

Figure 12. The VP building process.

As shown in Figure 12, using a connector, the DVAS is linked to the steering knuckle at the upper end and the IWM stator at the lower end. Since the axle is kept in steering synchronization with the IWM, there is only relative vertical motion between the two. The lower ASS end is fixed to the lower cross arm of the double cross arm; the control force is output through the ASS sensor. The proposed IVES structure only requires original chassis changes to the steering knuckle, while the springs and damping in the DVAS are integrated into one component. In other words, the structure requires minimal modification cost and is easy to integrate. Moreover, the built tire model is non-linear and can generate tangential forces and moments in the plane of the contact patch, as well as transient effects.

The mathematical model (MM) was validated using the VP model under ISO-B at 40 km/h.The proposed model was validated by comparing the four system responses. The responses of the MM model and the VP model of IVES are compared in Figure 13. Error statistics of the response variables is obtained to compare the two models, which is listed in Table 4. The modeling error is defined as:

$$\eta = \left| \frac{(RMS_{\mathrm{VP}} - RMS_{\mathrm{MM}})}{RMS_{\mathrm{VP}}} \right| \times 100\% \tag{51}$$

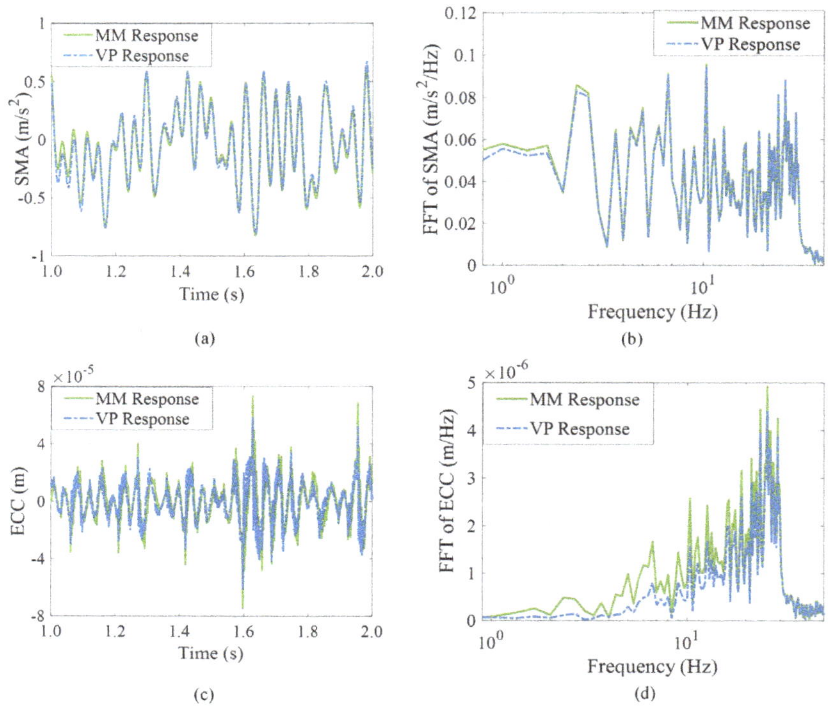

Figure 13. Responses of the MM and the VP: (**a**) time response of the SMA; (**b**) frequency response of the SMA; (**c**) time response of the ECC; (**d**) frequency response of the ECC.

Table 4. Modeling errors between the mathematical model and the VP model.

RMS	SMA (m/s^2)	ECC (10^{-5} m)	RS (10^{-3} m)	TD (10^{-3} mm)
MM	0.3217	1.444	4.222	4.318
VP	0.3089	1.323	4.142	4.039
Error η	4.12%	9.12%	1.9%	6.9%

In the time domain, the VP model response is smaller than that of the MM for both the SMA and the ECC, with the feature being more obvious in SMA. Regarding the FFT comparison, the difference is noticeable at lower frequencies. When considering differences in the time and frequency domains, those might be caused by the tire rubber properties and the nonlinearity of the DVAS connector and the double cross arm.

Comparative results shown in Table 4 indicate that responses used in both models have an error lower than 10%. Since the error is within the permitted range [56], it was concluded that the proposed model is valid.

5. Conclusions

In this paper, an IVES was developed aiming to improve the vertical dynamics performance of the IWMD EVs in both sprung and unsprung conditions while also considering the UEMF effects. The mathematical model of the IVES was established, containing the DVAS-ASS integrated model and the novel FCT model. Moreover, the UEMF model during the driving maneuvers was also considered in the IVES model and was based on the theoretically analyzed UEMF effects. A delay-dependent $H\infty$ controller was developed to improve the IVES performance by improving the ASS robustness to time delay. Finally, CATIA

ADAMS, and MATLAB/Simulink were used to create a virtual prototype environment needed to validate the accuracy and the practicability of the IVES.

Regarding the IVES performance, the numerical results obtained from the proposed mathematical model are consistent with the simulation results obtained from the virtual prototype. Compared to the passive suspension conditions, the RMS of sprung mass acceleration and the eccentricity are reduced via the IVES coordinated with the delay-dependent $H\infty$ controller by up to 67.6% and 38.9%, respectively. The ride comfort is improved and the IWM vibration is suppressed, compensating for the adverse effects of UEMF on the vehicle vertical dynamics. In future work, the real time experimental validation of the IVES applied to the IWMD EVs will be conducted.

Author Contributions: Project administration, L.G. and J.W.; resources, L.G. and X.Z.; supervision, H.Y. and X.Z.; validation, L.G.; visualization, Z.Z.; data curation, Z.Z. and J.W.; writing—original draft, Z.Z. and X.Z.; writing—review and editing, Z.Z. and J.W. All authors have read and agreed to the published version of the manuscript.

Funding: This research was funded by the National Natural Science Foundation of China (Grant No. 52202457).

Institutional Review Board Statement: Not applicable.

Informed Consent Statement: The data involved in this paper do not involve ethical issues.

Data Availability Statement: The datasets used and analyzed during the current study are available from the corresponding author on reasonable request.

Conflicts of Interest: All authors certify that they have no affiliations with or involvement in any organization or entity with any financial interest or non-financial interest in the subject matter or materials discussed in this manuscript.

References

1. Bilgin, B.; Liang, J.; Terzic, M.V.; Dong, J.; Rodriguez, R.; Trickett, E.; Emadi, A. Modeling and analysis of electric motors: State-of-the-art review. *IEEE Trans. Transp. Electrif.* **2019**, *5*, 602–617. [CrossRef]
2. Wu, J.; Wang, Z.; Zhang, L. Unbiased-estimation-based and computation-efficient adaptive MPC for four-wheel-independently-actuated electric vehicles. *Mech. Mach. Theory* **2020**, *154*, 104100. [CrossRef]
3. Luo, Y.; Tan, D. Research on the hub-motor driven wheel structure with a novel built-in mounting system. *Automot. Eng.* **2013**, *35*, 1105–1110.
4. Kim, K.T.; Hwang, K.M.; Hwang, G.Y.; Kim, T.J.; Jeong, W.B.; Kim, C.U. Effect of rotor eccentricity on spindle vibration in magnetically symmetric and asymmetric BLDC motors. In Proceedings of the 2001 IEEE International Symposium on Industrial Electronics Proceedings (ISIE 2001), Pusan, Republic of Korea, 12–16 June 2001; Cat. No. 01TH8570, Volume 2, pp. 967–972.
5. Ni, J.; Hu, J.; Xiang, C. A review for design and dynamics control of unmanned ground vehicle. *Proc. Inst. Mech. Eng. Part D J. Automob. Eng.* **2021**, *235*, 1084–1100. [CrossRef]
6. Peng, H.; Wang, W.; An, Q.; Xiang, C.; Li, L. Path tracking and direct yaw moment coordinated control based on robust MPC with the finite time horizon for autonomous independent-drive vehicles. *IEEE Trans. Veh. Technol.* **2020**, *69*, 6053–6066. [CrossRef]
7. Islam, R.; Husain, I. Analytical model for predicting noise and vibration in permanent-magnet synchronous motors. *IEEE Trans. Ind. Appl.* **2010**, *46*, 2346–2354. [CrossRef]
8. Tan, D.; Lu, C. The influence of the magnetic force generated by the in-wheel motor on the vertical and lateral coupling dynamics of electric vehicles. *IEEE Trans. Veh. Technol.* **2015**, *65*, 4655–4668. [CrossRef]
9. Liu, F.; Xiang, C.; Liu, H.; Han, L.; Wu, Y.; Wang, X.; Gao, P. Asymmetric effect of static radial eccentricity on the vibration characteristics of the rotor system of permanent magnet synchronous motors in electric vehicles. *Nonlinear Dyn.* **2019**, *96*, 2581–2600. [CrossRef]
10. Kim, T.J.; Hwang, S.M.; Park, N.G. Analysis of vibration for permanent magnet motors considering mechanical and magnetic coupling effects. *IEEE Trans. Magn.* **2000**, *36*, 1346–1350.
11. Qin, Y.; He, C.; Shao, X.; Du, H.; Xiang, C.; Dong, M. Vibration mitigation for in-wheel switched reluctance motor driven electric vehicle with dynamic vibration absorbing structures. *J. Sound Vib.* **2018**, *419*, 249–267. [CrossRef]
12. Jin, L.; Yu, Y.; Fu, Y. Study on the ride comfort of vehicles driven by in-wheel motors. *Adv. Mech. Eng.* **2016**, *8*, 1687814016633622. [CrossRef]
13. Xu, B.; Xiang, C.; Qin, Y.; Ding, P.; Dong, M. Semi-active vibration control for in-wheel switched reluctance motor driven electric vehicle with dynamic vibration absorbing structures: Concept and validation. *IEEE Access* **2018**, *6*, 60274–60285. [CrossRef]
14. Qin, Y.; Wang, Z.; Yuan, K.; Zhang, Y. Comprehensive analysis and optimization of dynamic vibration-absorbing structures for electric vehicles driven by in-wheel motors. *Automot. Innov.* **2019**, *2*, 254–262. [CrossRef]

15. Ma, Y.; Deng, Z.; Xie, D. Control of the active suspension for in-wheel motor. *J. Adv. Mech. Des. Syst. Manuf.* **2013**, *7*, 535–543. [CrossRef]
16. Nagaya, G.; Wakao, Y.; Abe, A. Development of an in-wheel drive with advanced dynamic-damper mechanism. *JSAE Rev.* **2003**, *24*, 477–481. [CrossRef]
17. Csekő, L.H.; Kvasnica, M.; Lantos, B. Explicit MPC-based RBF neural network controller design with discrete-time actual Kalman filter for semiactive suspension. *IEEE Trans. Control. Syst. Technol.* **2015**, *23*, 1736–1753. [CrossRef]
18. Shieh, M.Y.; Chiou, J.S.; Liu, M.T. Design of immune-algorithm-based adaptive fuzzy controllers for active suspension systems. *Adv. Mech. Eng.* **2014**, *6*, 916257. [CrossRef]
19. Sunwoo, M.; Cheok, K.C.; Huang, N. Model reference adaptive control for vehicle active suspension systems. *IEEE Trans. Ind. Electron.* **1991**, *38*, 217–222. [CrossRef]
20. Chen, S.A.; Wang, J.C.; Yao, M.; Kim, Y.B. Improved optimal sliding mode control for a non-linear vehicle active suspension system. *J. Sound Vib.* **2017**, *395*, 1–25. [CrossRef]
21. Badri, P.; Amini, A.; Sojoodi, M. Robust fixed-order dynamic output feedback controller design for nonlinear uncertain suspension system. *Mech. Syst. Signal Process.* **2016**, *80*, 137–151. [CrossRef]
22. Marzbanrad, J.; Zahabi, N. H∞ active control of a vehicle suspension system exited by harmonic and random roads. *Mech. Mech. Eng.* **2017**, *21*.
23. Sun, W.; Gao, H.; Yao, B. Adaptive robust vibration control of full-car active suspensions with electrohydraulic actuators. *IEEE Trans. Control. Syst. Technol.* **2013**, *21*, 2417–2422. [CrossRef]
24. Choi, S.B.; Han, S.S. H∞ control of electrorheological suspension system subjected to parameter uncertainties. *Mechatronics* **2003**, *13*, 639–657. [CrossRef]
25. Li, H.; Gao, H.; Liu, H. Robust quantised control for active suspension systems. *IET Control. Theory Appl.* **2011**, *5*, 1955–1969. [CrossRef]
26. Shao, X.; Naghdy, F.; Du, H. Reliable fuzzy H∞ control for active suspension of in-wheel motor driven electric vehicles with dynamic damping. *Mech. Syst. Signal Process.* **2017**, *87*, 365–383. [CrossRef]
27. Jing, H.; Wang, R.; Li, C.; Wang, J.; Chen, N. Fault-tolerant control of active suspensions in in-wheel motor driven electric vehicles. *Int. J. Veh. Des.* **2015**, *68*, 22–36. [CrossRef]
28. Shoukry, Y.; El-Shafie, M.; Hammad, S. Networked embedded generalized predictive controller for an active suspension system. In Proceedings of the 2010 American Control Conference, Baltimore, MD, USA, 30 June–2 July 2010; pp. 4570–4575.
29. Liu, M.; Zhang, Y.; Huang, J.; Zhang, C. Optimization control for dynamic vibration absorbers and active suspensions of in-wheel-motor-driven electric vehicles. *Proc. Inst. Mech. Eng. Part D J. Automob. Eng.* **2020**, *234*, 2377–2392. [CrossRef]
30. Wang, R.; Jing, H.; Yan, F.; Karimi, H.R.; Chen, N. Optimization and finite-frequency H∞ control of active suspensions in in-wheel motor driven electric ground vehicles. *J. Frankl. Inst.* **2015**, *352*, 468–484. [CrossRef]
31. Liu, M.; Gu, F.; Huang, J.; Wang, C.; Cao, M. Integration design and optimization control of a dynamic vibration absorber for electric wheels with in-wheel motor. *Energies* **2017**, *10*, 2069. [CrossRef]
32. Li, Z.; Zheng, L.; Gao, W.; Zhan, Z. Electromechanical coupling mechanism and control strategy for in-wheel-motor-driven electric vehicles. *IEEE Trans. Ind. Electron.* **2018**, *66*, 4524–4533. [CrossRef]
33. Jin, L.; Song, C.; Wang, Q. Evaluation of Influence of Motorized Wheels on Contact Force and Comfort for Electric Vehicle. *J. Comput.* **2011**, *6*, 497–505. [CrossRef]
34. Deur, J.; Asgari, J.; Hrovat, D. A 3D brush-type dynamic tire friction model. *Veh. Syst. Dyn.* **2004**, *42*, 133–173. [CrossRef]
35. Besselink, I.; Schmeitz, A.; Pacejka, H. An improved Magic Formula/Swift tyre model that can handle inflation pressure changes. *Veh. Syst. Dyn.* **2010**, *48*, 337–352. [CrossRef]
36. Zegelaar, P.; Pacejka, H. Dynamic tyre responses to brake torque variations. *Veh. Syst. Dyn.* **1997**, *27*, 65–79. [CrossRef]
37. Gipser, M. FTire: A physically based application-oriented tyre model for use with detailed MBS and finite-element suspension models. *Veh. Syst. Dyn.* **2005**, *43*, 76–91. [CrossRef]
38. Mao, Y.; Zuo, S.; Wu, X.; Duan, X. High frequency vibration characteristics of electric wheel system under in-wheel motor torque ripple. *J. Sound Vib.* **2017**, *400*, 442–456. [CrossRef]
39. Anwar, M.; Husain, I. Radial force calculation and acoustic noise prediction in switched reluctance machines. *IEEE Trans. Ind Appl.* **2000**, *36*, 1589–1597. [CrossRef]
40. Jokinen, T.; Hrabovcova, V.; Pyrhonen, J. *Design of Rotating Electrical Machines*; John Wiley & Sons: Hoboken, NJ, USA, 2013.
41. Zhu, W.; Pekarek, S.; Fahimi, B.; Deken, B.J. Investigation of force generation in a permanent magnet synchronous machine. *IEEE Trans. Energy Convers.* **2007**, *22*, 557–565. [CrossRef]
42. Xue, X.; Cheng, K.W.E.; Lin, J.; Zhang, Z.; Luk, K.; Ng, T.W.; Cheung, N.C. Optimal control method of motoring operation for SRM drives in electric vehicles. *IEEE Trans. Veh. Technol.* **2010**, *59*, 1191–1204. [CrossRef]
43. Ebrahimi, B.M.; Faiz, J.; Roshtkhari, M.J. Static-, dynamic-, and mixed-eccentricity fault diagnoses in permanent-magnet synchronous motors. *IEEE Trans. Ind. Electron.* **2009**, *56*, 4727–4739. [CrossRef]
44. Qin, Y.; Wang, Z.; Xiang, C.; Hashemi, E.; Khajepour, A.; Huang, Y. Speed independent road classification strategy based on vehicle response: Theory and experimental validation. *Mech. Syst. Signal Process.* **2019**, *117*, 653–666. [CrossRef]
45. Qin, Y.; Wei, C.; Tang, X.; Zhang, N.; Dong, M.; Hu, C. A novel nonlinear road profile classification approach for controllable suspension system: Simulation and experimental validation. *Mech. Syst. Signal Process.* **2019**, *125*, 79–98. [CrossRef]

46. *ISO 8608:1995*; Mechanical Vibration–Road Surface Profiles–Reporting of Measured Data. International Organization for Standardization: Geneva, Switzerland, 1995; Volume 8608.
47. Canudas-de Wit, C.; Tsiotras, P.; Velenis, E.; Basset, M.; Gissinger, G. Dynamic friction models for road/tire longitudinal interaction. *Veh. Syst. Dyn.* **2003**, *39*, 189–226. [CrossRef]
48. Ozerem, O.; Morrey, D. A brush-based thermo-physical tyre model and its effectiveness in handling simulation of a Formula SAE vehicle. *Proc. Inst. Mech. Eng. Part D J. Automob. Eng.* **2019**, *233*, 107–120. [CrossRef]
49. Guo, Z.; Wu, W.; Yuan, S. Longitudinal-vertical dynamics of wheeled vehicle under off-road conditions. *Veh. Syst. Dyn.* **2022**, *60*, 470–490. [CrossRef]
50. Karnopp, D. How significant are transfer function relations and invariant points for a quarter car suspension model? *Veh. Syst. Dyn.* **2009**, *47*, 457–464. [CrossRef]
51. Tseng, H.E.; Hrovat, D. State of the art survey: Active and semi-active suspension control. *Veh. Syst. Dyn.* **2015**, *53*, 1034–1062. [CrossRef]
52. Tuononen, A.; Hartikainen, L.; Petry, F.; Westermann, S. Parameterization of in-plane rigid ring tire model from instrumented vehicle measurements. In Proceedings of the 11th International Symposium on Advanced Vehicle Control (AVEC'12), Seoul, Republic of Korea, 9–12 September 2012; pp. 9–12.
53. Apkarian, P.; Tuan, H.D.; Bernussou, J. Continuous-time analysis, eigenstructure assignment, and H/sub 2/synthesis with enhanced linear matrix inequalities (LMI) characterizations. *IEEE Trans. Autom. Control.* **2001**, *46*, 1941–1946. [CrossRef]
54. Cottle, R.W. Manifestations of the Schur complement. *Linear Algebra Its Appl.* **1974**, *8*, 189–211. [CrossRef]
55. Guo, L.X.; Zhang, L.P. Robust H∞ control of active vehicle suspension under non-stationary running. *J. Sound Vib.* **2012**, *331*, 5824–5837. [CrossRef]
56. Guiggiani, M. *The Science of Vehicle Dynamics*; Springer: Dordrecht, The Netherlands, 2014; p. 15.

electronics

MDPI

Article

Data-Driven Intelligent Recognition of Flatness Control Efficiency for Cold Rolling Mills

Xiaomin Zhou [1,2,*], Liqi Li [1], Xinglong Ma [1] and Tao Xu [2]

[1] School of Mechanical Engineering, University of Science and Technology Beijing, Beijing 100083, China
[2] Shunde Innovation School, University of Science and Technology Beijing, Foshan 528399, China
* Correspondence: zhouxiaomin@ustb.edu.cn; Tel.: +86-1851-0257-595

Abstract: In the production process of strip tandem cold rolling mills, the flatness control system is important for improving the flatness quality. The control efficiency of actuators is a pivotal factor affecting the flatness control accuracy. At present, the data-driven methods to intelligently identify the flatness control efficiency have become a research hotspot. In this paper, a wavelet transform longitudinal denoising method, combined with a genetic algorithm (GA-WT), is proposed to handle the big noise of the measured data from each signal channel of the flatness meter, and Legendre orthogonal polynomial fitting is employed to extract the effective flatness features. Based on the preprocessed actual production data, the adaptive moment estimation (Adam) optimization algorithm is applied, to intelligently identify the flatness control efficiency. This paper takes the actual production data of a 1420 mm tandem cold mill as an example, to verify the performance of the new method. Compared with the control efficiency determined by the empirical method, the flatness residual MSE 0.035 is 5.4% lower. The test results indicate that the GA-WT-Legendre-Adam method can effectively reduce the noise, extract the flatness features, and achieve the intelligent determination of the flatness control efficiency.

Keywords: control efficiency; data noise reduction; Adam optimization; data-driven

Citation: Zhou, X.; Li, L.; Ma, X.; Xu, T. Data-Driven Intelligent Recognition of Flatness Control Efficiency for Cold Rolling Mills. *Electronics* **2023**, *12*, 875. https://doi.org/10.3390/electronics12040875

Academic Editor: Adão Silva

Received: 29 December 2022
Revised: 31 January 2023
Accepted: 7 February 2023
Published: 9 February 2023

1. Introduction

Cold-rolled strips are widely used in the fields of automobile manufacturing, aerospace, household appliance panels, and so on [1], because of their high geometric accuracy and great mechanical properties. With the development of technology, cold-rolled strips have become thinner and thinner, and the geometric shape and dimensional accuracy requirements are becoming higher and higher [2,3]. In order to improve the quality of the strips, modern rolling mills adopt an automatic gauge control system (AGC) to solve the gauge problem, and an automatic flatness control system (AFC) to solve the flatness problem. At present, the gauge control basically meets the requirements, but the flatness problem is becoming more and more complicated [4]. Due to the large number of actuators required for flatness control, and the strong coupling and hysteresis between the actuators [5], flatness control is still a great challenge in practice. The cold tandem rolling mill contains multi-stand flatness controls, and the final stand is a closed-loop feedback control. The control efficiency reflects the adjustment ability of the actuators to the flatness in a closed-loop feedback control. Distributing the flatness deviation to each actuator reasonably can improve the flatness control accuracy [6]. In actual production, the control efficiency of each actuator varies with different process parameters, especially, when the ratio of the width to thickness is larger, it is more unstable. The control efficiency is often obtained by an empirical method which has been used since setting up the factory. However, there are errors, inevitably, for specific working conditions, and the update speed is slow. Therefore, obtaining an accurate control efficiency is significant in the research on flatness control.

To the best of our knowledge, there are two main methods used to obtain the control efficiency: the experimental method and the finite element method [7]. The experimental method obtains the control efficiency by changing the setting value of the actuator one by one to obtain the change of the flatness. Although it is more accurate, the high cost and the limited working conditions make it infeasible. Compared with the experimental method, the finite element simulation has the advantage of being more economical, having a wider application, and more flexible simulation of various rolling conditions [8]. However, the finite element method needs many assumptions in the analysis of the rolling process, which leads to errors between the calculation results and the actual situations. Meanwhile, the calculation process is slow, so it is not suitable for the requirements of online real-time control [9]. Thanks to a large amount of rolling process data stored by steel companies, a lot of information related to the production process can be revealed with the development of big data and sensor technology [10]. However, the use of the rolling process data is not sufficient. Therefore, based on the rolling process data, mining the information with the goal of analyzing the relationship between the flatness and the process, provides not only a new way to improve the flatness control accuracy but also a new approach for the intelligent recognition of the flatness control efficiency.

The actual rolling production data has the characteristics of noise, outliers, deviations, and so on. In order to calculate the control efficiency, it is necessary to analyze and process the data. Thus, it is extremely important to reduce the noise of the production data. Wavelet transform, a key noise reduction method at home and abroad, has the advantages of multi-resolution, fast operation speed and small storage space. In addition, it has been widely used in the fields of image compression, data denoising, and edge detection.

Y. Kim et al. proposed cepstrum-assisted empirical wavelet transform (CEWT) to decompose the signal in the fault diagnosis of planetary gear trains. The CEWT solved the problem of empirical wavelet transform (EWT), which required physical understanding to isolate fault-related signals and improve performance of fault diagnosis [11]. F. Dengand et al. proposed a new peak detection algorithm based on continuous wavelet transform and image segmentation, which has made progress in identifying weak peaks, overlapping peaks, and removing false peaks [12].

M. Zolfaghari found that the electricity production data in the prediction is non-stationary and non-linear, and traditional forecasting methods display a poor robustness. Therefore, they proposed an AWT-LSTM-RF hybrid model, combining adaptive wavelet transform (AWT), long short-term memory network (LSTM), and random forest (RF) algorithms for the prediction. The results showed that the hybrid model of AWT-LSTM-RF is better than the benchmark model. At the same time, the wavelet transform algorithm for the treatment of the input data can improve the predictive ability of RF [13]. Denoising the actual data under the wavelet transform alone leads to signal distortion, therefore, the threshold of wavelet coefficients can be optimized in combination with the genetic algorithm, to remove noise as much as possible while retaining the characteristics of the signal. The flatness meter has dozens of sensor channels and these sensor channels can measure the flatness along the width direction at the same time. Figure 1 shows the measurement signal of the strip flatness coil. Because of the noise interference in the actual signal, the characteristics of the flatness cannot be expressed directly. Therefore, data processing is required. Simply denoising each signal channel can only solve the noise problem of a single sensor channel, and the strip flatness represented by the width direction sensor still has no obvious strip flatness features. So it is necessary to fit in the width direction with a Legendre orthogonal polynomial to obtain the strip flatness represented by the width direction sensor signal. Based on the strip flatness features extracted after denoising, Legendre fitting, and the corresponding actual production process data, an optimized algorithm is used to intelligently identify the strip flatness control effect. The Adam optimization algorithm is an extension of the stochastic gradient descent method. It has the advantages of simplicity, directness, high efficiency, small memory, and handling sparse gradients in solving non-convex optimization problems.

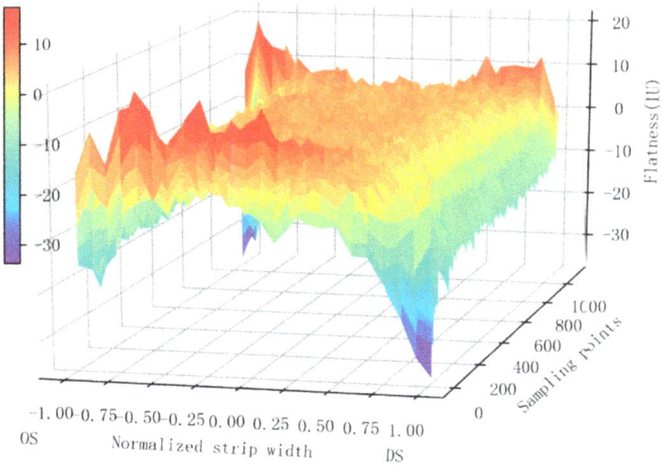

Figure 1. Three−dimensional strip flatness.

This work chooses the Adam optimization method to identify the flatness control efficiency. Based on the actual production process data of a tandem cold rolling mill, the influence of the different actuators on the flatness is analyzed. The GA-WT-Legendre-Adam intelligent identification method for flatness adjustment and control effects is proposed and compared with the control efficiency determined by the empirical method. The analysis results show that the method proposed in this paper can effectively reduce the influence of noise, and the flatness residual MSE calculated by the recognized flatness control efficiency is reduced by 5.4% compared with the empirical method. This noise reduction method can use the industrial data of online production to realize the online identification of the flatness control efficiency.

2. Flatness Closed-Loop Control System

A cold tandem rolling mill consists of a five-stands six-high UCM (Universal Crown Mill). Among them, the first four stands use a preset model to make preliminary adjustments to the thickness and flatness of the strip. The fifth stand uses a closed-loop feedback control strategy to adjust the flatness, which is the most critical part in the flatness control system. The flatness control system is shown in Figure 2. The flatness closed-loop feedback control system is established in the rolling stabilization stage. The closed-loop feedback control model calculates the adjustment of the actuator through the deviation between the measured flatness and the target flatness. Then a control signal is sent to the actuator and the new flatness deviation is detected at the same time. Continuous dynamic adjustments for the actuator in real time are essential for obtaining a stable and good flatness [14,15]. At present, the five-stands cold tandem rolling mill includes intermediate roll bending (IRB), intermediate roll shifting (IRS), work roll bending (WRB), tilting and segment cooling, and other actuators [16,17]. Among them, tilting can eliminate the first order flatness deviation. WRB, IRB, and IRS can eliminate the secondary and the fourth order flatness deviations. Elimination of the higher order flatness deviation requires segmented cooling [18]. The control performance of each type of actuator can be reflected on the control efficiency curve. The general segmented cooling of work rolls cannot be solved by formulas, and the response speed is slow. It is often used to eliminate the remaining flatness deviation after the adjustment of other actuators. Therefore, this paper only takes the influence of tilting, WRB, IRB, and IRS into account.

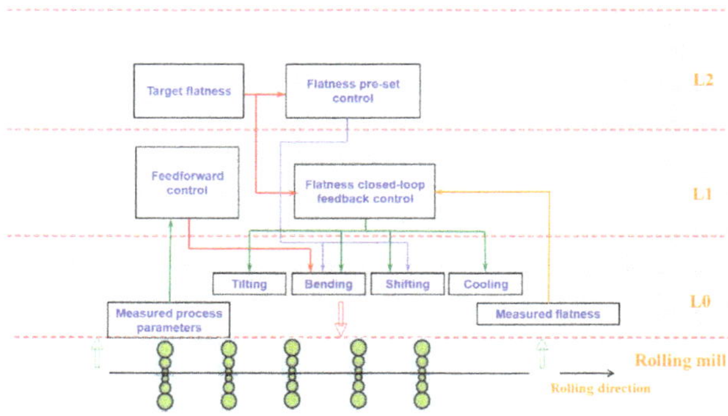

Figure 2. Flatness control system.

The basis of the flatness control is the control efficiency, which can effectively reflect the control performance of the actuators on the flatness. Because of its simplicity and fast calculation process, it can play a key role in a unit with a variety of control methods. The control efficiency can be optimized by making full use of the rolling production process data to obtain a more precise flatness control [19]. The control efficiency, described by the following Formula (1), means the change of the flatness deviation caused by the change of each unit of the actuator:

$$eff_{ij} = \frac{\Delta flat_{ij}}{\Delta Act_j} \tag{1}$$

where eff_{ij} is the actuator efficiency of the jth actuator in the ith segment of the width direction, ΔAct_j is the adjustment of the jth actuator and $\Delta flat_{ij}$ is the flatness change caused by the jth actuator in the ith segment of the width direction.

The multi-variable optimal flatness control algorithm, based on control efficiency, studies the effect of the actuator on the flatness deviation of each measurement section in the width direction. The control amount of each actuator can be solved in connection with the flatness deviation of each measurement section to get the optimal control for the entire system. Therefore, as long as the control efficiency of the flatness is obtained, the adjustment of each actuator under the optimal flatness control can be calculated using Formula (2), to guarantee the minimum value of the performance index J.

$$J = \sum_{i=1}^{n} \left[Tarflat_i - \left(\sum_{j=1}^{m} eff_{ij} \cdot \Delta Act_j \right) \right]^2 \tag{2}$$

$Tarflat_i$ is the target flatness deviation in the ith segment of the width direction, m is the number of flatness control technologies, and n is the number of flatness measurement channels in the width direction.

Among the required actuators, the response speed of the tilting is the fastest, which mainly eliminates the asymmetric flatness deviation, so the control strategy mainly refers to the control of the roll bending and roll shifting. Due to the large number of different control methods, various control goals and the coupling effects of them, choosing either the sequential or parallel calculation needs to be considered for adjustment in the flatness control. The cold tandem rolling mill that provides the data source for this paper adopts a sequential solution strategy, also known as a relay solution, which is based on the least squares principle of the priority sequence table. The execution mechanism is calculated hierarchically based on the sequence in the priority sequence table. The objective function of an executive agency can be expressed as

$$J_j = (Rflat_{ij} - \Delta Act_j \times eff_{ij})^2 \tag{3}$$

where J_j is the objective function of the jth actuator; $Rflat_{ij}$ is the remaining flatness deviation after the adjustment of the first $j-1$ actuators.

In order to minimize the objective function J_j, the partial derivative of the objective function is calculated, as in Equation (4), to obtain the adjustment of the jth actuator.

$$\frac{\partial J_j}{\partial eff_j(x)} = 0 \tag{4}$$

The priority is determined by the sensitivity of the actuator. Generally, it has the sequence of WRB, IRB, and IRS. The sequential solution strategy makes the best use of the actuators to eliminate the flatness deviation. When the front actuator cannot completely eliminate the flatness deviation, the remaining flatness deviation is eliminated by the latter actuator. Because this kind of control strategy calculates the control quantity of each actuator in sequence, according to the artificially prescribed sequence, there is no unsolvable situation. It successfully avoids the matrix solving process and the limit problem of the actuator, which are not suitable for online and real-time use. From the above calculation formula, it is obvious that accurate control efficiency is the core content of the flatness closed-loop control system, which is different from the empirical method that has a long update cycle and is not suitable for specific working conditions. The data-driven method can dynamically obtain the flatness control functions of different steel specifications from the data more accurately and with a greater flexibility, thereby the accuracy of the closed-loop feedback control of the current steel specifications can be improved.

3. Establishment and Test of Noise Reduction Model

Using actual production data to establish a data-driven model for the intelligent recognition of the flatness control efficiency can reflect the real performance of the flatness control mechanism, and realize the adjustment of the control efficiency with the change of working conditions. However, the data actually collected at the production site contains noise, has a wide range of fluctuations, and thus the true situation of the data is obscured. It is necessary first to extract meaningful flatness features to establish a data-driven intelligent identification model of flatness control efficiency.

Wavelet transform is a time-frequency localization analysis method, in which the wavelet window area is constant and the time and frequency domain window ranges are variable. It solves the shortcoming of the short-time Fourier transform, which adopts a fixed sliding window function and cannot change the time resolution. Wavelet transform uses a short-time window when analyzing high-frequency signals and a long-time window when analyzing low-frequency signals. This achieves the adaptive change of the time-frequency window and has the abilities of local analysis and refinement.

The wavelet of continuous wavelet transform is a short-duration time function. Suppose that $\psi(t)$ is an integrable function and $\psi(t) \in H^2(R)$. $\psi(t)$ is a basic wavelet or wavelet mother function, if the Fourier transform satisfies:

$$C_\psi = \int_{-\infty}^{+\infty} \left|\hat{\psi}(\omega)\right|^2 w^{-1} d\omega < \infty \tag{5}$$

By shifting and scaling the wavelet $\psi(t)$, a set of wavelet basis functions $\{\psi_{a,b}(t)\}$ can be obtained:

$$\psi_{a,b}(t) = |a|^{-\frac{1}{2}} \psi\left(\frac{t-b}{a}\right) \tag{6}$$

where $a, b \in R$ and $a > 0$. b is the translation factor indicating the position where the function is translated along the t axis and a is the scale factor of a wavelet basis function.

For a signal $f(t) \in L^2(R)$, its continuous wavelet transform is:

$$W_f(a,b) = \langle f, \psi_{a,b} \rangle = |a|^{-\frac{1}{2}} \int_R f(t) \psi\left(\frac{t-b}{a}\right) dt \tag{7}$$

where $W_f(a,b)$ is the coefficient after wavelet transform of the signal $f(t)$. Through the inverse transform of $W_f(a,b)$, we can obtain:

$$f(t) = \frac{1}{C_\psi} \int_{R^+} \int_{R} \frac{1}{a^2} W_f(a,b) \psi \frac{t-b}{a} dadb \tag{8}$$

In practical applications, the wavelet transform needs to be discretized. Generally, the scale factor and translation factor of the continuous wavelet transform are discretized. The scale factor is $a = a_0^j$ and the translation factor is $b = kb_0 a_0^j$, $j,k \in Z$. The step value is a fixed value other than 1, and usually $a_0 > 1$, then the function can be written as

$$\psi_{j,k}(t) = a_0^{-\frac{j}{2}} \psi\left(\frac{t - kb_0 a_0^j}{a_0^j}\right) = a_0^{-\frac{j}{2}} \psi\left(\int_{-\infty}^{+\infty} a_0^{-j} t - kb_0\right) \tag{9}$$

The corresponding discrete wavelet transform can be written as

$$(W_\psi f)(a,b) = \langle f, \psi_{a,b} \rangle = a^{-\frac{j}{2}} \int_{-\infty}^{+\infty} f(t) \overline{\psi\left(a_0^{-j} t - kb_0\right) dt} \tag{10}$$

Because the wavelet transform has the characteristics of multi-resolution, it can effectively remove the noise while preserving the characteristic information of the original signal as much as possible. The formula for wavelet denoising is as follows:

$$s(n) = f(n) + \sigma e(n) \tag{11}$$

where $s(n)$ is the actual noisy signal, $f(n)$ is the effective signal, n is the sampling time, σ is the noise intensity and $e(n)$ is the noise component.

Wavelet denoising requires processes such as multi-layer decomposition of noisy signals, threshold processing of wavelet coefficients, and reconstruction of the wavelet inverse transform. The more decomposition layers of a noisy signal there are, the easier it is to eliminate the noise signal, but the error is greater after reconstruction. So it is very important to choose an appropriate number of decomposition layers. The appropriate number of decomposition layers can be calculated using Formula (12):

$$max\_level = \log_2\left(l_{data} / l_{filter}\right) \tag{12}$$

where max_level is the number of decomposition levels, l_{data} is the length of the data, and l_{filter} is the length of the filter.

The processing of the wavelet coefficient threshold is a key factor to improve the effect of wavelet denoising. Common threshold functions include the hard threshold and the soft threshold. Since the soft threshold function can shrink large coefficients, it can reduce the discontinuity problems in the hard threshold noise reduction process. At the same time, it has a good adaptability and an excellent noise reduction effect. Therefore, the soft threshold function is selected in this paper as follows:

$$w_\lambda = \begin{cases} [\mathrm{sgn}(w)](|w| - \lambda) & |w| \geq \lambda \\ 0 & |w| < \lambda \end{cases} \tag{13}$$

where w is the wavelet coefficient, λ is the threshold, and w_λ is the wavelet coefficient after threshold processing.

From (13), it can be found that, how to choose an appropriate threshold is the key to wavelet denoising. In order to obtain the optimal wavelet coefficients, this paper combines a genetic algorithm to optimize the threshold. After the wavelet coefficients are decomposed by the wavelet transform, the genetic algorithm is used to select an appropriate threshold remove the noise in the high frequency, and retain the useful information. Then the wavelet inverse transform is reconstructed to obtain the noise-reduced data. Figure 3 shows the noise reduction process of GA-WT (Genetic Algorithm-Wavelet).

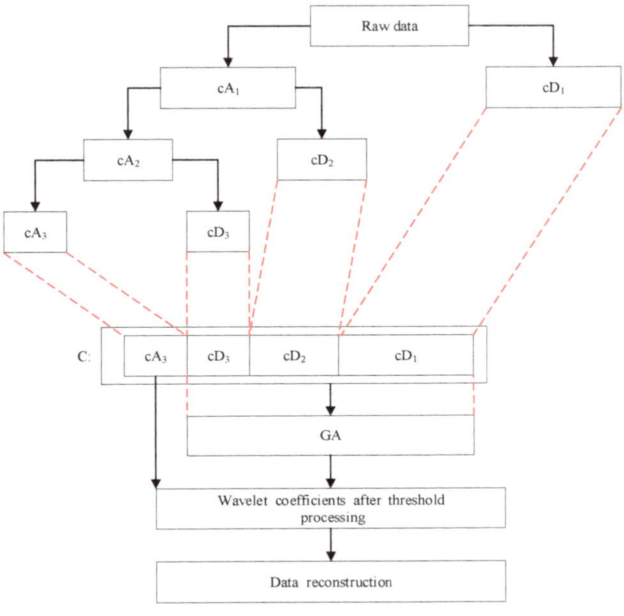

Figure 3. GA-WT noise reduction process.

To establish the GA-WT model, it is necessary to select an appropriate wavelet basis function. In this paper, the db8 basis function is selected, and the number of decomposition layers calculated by Formula (12) is six. The low-frequency wavelet coefficients are reserved, and the optimal threshold is selected by the genetic algorithm for high-frequency wavelet coefficients. The initial population of the genetic algorithm is 200, the crossover probability is 0.8, the number of iterations is 50, and the mutation probability is 0.003.

In order to verify the performance of the model, a standard data set is established. White noise is added to obtain simulation data. The commonly used EEMD (Ensemble Empirical Mode Decomposition) noise reduction method, and the wavelet transform method without genetic algorithm optimization, are compared with the GA-WT method proposed in this paper. The Signal to Noise Ratio (SNR) and Root Mean Square Error (RMSE) are used to measure the noise reduction performances of these three methods. The SNR is defined as the ratio of the effective component of the reconstructed signal to the noise. A larger ratio indicates a better noise reduction performance. The calculation formula is:

$$\mathrm{SNR} = 10\lg\left(\frac{\sum\limits_{n=1}^{N} x^2(n)}{\sum\limits_{n=1}^{N} [x(n) - y(n)]^2}\right) \tag{14}$$

where $y(n)$ is the original signal, $x(n)$ is the signal after denoising, and N is the signal length.

The RMSE reflects the similarity between the reconstructed signal and the original signal. A smaller value of RMSE indicates that the reconstructed signal is closer to the original signal. It is defined as:

$$\text{RMSE} = \sqrt{\frac{\sum_{n=1}^{N} [x(n) - y(n)]^2}{N}} \tag{15}$$

According to the GA-WT model, the appropriate wavelet basis function needs to be selected firstly. This paper chooses the db8 basis function, and the number of decomposition layers is 6, from Formula (12). The low frequency part is retained, and the genetic algorithm is used to select the optimal soft threshold for the high frequency part. Figure 4 shows the comparison of these three methods before and after denoising. It is obvious that these three methods can basically restore the signal trend. In detail, the wavelet transform based on the genetic algorithm performs better and reconstructs the original data more precisely. The unoptimized wavelet transform and EEMD fluctuate greatly, and the noise is not removed as much as possible. Even some valid information is deleted at the same time.

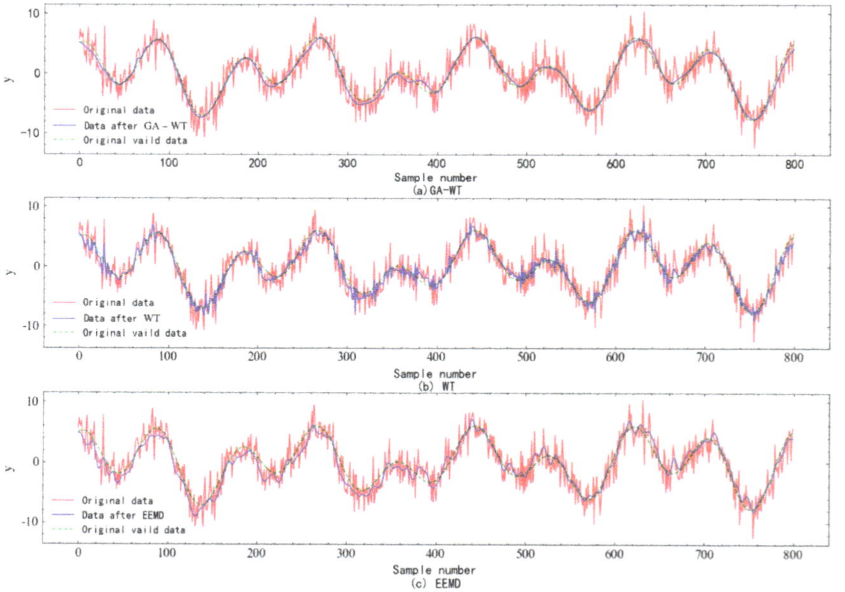

Figure 4. Comparison of the three methods.

The SNR comparison of these three methods is shown in Table 1. The SNR of GA-WT is 18.98 db, which is higher than the 14.41 db of the unoptimized wavelet transform and 12.42 db of EEMD. At the same time, the RMSE of GA-WT is 0.39 IU. This is lower than 0.67 IU of the wavelet transform without optimization and 0.87 IU of EEMD. It shows that the noise reduction performance of GA-WT is the best, so the GA-WT method is chosen in this paper to reduce the noise of the measured flatness data.

Table 1. Comparison of the three methods.

Method	SNR/db	RMSE/IU
GA-WT	18.98	0.39
WT	14.41	0.67
EEMD	12.42	0.87

4. Intelligent Recognition of Flatness Control Efficiency Based on Data-Driven Model

4.1. Data Preprocessing

Modern rolling mills are equipped with a variety of actuators with different control capabilities. This paper mainly studies the control effects of tilting, WRB, IRB, and IRS. The experimental data is the actual data produced by a 1420 mm UCM cold tandem rolling mill. The steel grade is AQ0510B1, the strip width is 1150 mm, and the exit thickness is 0.8 mm. A total of 1514 sampling points are selected in the stable stage of the rolling process, and each sampling point includes variables such as 40 sections of flatness measurement data, rolling force, tilting, roll bending force, and roll shifting, etc. After a preliminary observation of the data, it is found that the width direction of the flatness fluctuates sharply, so it is difficult to extract the effective flatness features. There exists noise in the rolling direction of each signal channel, so GA-WT is adopted for each signal channel in the rolling direction of the strip to obtain the flatness after noise reduction. The denoising flatness data of some certain signal channels, such as the 29th channel, which is near the strip edge channel, and the 12th channel, which is near the strip center channel, are shown in Figure 5.

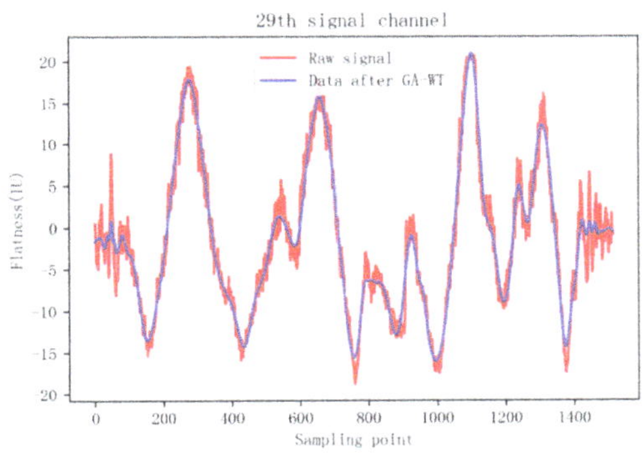

(**a**) Denoising effect of the 29th channel

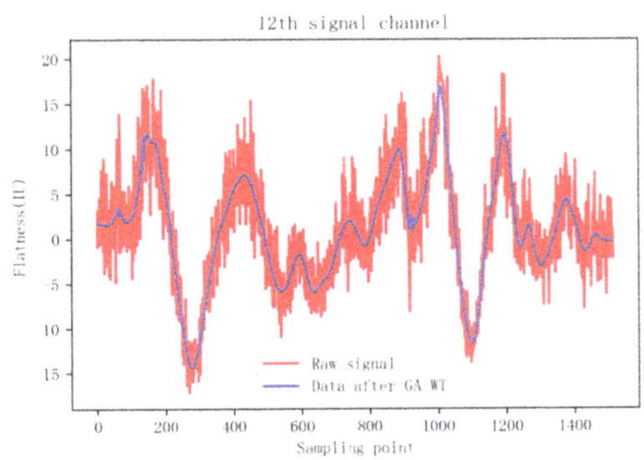

(**b**) Denoising effect of the 12th channel

Figure 5. Denoising effect of certain signal channels.

After the noise reduction, the fluctuation of the flatness composed of all signal channels in the width direction at the same time, is reduced. However, the flatness features are still not obvious. In general, Legendre polynomials or Chebyshev polynomials are used to describe the shape defects. The integral value of the even-degree polynomial of the Chebyshev polynomial is not zero along the strip width direction, which cannot satisfy the self-equilibrium condition of the residual stress of the strip [20]. The Legendre polynomial can satisfy the self-equilibrium condition of the residual stress in the plate width direction.

Therefore, it is necessary to perform Legendre orthogonal polynomial fitting on the flatness in the width direction, which can be effective to extract the flatness features of the strip. Due to the limitation of the control ability of the actuator, the fourth-order Legendre orthogonal polynomial is used for fitting, which can remove the influence of higher-order terms. The Legendre orthogonal polynomial is shown as:

$$y_1 = \pm x \tag{16}$$

$$y_2 = \pm\left(\frac{3}{2}x^2 - \frac{1}{2}\right) \tag{17}$$

$$y_3 = \pm\frac{1}{2}\left(5x^3 - 3x\right) \tag{18}$$

$$y_4 = \pm\frac{1}{8}\left(35x^4 - 30x^2 + 3\right) \tag{19}$$

where x is the sampling points in the width direction of the strip, y_1 is the first order flatness, y_2 is the second order flatness, y_3 is the third order flatness, and y_4 is the fourth order flatness.

Figure 6 shows that the original signal width direction is severely oscillating, which masks the characteristics of the flatness shape. After longitudinal denoising, the oscillation amplitude is weakened, but there are still fluctuations. Then after the Legendre orthogonal polynomial fitting, the data transition is smoother, in line with the current control performance of the actuator. At the same time, the influence of the high-order shape deviation is excluded, so the flatness control efficiency can be calculated by a Legendre orthogonal polynomial.

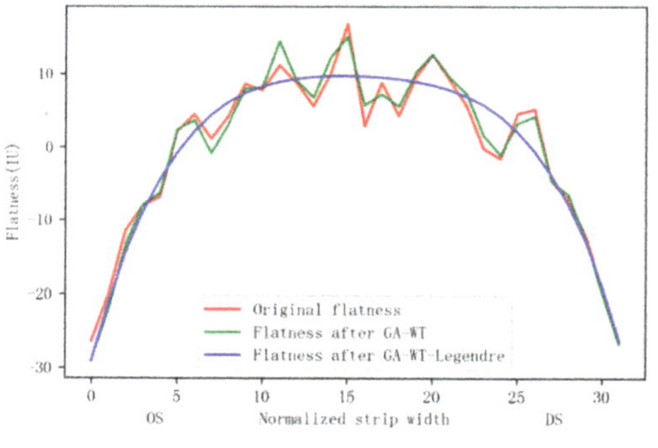

Figure 6. Comparison of the three flatnesses.

4.2. Optimization Method

Gradient descent is one of the most commonly used methods to optimize model parameters, and Adam is an optimization method that can be used to improve the gradient descent method. The Adam algorithm has a fast calculation speed, uses the momentum method to select the optimization direction, and uses the RMSProp algorithm to select the learning rate [21]. With the idea of moment estimation, the Adam optimization algorithm dynamically adjusts the learning rate by modifying the first-order and second-order moment estimation of the gradient. The Adam algorithm calculates the current time gradient as follows:

$$g_t = \nabla J(w) \tag{20}$$

where $J(w)$ is the objective function to be optimized.

The first moment estimation at the current moment is shown as follows:

$$M_t = \beta_1 \cdot M_{t-1} + (1 - \beta_1) \cdot g_t \tag{21}$$

where t is the time step, M_{t-1} is the first moment estimate at the previous moment, and $\beta_1 \in [0,1)$ is the hyperparameter, which is the exponential decay rate of the instantaneous estimation.

The second moment at the current moment is estimated as follows:

$$V_t = \beta_2 \cdot V_{t-1} + (1 - \beta_2) \cdot {g_t}^2 \tag{22}$$

where V_{t-1} is the second moment estimate at the previous moment and $\beta_2 \in [0,1)$ is the hyperparameter, which is the exponential decay rate of the instantaneous estimation.

The second-order moment estimation is actually calculated according to the average value of the exponential movement. The second-order moment estimation is initialized to a 0 vector, that is, at the beginning of the iteration, V_t is close to 0, which requires the estimation of the second-order moment. The moment uses the exponentially averaged original moment of the gradient squared at the decay rate β_2, that is, the centerless variance. Let $g_1, g_2 \ldots, g_t$ be the gradient of the subsequent time steps, then Equation (22) can be written as:

$$V_t = (1 - \beta_2) \sum_{i=1}^{t} \beta_2^{t-i} \cdot {g_i}^2 \tag{23}$$

Then the relationship between the expected value $\mathbb{E}[V_t]$ of the second-order moment estimation at the time step and the expected value of the second-order moment $\mathbb{E}[g_t^2]$ is the expectation on both sides of Equation (23):

$$\begin{aligned} \mathbb{E}[V_t] &= \mathbb{E}\left[(1 - \beta_2) \sum_{i=1}^{t} \beta_2^{t-i} \cdot {g_i}^2\right] \\ &= \mathbb{E}[g_t^2] \cdot (1 - \beta_2) \sum_{i=1}^{t} \beta_2^{t-i} + \zeta \\ &= \mathbb{E}[g_t^2] \cdot (1 - {\beta_2}^t) + \zeta \end{aligned} \tag{24}$$

When the true second-order moment $\mathbb{E}[g_t^2]$ is a constant value, ζ is 0. Otherwise, the exponential decay rate β_2 is usually selected by adding the past moment to the current moment gradient with a small weight, so ζ is a minimum value. In the case of sparse gradients, in order to reliably estimate the second moment, it is necessary to choose a smaller β_2 to average many gradients. Since β_2 is small, the lack of an initialization bias correction will lead to a much larger initial step size.

Then the deviation correction of the second-order moment estimation V_t is shown as follows:

$$\hat{V}_t = \frac{V_t}{1 - \beta_2^t} \tag{25}$$

In the same way, the deviation correction of the first-order moment estimation M_t is shown in Equation (26):

$$\hat{M}_t = \frac{M_t}{1 - \beta_1^t} \tag{26}$$

Update the control amount by Formula (27):

$$w_t = w_{t-1} - \eta \frac{\hat{M}_t}{sqr(\hat{V}_t) + \varepsilon} \tag{27}$$

where $\eta \in (0,1]$ is the learning rate and ε is a very small constant to avoid the situation where the second moment estimation V_t is zero during the calculation process.

4.3. *Intelligent Recognition of Flatness Control Efficiency*

The objective function that defines the intelligent recognition of the flatness control efficiency is shown in Equation (28):

$$J(w) = \frac{\sum_{i=1}^{n} (Tarflat_i - \sum_{j=1}^{4} \Delta Act_j \cdot eff_{ij})^2}{2n} \tag{28}$$

In order to minimize the objective function $J(w)$, calculate its gradient at the current position:

$$g_t = \frac{\partial J(w)}{\partial \Delta Act} = -\frac{\begin{bmatrix} w_1 \\ w_2 \\ w_3 \\ w_4 \end{bmatrix} \cdot (Tarflat - \sum_{j=1}^{4} \Delta Act_j \cdot eff_j)^T}{n} \tag{29}$$

where w_1, w_2, w_3, w_4 are respectively the adjustment amounts of tilting, WRB, IRB, and IRS.

Figure 7 shows the results using the Adam algorithm and the empirical method. The results of the two methods both show that the tilting has the adjustment ability of the asymmetric term; and the work roll bending roll, the middle roll bending roll, and the middle roll shifting roll have the adjustment ability of the symmetric term.

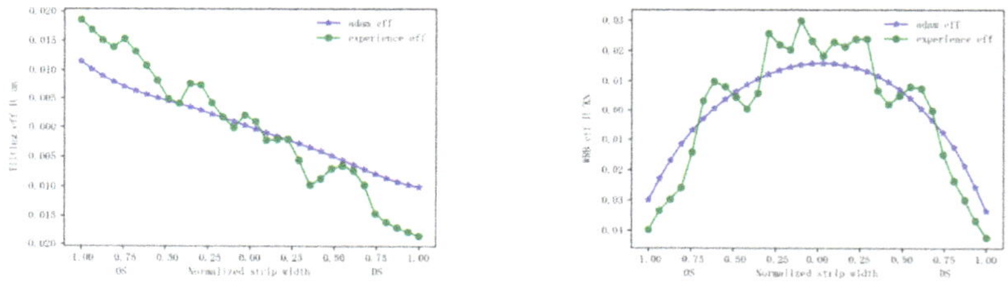

(**a**) Control efficiency of tilting (**b**) Control efficiency of WRB

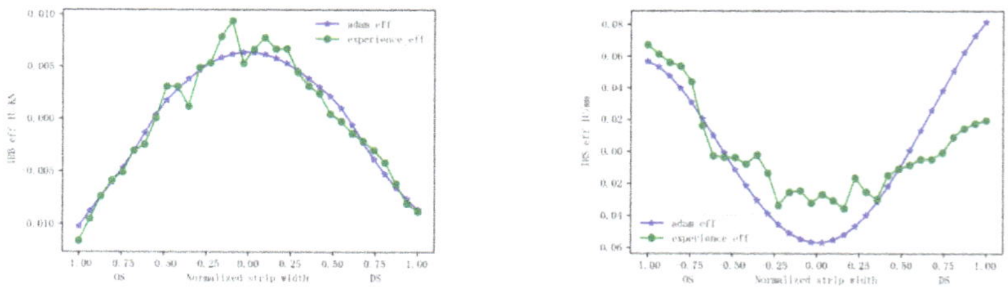

(**c**) Control efficiency of IRB (**d**) Control efficiency of IRS

Figure 7. Comparison of the two control efficiencies.

Substituting the control efficiency of the two methods into the control system, calculate the flatness residual after flatness control and defines the MSE of the flatness residual as shown in Equation (30):

$$\text{MSE} = \frac{\sum_{k=1}^{m} \left(\frac{\sum_{i=1}^{n} \left(Tarflat_i - \sum_{j=1}^{4} w_j \cdot eff_{ij} \right)^2}{n} \right)}{m} \tag{30}$$

where m is the number of sampling points with length, n is the number of flatness meter sensors in the width direction, and the other parameters remain unchanged.

The flatness residuals obtained by the two methods are shown in Figure 8. It can be seen that the flatness residuals calculated by the two methods are very small, especially, the control efficiency of Adam's method has smaller flatness residuals at certain sampling points. The MSEs of the flatness residuals are shown in Table 2. It can be seen that the MSE of the flatness deviation calculated by the control efficiency of the Adam algorithm is 0.035, and the MSE of the flatness deviation calculated by the control efficiency of the empirical method is 0.037. The MSE of the proposed method is reduced by 5.4% compared to the empirical method, so the regulation effect based on the Adam algorithm is more accurate.

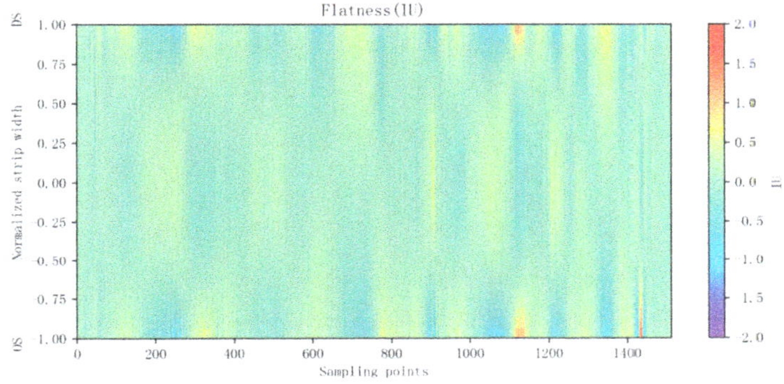

(**a**) Error under Adam method

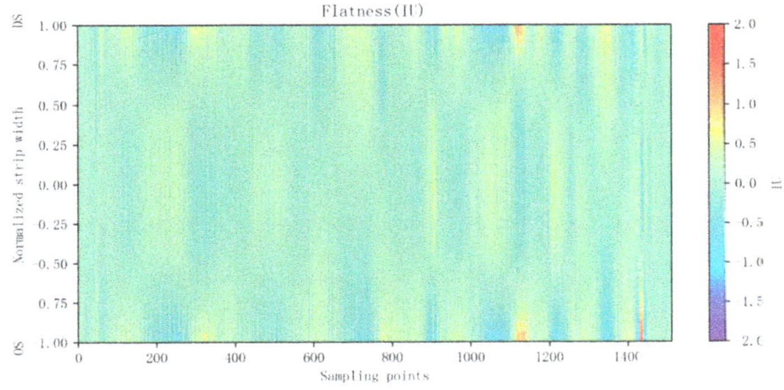

(**b**) Error under the empirical method

Figure 8. Error under the two methods.

Table 2. MSE under the two methods.

Method	MSE
Adam	0.035
Empirical	0.037

5. Conclusions

(1) This paper employs a GA-WT denoising method to deal with the large measurement noise in each signal channel of the flatness meter. Compared with the unprocessed wavelet transform, the SNR of GA-WT is increased by 31.7% and the RMSE is reduced by

41.8%. Compared with EEMD, the SNR of GA-WT is increased by 52.8% and the RMSE is reduced by 55.2%.

(2) GA-WT is used to perform longitudinal denoising in the strip length direction for each signal channel of the flatness meter. Legendre orthogonal polynomial fitting is used on all the flatness signal channels in the width direction, after longitudinal denoising. The method proposed can extract the flatness features, and reflect the real flatness, which can be controlled by the flatness control actuators.

(3) Based on the processed flatness data and the adjustment of the actuator, compared with the empirical method, the control efficiency calculated by the Adam algorithm is 0.035. This is 5.4% lower than the empirical method, which indicates that the GA-WT-Legendre-Adam method can effectively extract flatness features and obtain a more accurate control efficiency.

(4) Production data is characterized by high noise and low quality. In the future, noise reduction and calculation accuracies will be improved further by improving the quality of the actual production data.

Author Contributions: Conceptualization, X.Z.; methodology, X.Z. and L.L.; software, X.Z.; validation, X.Z., L.L. and X.M.; formal analysis, X.Z.; investigation, L.L. and T.X.; resources, X.Z.; data curation, X.Z.; writing—original draft preparation, X.Z.; writing—review and editing, X.Z. and L.L.; visualization, X.Z.; supervision, X.M.; project administration, X.Z. and T.X.; funding acquisition, X.Z. All authors have read and agreed to the published version of the manuscript.

Funding: This research was funded by the Scientific and Technological Innovation Foundation of Foshan (No. BK22BE014).

Data Availability Statement: Due to the nature of this research, participants of this study did not agree for their data to be shared publicly, so supporting data is not available.

Acknowledgments: We sincerely thank the reviewers for their comments, which can help us to revise the paper better.

Conflicts of Interest: The authors declare no conflict of interest.

References

1. Jia, D.; Li, S.; Mi, N.; Shen, Z.; Li, X. Core technology of process control system for 1250 mm five stands 6-high tandem cold mill. *Steel Roll.* **2020**, *37*, 71–76.
2. Huang, C.; Zhao, M. Research progress on cold-rolled strip shape recognition and control. *J. Iron Steel Res.* **2013**, *25*, 1–7.
3. Liu, L.; Han, J.; He, Y. Development of cold rolling flatness control theory. *J. Iron Steel Res.* **1997**, *9*, 51–54.
4. Wang, P.; Zhang, D.; Liu, J.; Wang, J.; Yu, X. Analysis and improvement for flatness control system of 1450 tandem cold mill. *China Metall.* **2009**, *19*, 31–35.
5. Li, X.; Wang, Q.; Zhang, Y.; Zhang, D. Present status and future prospects of strip flatness control based on elastic-plastic finite element. *Steel Roll.* **2020**, *37*, 1–11.
6. Song, L.; Shen, M.; Yang, L.; Liu, J.; Wang, J.; Chen, X. Shape control dimensionality reduction efficiency inherited regulation method of cold rolling wide strip. *Iron Steel* **2016**, *51*, 70–75.
7. Huang, L.; Chen, X.; Zhang, Q.; Chen, S.; Wei, C.; Wu, B.; Wang, J. Study on the Flatness Controlling Actuator's Efficiency of Cold Rolling Mill. *Metall. Equip.* **2000**, *1*, 4–7.
8. Liu, X. The Simulation Research of Shape Control Efficiency Function on Cold Rolling Mill. Master's Thesis, Yanshan University, Qinhuangdao, China, 2015.
9. Wang, H. Model Optimization and Simulation Software Development of Flatness Control System for Cold Rolled Strip. Master's Thesis, Yanshan University, Qinhuangdao, China, 2020.
10. Wang, X.; Yan, Y. Promoting the technical progress of steel industry with intelligent manufacturing. *Metall. Autom.* **2019**, *43*, 1–5.
11. Kim, Y.; Ha, J.M.; Na, K.; Park, J.; Youn, B.D. Cepstrum-assisted empirical wavelet transform (CEWT)-based improved demodulation analysis for fault diagnostics of planetary gearboxes. *Measurement* **2021**, *183*, 109796. [CrossRef]
12. Deng, F.; Li, H.; Wang, R.; Yue, H.; Zhao, Z.; Duan, Y. An improved peak detection algorithm in mass spectra combining wavelet transform and image segmentation. *Int. J. Mass Spectrom.* **2021**, *465*, 116601. [CrossRef]
13. Zolfaghari, M.; Golabi, M.R. Modeling and predicting the electricity production in hydropower using conjunction of wavelet transform, long short-term memory and random forest models. *Renew. Energy* **2022**, *170*, 1367–1381. [CrossRef]
14. Liang, X.; Xu, J.; Wang, G.; Liu, X. Close loop feedback control principle of cold rolling strip flatness and its application effect. *Metall. Autom.* **2006**, *30*, 36–39.

15. He, X.; Wang, C. Study on closed-loop shape control strategy of cold colling. *Metall. Autom.* **2018**, *42*, 44–48.
16. Pin, G.; Francesconi, V.; Cuzzola, F.A.; Parisini, T. Adaptive task-space metal strip-flatness control in cold multi-roll mill stands. *J. Process Control* **2013**, *23*, 108–119. [CrossRef]
17. Bemporad, A.; Bernardini, D.; Cuzzola, F.A.; Spinelli, A. Optimization-based automatic flatness control in cold tandem rolling. *J. Process Control* **2010**, *20*, 396–407. [CrossRef]
18. Wang, X.; Yang, Q. Research on Asymmetric Shape Control Character and Generalized Overall Shape Control Strategy for Universal Crown Mill. *Chin. J. Mech. Eng.* **2012**, *48*, 58–65. [CrossRef]
19. Sun, J.; Shan, P.; Peng, W.; Zhang, D. Cold-rolled flatness actuator efficiency coefficient obtaining based on data noise reduction. *Iron Steel* **2021**, *56*, 67–74.
20. Peng, Y. Theoretical Studies and Engineering Application of Shape Preset Control for HC Cold Mill Based on Strip Element Method. Ph.D. Thesis, Yanshan University, Qinhuangdao, China, 2000.
21. Kingma, D.P.; Ba, J. Adam: A method for stochastic optimization. In Proceedings of the 3rd International Conference for Learning Representations, San Diego, CA, USA, 7–9 May 2015.

electronics

Article

Investigation on Battery Thermal Management Based on Enhanced Heat Transfer Disturbance Structure within Mini-Channel Liquid Cooling Plate

Renzheng Li [1,2,*], Yi Yang [3], Fengwei Liang [4], Jichao Liu [5,6] and Xinbo Chen [1,2]

1 School of Automotive Studies, Tongji University, Shanghai 201804, China
2 Clean Energy Automotive Engineering Center, Tongji University, Shanghai 201804, China
3 School of Automotive and Traffic Engineering, Jiangsu University, Zhenjiang 212013, China
4 School of Mechanical Engineering, University of Science and Technology Beijing, Beijing 100083, China
5 Jiangsu Advanced Construction Machinery Innovation Center Ltd., Xuzhou 221000, China
6 School of Materials and Physics, China University of Mining and Technology, Xuzhou 221116, China
* Correspondence: 1911070@tongji.edu.cn

Abstract: The battery thermal management system plays an important role in the safe operation of a lithium-ion battery system. In this paper, a novel liquid cooling plate with mini-channels is proposed and is improved with disturbance structures. First, an accurate battery heat generation model is established and verified by experiments. The error is less than 4%, indicating the heat generation power is reliable. Then, five designs are proposed first to determine the suitable number of disturbance structures, and plan 3 with five disturbance structures shows a satisfying performance in heat dissipation and flow field. Moreover, four layout plans are proposed, namely uniform, interlaced, thinning, and gradually denser distribution. Results show that plan 5 (uniform) achieves the best performance: the maximum average temperature is 36.33 °C and the maximum average temperature difference is 0.16 °C. At last, the orthogonal experiment and range analysis are adopted to optimize the structure parameters. Results show that the best combination is space between adjacent disturbance structures d1 = 20 mm, length d2 = 5 mm, width d3 = 1.5 mm, and tilt angle $\beta = 60°$.

Keywords: lithium-ion battery; thermal management; liquid cooling; mini-channel; disturbance structure

Citation: Li, R.; Yang, Y.; Liang, F.; Liu, J.; Chen, X. Investigation on Battery Thermal Management Based on Enhanced Heat Transfer Disturbance Structure within Mini-Channel Liquid Cooling Plate. *Electronics* **2023**, *12*, 832. https://doi.org/10.3390/electronics12040832

Academic Editor: Pablo García Triviño

Received: 31 December 2022
Revised: 16 January 2023
Accepted: 3 February 2023
Published: 7 February 2023

1. Introduction

With the rapid development of transportation electrification, lithium-ion batteries are considered an ideal power source because of their high energy density, low self-discharge rate, and long life [1]. However, lithium-ion batteries have unavoidable defects and need to be operated in a suitable but narrow temperature interval [2]. The best temperature range for lithium-ion batteries is 25–40 °C [3]. During the charging and discharging process, lithium-ion batteries generate a lot of heat, the available energy will be limited, and the battery life will decrease more quickly if the temperature is beyond the range [4]. In addition, hundreds and even thousands of batteries are connected in parallel and series to form the battery systems in practical electric vehicles [5]. According to existing research and engineering experience, the temperature difference of all the batteries in the system should be kept below 5 °C to maintain a stable and safe operation [6]. Therefore, an efficient battery thermal management system (BTMS) is necessary for the battery system to maintain the temperature in the suitable range.

There are many different ways to classify the battery thermal management systems. According to whether external energy is consumed, the BTMSs can be divided into three categories, namely active, passive, and hybrid [7–9]. Another common way is based on the type of cooling medium, and the BTMSs can be divided into five categories, namely air cooling, liquid cooling, phase change material (PCM), heat pipe, and hybrid cooling [10–14]. A structural ventilation path or a few fans are used in air cooling and the design is simple. This cooling way is widely used in the early time and small battery systems for its energy efficiency, low cost, and long-lasting reliability and durability [15]. Saw et al. [16] utilized the CFD method to analyze the air cooling of a battery pack comprising 38,120 cells. The simulation results demonstrate that an increase in the cooling airflow rate will increase the heat transfer coefficient and pressure drop. Chen et al. [17] proposed an improved air cooling design with Z-type flow and adjustable battery pacing. Results show that the new design achieved better heat dissipation. Although air cooling has many advantages, it cannot meet the increasing demand for heat dissipation due to the low thermal conductivity of air [18]. The BTMS based on PCM dissipates heat by storing the energy with the latent heat. When the temperature increases, the PCM receives heat, and the state changes. When the temperature decreases to the phase change point, heat is released again and the material returns to the original state again [19,20]. In recent years, PCM cooling has been of great interest due to its significant cooling performance. Huang et al. [21] proposed a novel composite PCM to serve BTMSs. Results show that the material can reduce the contact thermal resistance and improve the cooling performance. Luo et al. [22] proposed a new PCM consisting of paraffin with dual-phase transition ranges, expanded graphite, and epoxy resin. Results show that the PCM can achieve great cooling performance under a 4C discharging rate. However, a significant problem is that a large amount of PCM is needed to achieve the ideal heat dissipation performance, which adds too much weight to the battery system and makes the energy density drop a lot. In addition, leakage cannot be avoided when the PCM changes to the liquid state. Heat pipes essentially use the phase change of materials to dissipate heat as well [23,24]. Due to the outstanding heat transfer performance, heat pipes with PCM are commonly studied for BTMSs, and this combination provides better heat dissipation performance than the above several ways [25]. However, the complex structure, high cost, and low energy density are still inevitable problems limiting its application in electric vehicles.

Compared to the above several cooling ways, liquid cooling has many advantages, such as high heat dissipation performance, high engineering applicability, and high energy density [26]. By adding tubes or cooling plates around the batteries, the coolant is driven by pumps and flows along the paths to dissipate the heat [27,28]. In addition, another form of liquid cooling is submerging the batteries directly in the coolant, and this way can significantly improve power consumption and temperature uniformity [29,30]. However, direct liquid cooling needs complex sealing structures, and its reliability is hard to maintain in real applications. Therefore, much attention is paid to indirect liquid cooling. Zhou et al. [31] designed a cold plate with half-helical ducts and compared its performance with the jacket liquid cooling. Results show that half-helical ducts are more efficient. Rao et al. [32] proposed a novel form of liquid cooling in which the contact surface is variable. Results show that the maximum temperature of the battery can be efficiently controlled under 40 °C. Shang et al. [33] proposed a novel liquid-cooled BTMS with changing contact surface to improve the heat dissipation performance. Results show that the battery temperature is proportional to the inlet temperature. Under the requirement of the increased energy density of the battery system, more attention has been paid to liquid cooling plates with mini-channels due to their tight size, high heat transfer efficiency, and low weight [34,35] S. Panchala et al. [36,37] developed a new CFD model to research the heat flow field of a mini-channel liquid cooling plate; they investigated the temperature and velocity distribution of liquid-cooled LiFePO4 thermal management and revealed that increasing the battery charge–discharge rate will cause the battery temperature to rise rapidly. Huang et al. [38] applied the streamlined concept to the design of the micro-channels, and they found that the streamlined structure can reduce flow resistance and strengthen heat dissipation Liu et al. [39] researched the effect of different types of coolant on the cooling efficiency of a micro-channel liquid cooling plate, and the results showed that water achieved the best cooling effect; then, nanofluids were added to the water, and the authors found that nanoparticle addition can relieve the temperature rise of the battery. According to the existing research, improving the flow field of the mini-channel is an effective way, but most research focuses on the channel shape design. Improving the flow state within the original channel is still an effective way. Contributions of this paper can be shown in the following aspects:

1. Novel mini-channel design with disturbance structure: A novel mini-channel liquid cooling plate with is proposed with improved disturbance structures. High heat dissipation and temperature uniformity are achieved based on the BTMS.
2. Efficient mini-channel structure optimization method: Orthogonal experiment design especially for the disturbance structure is proposed to improve the designing process significantly.

2. Battery Heating Generation Power Experiment

2.1. Battery Property

In this paper, a commercial 40 Ah lithium-ion battery is selected. The heating generation power of the battery is obtained based on the simulation first, and then experiments are conducted to verify the simulation results. The battery properties are provided by the manufacturer, as shown in Table 1.

Table 1. Battery properties.

Content	Parameter
Battery type	NCM
Nominal capacity	40 Ah
Nominal voltage	3.6 V
Weight	825 g ± 10 g
Size	28 mm × 148 mm × 93 mm
DC internal resistance	≤2.0 mΩ
AC internal resistance	≤1.0 mΩ
Density	2140 kg/m^3
Thermal conductivity	λ_z = 1.5 W/(m·K) λ_y = 20.6 W/(m·K) λ_z = 20.6 W/(m·K)
Heat capacity	1030 J/(kg·K)

2.2. Battery Model and Verification

In the charging and discharging process of lithium-ion batteries, heat is generated along with the internal electrochemical reactions. The heat can be divided into four parts, namely reaction heat, polarization heat, Joule heat, and heat of side reaction [17]. The expression is shown as Equation (1).

$$Q = Q_r + Q_p + Q_j + Q_s \tag{1}$$

where Q_r means reaction heat, Q_p means polarization heat, Q_j means Joule heat, and Q_z means the heat of side reaction. In addition, the battery temperature will not exceed 50 °C, so the reaction heat takes a very small part and side reactions will not happen under the temperature. Therefore, the heating generation expression can be simplified as Equation (2).

$$Q = Q_j + Q_p = I^2R = I^2(R_j + R_p)t \tag{2}$$

where R_j means Joule resistance, R_p means polarization resistance, t means reaction time, and I means current.

Based on the above analysis, the Bernardi battery model is adopted in the simulation, and its expression is shown in Equation (3) [29].

$$P = I\left[(U_{oc} - U) + T\frac{\partial U_{oc}}{\partial T}\right] \tag{3}$$

where P means the battery heating generation power, U_{oc} means the battery open circuit voltage, U means the terminal voltage, T is the battery temperature, and I is the charging and discharging current.

To verify the accuracy of the battery model, the temperature rise experiment was conducted. Two PT100 temperature sensors were mounted on the side faces of the battery as shown in Figure 1. The battery temperature during the discharging process at 1 C, 2 C, and 3 C is assumed as the average value of the two sensors. Figure 2 shows the simulation and experiment results; it can be seen that the temperature rising rate and the maximum temperature increase obviously with the increase in the discharging rate. In addition, the simulation temperature is a little higher than that in the experiment because the heat exchange between the battery and ambient air cannot be avoided completely in the experiment like the setup in the simulation. The maximum temperature difference between the simulation and experiment is 0.38 °C, 0.95 °C, and 2.08 °C at 1 C, 2 C, and 3 C, respectively. The error is less than 4%, indicating that the model is accurate. In addition, the heating generation power is calculated according to Equation (4) based on the experimental results, as shown in Table 2.

$$P = \frac{Q_t}{t} = \frac{cm\Delta T}{t} \tag{4}$$

where Q_t means the heat generated by the battery, c means the heat capacity of the battery, m means battery weight, ΔT means temperature rise in the discharging process, and t means the discharging time.

Table 2. Heating generation power of 40 Ah battery at different discharge rates.

Discharging Rate	1 C	2 C	3 C
Time (s)	3600	1800	1200
Heating generation power (W)	3.24	11.48	22.18

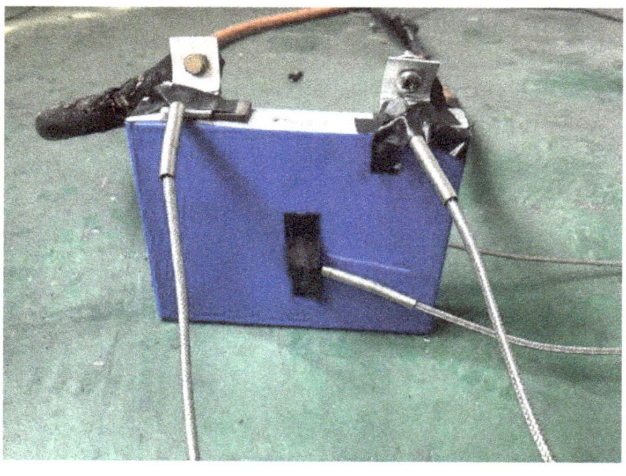

Figure 1. Experiment setup.

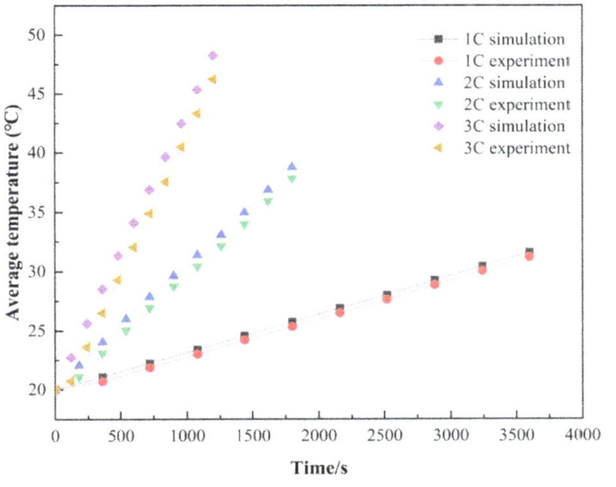

Figure 2. Battery model verification.

2.3. System Model

In practice, battery packs usually contain batteries, heat dissipation components, mechanical components, and control modules. However, the heat exchange mainly occurs between the heat dissipation components and the battery, and the influence of the remaining components is small. In order to optimize the computational performance, a battery module model consisting of 12 batteries and 6 mini-channel liquid cooling plates was built for analysis in this study, as shown in Figure 3. The batteries are numbered from 1 to 12. In addition, thermal conductive sheets are set between adjective batteries and mini-channel liquid cooling plates to decrease the contact thermal resistance. The thickness is 2 mm and the thermal conductivity is 2 W $\times$ K^{-1} $\times$ m^{-1}. In the primary design, 2 parallel 3 tandem flow channels are used; the width is 7.6 mm and the height is 2 mm for one flow channel. In addition, the material of the plate is set as aluminum in all the simulations, and the properties are the defaults in the software. The other structural parameters are shown in Figure 3 as well.

In practical engineering applications, adding disturbance design to change the original velocity field of the fluid to make a sudden change in the flow direction and produce secondary flow, or adding fins, manifolds, and other microstructures on the flow channel to change the fluid boundary layer and flow state, can improve the heat dissipation performance without generating additional energy consumption. In this paper, the disturbance structure is applied to the design of the mini-channel liquid cooling plate, as shown in Figure 4. The structure parameters include the number of the disturbance structure within a single channel N, the spacing between two adjacent disturbance structures d1, the length of one disturbance structure d2, the width of one disturbance structure as well as the depth of the cavity d3, and the tilt angle of the cavity β.

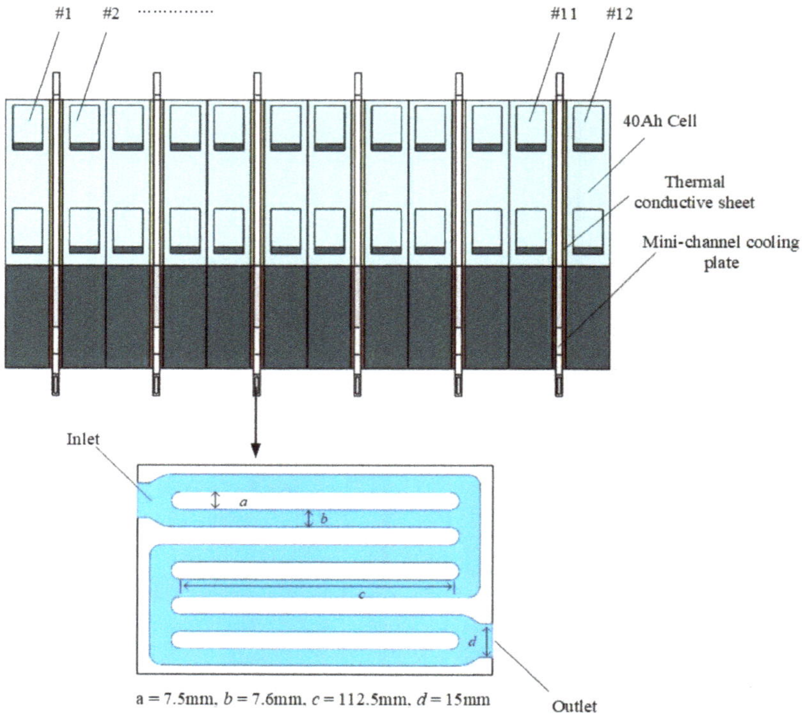

Figure 3. Design of the mini-channel liquid-cooling battery module.

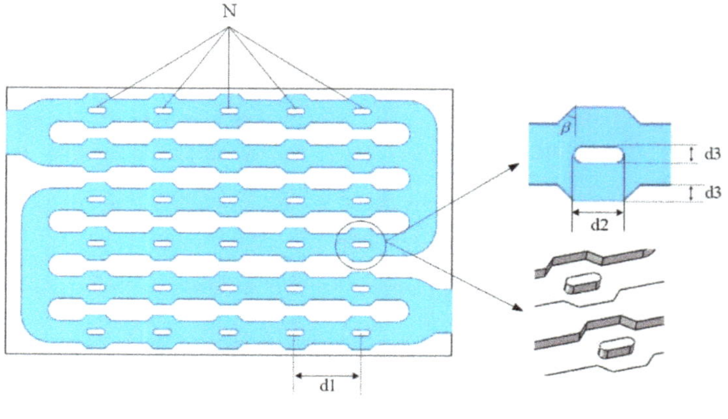

Figure 4. Design of the disturbance structure.

2.4. Computational Fluid Control Model

For the proposed mini-channel liquid cooling battery module, the flow and heat transfer process meet the conservation of mass, momentum, and energy equations [21]. The detailed expressions are as follows:

Mass conservation equation:

$$\frac{\partial \rho}{\partial t} + div(\rho u) = 0 \quad (5)$$

where ρ means density, u means velocity vector, and t means time.

Momentum conservation equation:

$$\frac{\partial(\rho u_x)}{\partial t} + div(\rho u_x u) = div(\mu grad u_x) - \frac{\partial p}{\partial x} + R_x \tag{6}$$

$$\frac{\partial(\rho u_y)}{\partial t} + div(\rho u_y u) = div(\mu grad u_y) - \frac{\partial p}{\partial y} + R_y \tag{7}$$

$$\frac{\partial(\rho u_z)}{\partial t} + div(\rho u_z u) = div(\mu grad u_z) - \frac{\partial p}{\partial z} + R_z \tag{8}$$

where μ denotes dynamic viscosity; u_x, u_y, and u_z denote the components of u in x, y, and z directions; p denotes pressure on the computational domain; R_x, R_y, and R_z denote generalized source terms.

Energy conservation equation:

$$\frac{\partial(\rho T)}{\partial t} + div(\rho u T) = div(\frac{\lambda}{C} grad T) + R_T \tag{9}$$

where λ denotes thermal conductivity, C denotes specific heat capacity, and R_T denotes heat source.

Mathematical formula of battery temperature field:

$$\rho c \frac{\partial T}{\partial t} = \lambda_x \frac{\partial^2 T}{\partial x^2} + \lambda_y \frac{\partial^2 T}{\partial y^2} + \lambda_z \frac{\partial^2 T}{\partial z^2} + q \tag{10}$$

where q means heat generation per unit volume of battery and λ_x, λ_y, and λ_z denote the thermal conductivity in x, y, and z directions, respectively.

Turbulence equation:

Turbulent flow phenomena exist in nature and in various engineering fields. Convective heat transfer during turbulent flow is a common method of heat transfer in engineering. Applying the standard model is currently the main method for solving the flow and heat transfer [26,27]. In this paper, the turbulence is considered according to Equation (11).

$$\rho \frac{\partial k}{\partial t} + \rho u_j \frac{\partial k}{\partial x_j} = \frac{\partial}{\partial x_j}\left[\left(\eta + \frac{\eta_t}{\sigma_k}\right)\frac{\partial k}{\partial x_j}\right] + \eta_t \frac{\partial u_i}{\partial x_j}\left(\frac{\partial u_i}{\partial x_j} + \frac{\partial u_j}{\partial x_i}\right) - \rho\varepsilon \tag{11}$$

where ε is the dissipation rate, and its control equation is shown as follows:

$$\rho \frac{\partial \varepsilon}{\partial t} + \rho u_k \frac{\partial \varepsilon}{\partial x_k} = \frac{\partial}{\partial x_k}\left[\left(\eta + \frac{\eta_t}{\sigma_\varepsilon}\right)\frac{\partial \varepsilon}{\partial x_k}\right] + \frac{c_1}{k}\eta_t \frac{\partial u_i}{\partial x_j}\left(\frac{\partial u_i}{\partial x_j} + \frac{\partial u_j}{\partial x_i}\right) - c_2\rho\frac{\varepsilon^2}{k} \tag{12}$$

2.5. Boundary Condition

The flow of the coolant fluid in the entire computational domain is set as a steady-state, steady incompressible turbulent flow; all walls are considered hydraulically smooth. K-epsilon standard turbulence equation and standard wall function are used in simulation calculation; the non-coupling implicit algorithm, second-order solution accuracy, and the energy equation are selected as well. Stop criteria are the maximum number of internal iteration steps and maximum physical time.

In addition, the heating generation power of batteries at different discharging rates has been determined by experiments. The volume flow inlet boundary condition is considered, and the value is set as 6 L/min; the pressure outlet boundary condition is considered, and the pressure value is standard atmospheric pressure. The ambient temperature is set as 27 °C. The heat exchange between the batteries and the air is considered according to Newton's law of cooling, and the expression is shown in Equation (13).

$$Q_e = hA(T_b - T_a) \tag{13}$$

where Q_e means the exchanged heat, h is the heat transfer coefficient, A means the surface area of the batteries, T_b means the battery temperature, and T_a means ambient temperature.

2.6. Grid independence Validation

In this paper, the hexahedral mesh is used for mesh construction in the simulation. To ensure computation accuracy, mesh densification is used for the fluid area, and the basic size is 0.2 mm. Shapes of the cooling plate and thermal conductive sheets are simple and regular, which have low requirements for the density of the grid, so larger grids can be used. The basic size is 1 mm. In order to ensure the accuracy of the calculation, the grid independence validation of the numerical model is conducted. Figure 5 shows the simulation results of the highest temperature and maximum temperature difference of the battery module under a 1 C discharging rate with different grid numbers. It can be seen that when the number of grids exceeds 8.4×10^5, the calculation results tend to be stable, which means that the simulation results are not affected by the number of grids. Therefore, the mesh number of 8.4×10^5 is adopted.

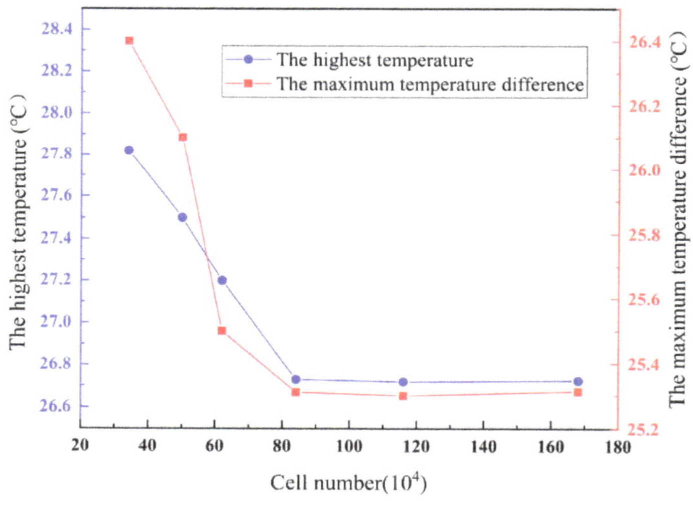

Figure 5. Grid dependency test.

3. The Influence of Disturbance Structure Arrangement on Enhanced Heat Transfer Performance

3.1. Influence of Disturbance Structure Number

In the liquid-cooled battery thermal management system, the coolant needs to be driven by an onboard water pump to circulate the flow, which requires additional system energy consumption and has an impact on the range of the whole vehicle. The pressure drop and friction factor in the flow channel are the main factors affecting energy consumption, and a larger pressure drop and friction factor indicate a higher pump power required The friction factor is the group of uncaused times when the fluid flows in the flow channel, and its expression is given in Equation (14).

$$f = \frac{\Delta P}{\frac{\rho u^2}{2} * \frac{L}{d_H}} \tag{14}$$

where f means the friction factor; ΔP is the pressure drop of the fluid in the channel, Pa; u is the fluid flow rate, m/s; L is the length of the channel, m; and d_H is the hydraulic diameter of the flow section, m.

According to the heat transfer theory, it is known that the increase in heat transfer area can enhance heat transfer efficiency. Adding the disturbance structure inside the mini-channel can change the flow state and increase the contact area with the coolant, which promotes the rapid heat transfer between the coolant and the mini-channel cooling plate. However, the disturbance structure also causes external pressure loss and frictional resistance. In order to study the influence of disturbance structure number, five design plans are compared, as shown in Figure 6. The plans include one, three, five, and seven disturbance structures. The original plan has none. The structure parameters of the disturbance in different plans are all length (mm) × width (mm) × height (mm) × tilt angle = 4 × 2 × 2 × 45°. In addition, the simulation conditions are all set as discharging at 3 C.

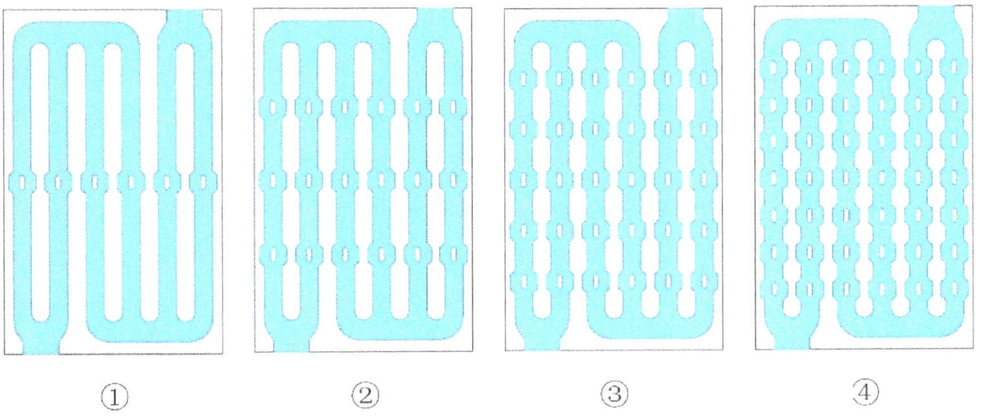

Figure 6. Different numbers of disturbance structures.

The average temperature distribution of the 12 batteries in the module at a 3 C discharge rate is shown in Figure 7. The average temperature of each battery decreases after disturbance structures are added to the mini-channel, and the decreasing trend increases when the disturbance structure number increases. When there are seven disturbance structures, the highest battery temperature is 36.18 °C, which is 0.63 °C lower than the initial plan, and the most obvious improvement in heat dissipation is achieved at this time. In addition, the addition of disturbance structures obviously changes the average temperature distribution of the batteries in the module. In the initial plan (no disturbance structure), the highest temperature occurs at the #4 battery, and the module temperature distribution shows the pattern of the temperature of the front batteries being higher than that of the back batteries. However, the temperature difference trend decreases a little when disturbance structures are added, and the temperature of all batteries decreases obviously when the number increases.

In addition to the average temperature of each battery, the maximum temperature of each battery is also monitored in the simulation. Figure 8 shows the maximum temperature and temperature difference of the battery module under different numbers of disturbances. With the addition of the disturbance structure, the maximum temperature and temperature difference both show a significant decrease compared to the initial plan. The lowest value appears when there are seven disturbance structures: the maximum temperature is 39.51 °C and the maximum temperature difference is 11.38 °C. Compared to the initial plan, the two values decrease by 0.75 °C and 0.41 °C, respectively. The phenomenon indicates that the flow field inside the mini-channel changes and the heat transfer is enhanced obviously, which further improves the temperature rise and uniformity. Figure 9 shows the effect of different numbers of disturbance structures on the performance of the flow channel. It can be seen that disturbance structures increase the pressure drop and friction factor. The increase is small when the number of the disturbance structures increases from zero to five, but when the number of disturbance structures increases from five to seven, the pressure drop and friction factor are 7445.76 Pa and 1.25, which increase 14% and 31%, respectively, which indicates that the energy consumption is obvious. Considering the heat dissipation performance and energy consumption, plan 3 (five disturbance structures in the mini-channel) is adopted in the following analysis.

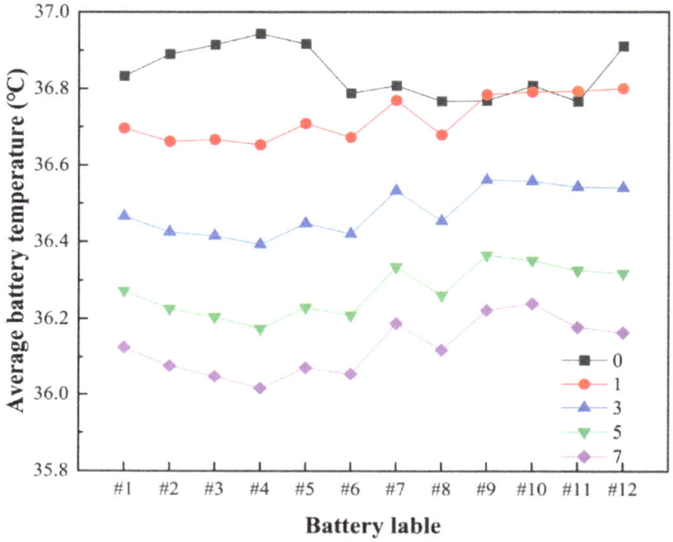

Figure 7. Battery temperature under different numbers of disturbance structures.

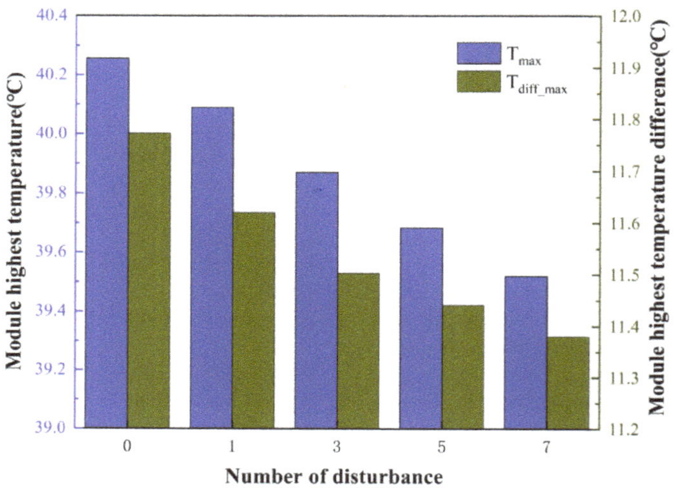

Figure 8. Module temperature under different numbers of disturbance structures.

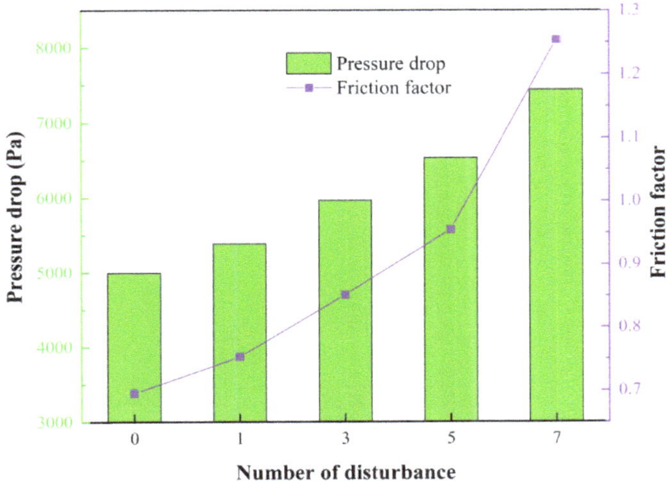

Figure 9. Channel performance under different numbers of disturbance structures.

3.2. Influence of Disturbance Structure Layout

Different distributions of the disturbance structures within the flow channel also affect the flow field of the coolant, which may improve the heat dissipation and the flow channel performance. Based on plan 3, four different distribution plans, numbered 5 to 8, are designed as shown in Figure 10. Plan 5 means the uniform distribution, plan 6 means the interlaced distribution, plan 7 means the gradually thinning distribution, and plan 8 means the gradually denser distribution. The structure parameters of the disturbance structure are the same as the initial value, and the simulation condition is still under the 3 C discharging rate.

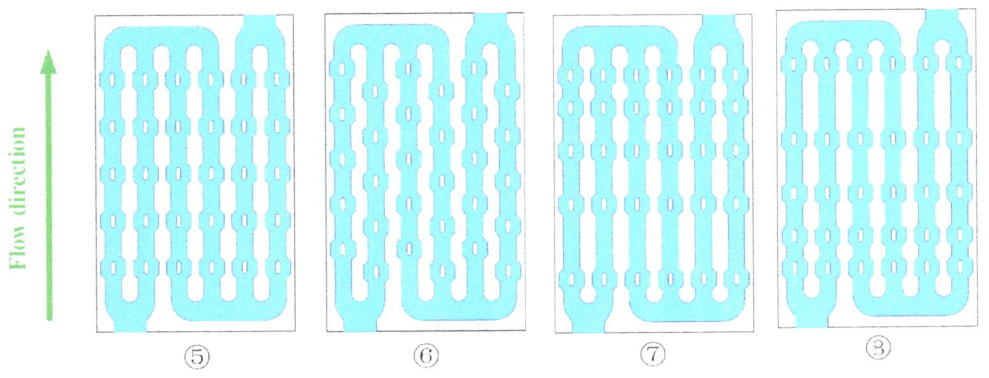

Figure 10. Different distributions of disturbance structures.

The effect of different distributions of the disturbance structures in the mini-channel on the average temperature of each battery is shown in Figure 11. The average temperature distribution of each battery is basically the same in the four plans, in which the average temperature of each battery is almost the lowest in plan 5 and is the highest in plan 8, which indicates the best and worst heat dissipation performance. In plan 5, the average temperature of battery #7 is the highest at 36.33 °C, and that of battery #4 is the lowest at 36.17 °C. The maximum average temperature difference is 0.16 °C, which indicates that the thermal balance performance is significant.

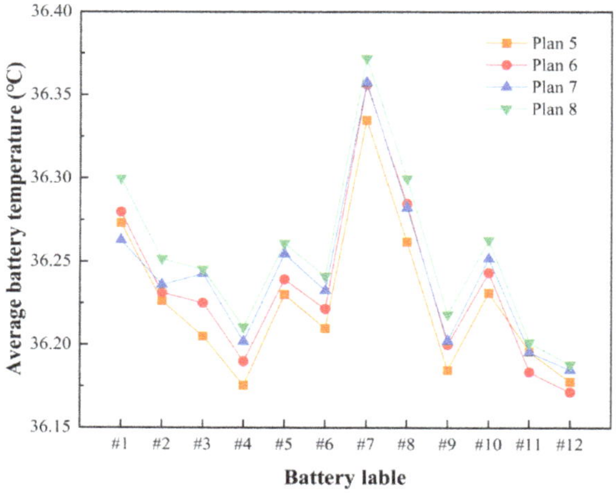

Figure 11. Average cell temperature under different distributions.

Figure 12 shows the maximum temperature and the maximum temperature difference under different distributions of disturbance structures. From the figure, it can be seen that the maximum temperature of the battery module in plan 5 is 36.68 °C, and the maximum temperature difference of the module in the remaining three plans is smaller but significantly higher than that of plan 5 (11.44 °C), which indicates that plan 5 has a greater advantage in temperature control. Figure 13 shows the influences of different distribution plans on the channel performance. The pressure drop and friction factor inside the mini-channel of plan 6 are the largest among the four plans, reaching 6743.98 Pa and 0.98, respectively, which indicates that the flow resistance is large and the operating power consumption of the pump is high. The difference in pressure drop between plan 5 and plan 7 is small; both are significantly lower than the remaining two plans, and while comparing the friction factors of the two plans, it can be seen that the friction factor of the mini-channel in the form of uniform distribution is 0.95, which is a larger decrease compared to plan 7. Therefore, the best performance of enhanced heat transfer within the mini-channel is achieved when five pieces of disturbance structures are uniformly distributed in each flow channel, and the following analysis of disturbance structure parameter optimization is carried out on this basis.

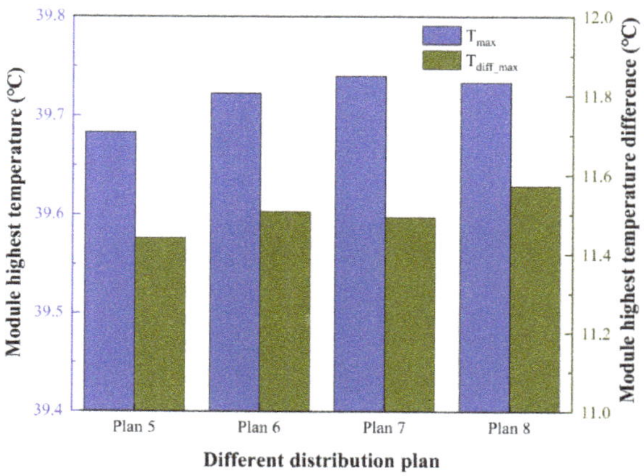

Figure 12. Module temperature under different distribution plans.

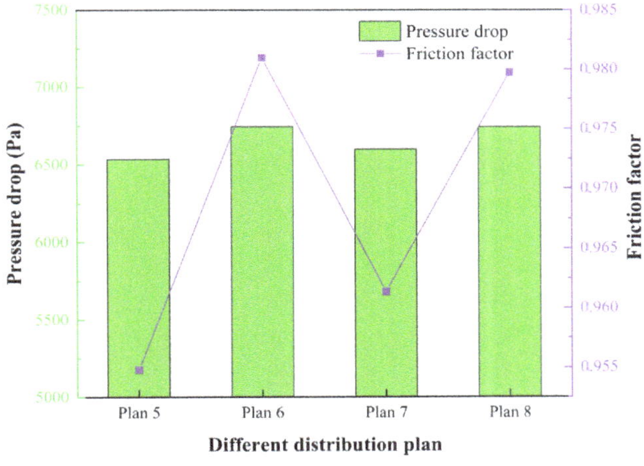

Figure 13. Channel performance under different distribution plans.

The coolant flow traces inside the mini-channel in the form of uniform distribution in plan 5 are shown in Figure 14. The coolant flow rate is relatively evenly distributed, with an average flow rate of about 0.8 m/s. As can be seen from the figure, the existence of the disturbance structure changes the original uniform flow field distribution. The flow velocity is relatively stable before the disturbance structure. However, after flowing through the disturbance structure, the growth of the fluid boundary layer is promoted, a new boundary layer is continuously generated in the middle region of the fluid, and even low-speed secondary flow and reflux will appear in the local region; the coolant is continuously disturbed and guided to the cavities on both sides, and the flow state of the fluid is abruptly changed so that its flow is constantly close to the wall of the flow channel, which increases the turbulence and accordingly enhances convective heat transfer in the mini-channel.

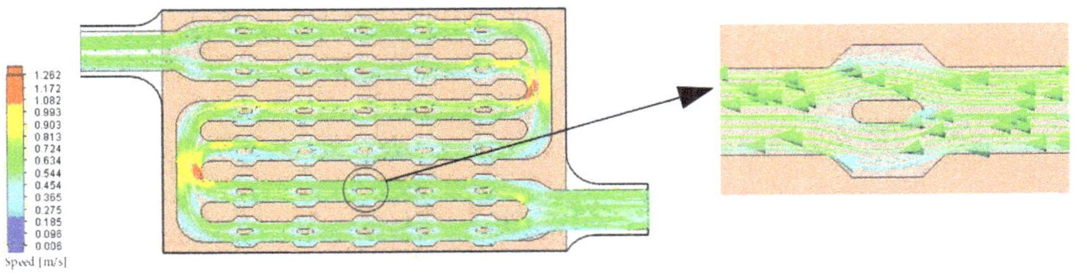

Figure 14. Fluid trace inside mini-channel in plan 5.

Figure 15 shows the change in the velocity vector of the coolant before and after flowing through the disturbance structure. When approaching the disturbance structure, the middle region of the fluid already starts to flow to both sides of the flow channel and creates a new boundary layer in the middle region. When reaching the middle region of the disturbance structure, the flow of fluid to both sides becomes more intense, and there is even backflow and secondary flow on both sides of the wall of the disturbed fluid. After passing through the disturbance structure, the fluids on both sides converge to the middle region again, and the boundary layer in the middle region gradually disappears. The process repeats at each disturbance structure, which promotes heat exchange with the wall during the lateral flow, thus enhancing the heat dissipation performance.

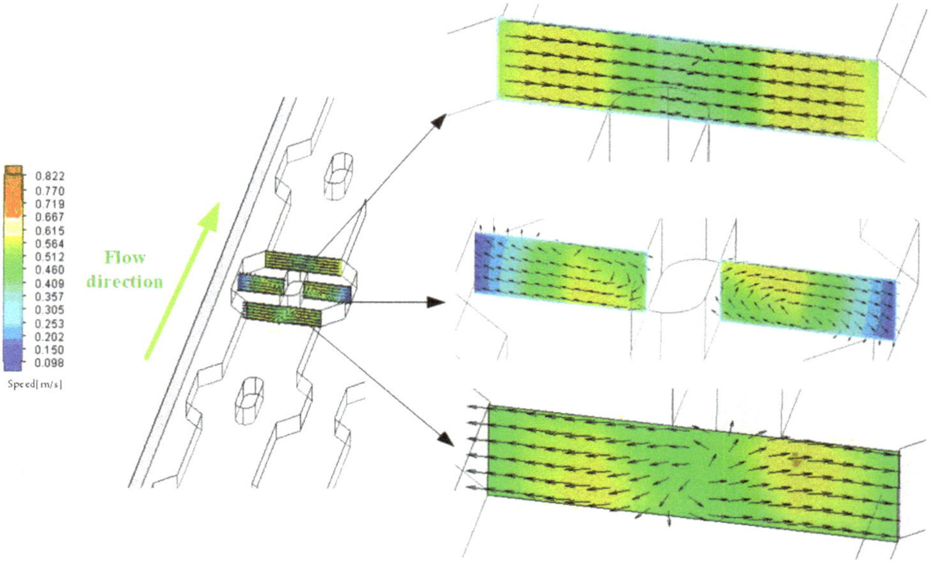

Figure 15. Cross-sectional view of flow velocity at different locations.

4. Parameter Optimization of the Disturbance Structure

4.1. Orthogonal Experiment Design

Changing the parameters of the disturbance structure will directly affect the heat dissipation performance of the mini-channel liquid cooling plate, and the processes of reconstructing the finite element model and conducting numerical simulation will be repeated many times if the parameters are changed one by one. This way may be overly demanding in terms of computation resources and time. In order to improve computational efficiency, the orthogonal test method is used to design the parameter optimization process [30]. Based on plan 5, the design parameters of the disturbance structure include space between adjacent disturbance structures d1, length d2, width d3, and tilt angle β. The factors and levels for this orthogonal experiment are shown in Table 3.

Table 3. Experimental factors and levels.

Level	Factor			
	d1 (mm)	d2 (mm)	d3 (mm)	β (°)
L1	14	3	1.5	15
L2	16	4	2	30
L3	18	5	2.5	45
L4	20	6	3	60

According to the test factors and levels in Table 3, the L_{16} (4^4) orthogonal test table was designed to carry out the orthogonal analysis. Adjacent disturbance structures d1, length d2, width d3, and tilt angle β were selected as the factors. The maximum temperature $T_{\max}$, maximum temperature difference T_{diff}, and friction factor f were selected as the evaluation indexes of the heat dissipation performance. Numerical simulations were conducted for each scheme under the same conditions: 3 C discharging rate, 25 °C the coolant inlet temperature, and 400 L/h coolant flow rate. The detailed scheme of the orthogonal experiment and results are shown in Table 4.

Table 4. Orthogonal test and simulation results.

Number	Factor				Evaluation Index		
	d1 (mm)	d2 (mm)	d3 (mm)	β (°)	T_{max} (°C)	T_{diff} (°C)	f
1	L1 (14)	L1 (3)	L1 (1.5)	L1 (15)	39.8433	11.4737	1.0430
2	L1 (14)	L2 (4)	L2 (2)	L2 (30)	39.8168	11.5788	1.1270
3	L1 (14)	L3 (5)	L3 (2.5)	L3 (45)	39.7650	11.5139	1.0949
4	L1 (14)	L4 (6)	L4 (3)	L4 (60)	39.7281	11.5574	0.9870
5	L2 (16)	L1 (3)	L2 (2)	L3 (45)	39.8034	11.4672	1.0533
6	L2 (16)	L2 (4)	L1 (1.5)	L4 (60)	39.7154	11.4105	0.9154
7	L2 (16)	L3 (5)	L4 (3)	L1 (15)	39.6499	11.5349	1.6825
8	L2 (16)	L4 (6)	L3 (2.5)	L2 (30)	39.7101	11.5225	1.2705
9	L3 (18)	L1 (3)	L3 (2.5)	L4 (60)	39.6481	11.4087	0.9787
10	L3 (18)	L2 (4)	L4 (3)	L3 (45)	39.7109	11.4318	1.1623
11	L3 (18)	L3 (5)	L1 (1.5)	L2 (30)	39.7883	11.4652	1.0405
12	L3 (18)	L4 (6)	L2 (2)	L1 (15)	39.7728	11.4790	1.2455
13	L4 (20)	L1 (3)	L4 (3)	L2 (30)	39.6904	11.4237	1.4371
14	L4 (20)	L2 (4)	L3 (2.5)	L1 (15)	39.7117	11.4396	1.3585
15	L4 (20)	L3 (5)	L2 (2)	L4 (60)	39.6155	11.3776	0.9546
16	L4 (20)	L4 (6)	L1 (1.5)	L3 (45)	39.6868	11.4347	0.9958

4.2. Range Analysis of Orthogonal Experiment Results

The range analysis method includes two steps of calculation and judgment. From the orthogonal test simulation results in Table 4, the sum of the factors K_i in each column can be calculated, and thus the average value k_i of K_i can be calculated as well. Based on the average value k_i, the range value of each factor for the evaluation indexes can be obtained according to Equation (15).

$$R_j = \max\{k_1, k_2, k_3, k_4\} - \min\{k_1, k_2, k_3, k_4\} \quad (15)$$

Therefore, the sensitivity of each structural parameter of the disturbance structure to different evaluation indexes can be visualized by the range value R derived from the orthogonal experiment. A larger R value indicates a greater sensitivity, i.e., a greater influence of the factor on the evaluation index. The results of the range analysis are shown in Table 5.

According to the results in Table 5, the sensitivity of each structural parameter of the disturbance structure is in the order of $d_1 > \beta > d_3 > d_2$ for the evaluation index maximum temperature, and the detailed values are d_1 = 20 mm, d_2 = 5 mm, d_3 = 3 mm, and β = 60°; for the maximum temperature difference, the order is $d_1 > \beta > d_2 > d_3$, and the detailed values are d_1 = 20 mm, d_2 = 3 mm, d_3 = 1.5 mm, and β = 60°; for the friction factor, the order is $\beta > d_3 > d_1 > d_2$, and the detailed values are d_1 = 14 mm, d_2 = 6 mm, d_3 = 1.5 mm, and β = 60°.

With d_1 as the horizontal coordinate and k as the vertical coordinate, the trend of the influence of the disturbance structure spacing on each evaluation index can be obtained, as shown in Figure 16. The maximum temperature difference of the battery module gradually decreases as the spacing of the disturbance structure increases, because the larger spacing indicates that the disturbance structure is more uniformly distributed inside the mini-channel, and the plate can evenly dissipate the heat to improve the temperature uniformity. In addition, the maximum temperature and the friction factor show an opposite trend, which indicates that the two indexes are negatively correlated. In this section, the maximum temperature and the maximum temperature difference achieve the best value when d_1 = 20 mm, and this value is selected as the final parameter, although the friction factor is not optimal. Here, the heat dissipation performance is considered first.

Table 5. Range analysis results.

Index	Parameter	Factor			
		d1 (mm)	d2 (mm)	d3 (mm)	β (°)
T_{max}	k_1	39.7883	39.7463	39.7585	39.7444
	k_2	39.7197	39.7387	39.7521	39.7514
	k_3	39.7323	39.7047	39.7101	39.7415
	k_4	39.6761	39.7245	39.6948	39.6768
	R	0.1122	0.0416	0.0637	0.0746
	Sensitivity	$d_1 > \beta > d_3 > d_2$			
	Best solution	d_1 = 20 mm, d_2 = 5 mm, d_3 = 3 mm, β = 60°			
T_{diff}	k_1	11.5310	11.4433	11.4427	11.4818
	k_2	11.4838	11.4652	11.4757	11.4976
	k_3	11.4462	11.4729	11.4712	11.4619
	k_4	11.4189	11.4984	11.4878	11.4386
	R	0.1121	0.0551	0.0451	0.059
	Sensitivity	$d_1 > \beta > d_2 > d_3$			
	Best solution	d_1 = 20 mm, d_2 = 3 mm, d_3 = 1.5 mm, β = 60°			
f	k_1	1.0630	1.1280	0.9987	1.3324
	k_2	1.2304	1.1408	1.0951	1.2188
	k_3	1.1068	1.1931	1.2323	1.0766
	k_4	1.1865	1.1247	1.3172	0.9589
	R	0.1674	0.0684	0.3185	0.3735
	Sensitivity	$\beta > d_3 > d_1 > d_2$			
	Best solution	d_1 = 14 mm, d_2 = 6 mm, d_3 = 1.5 mm, β = 60°			

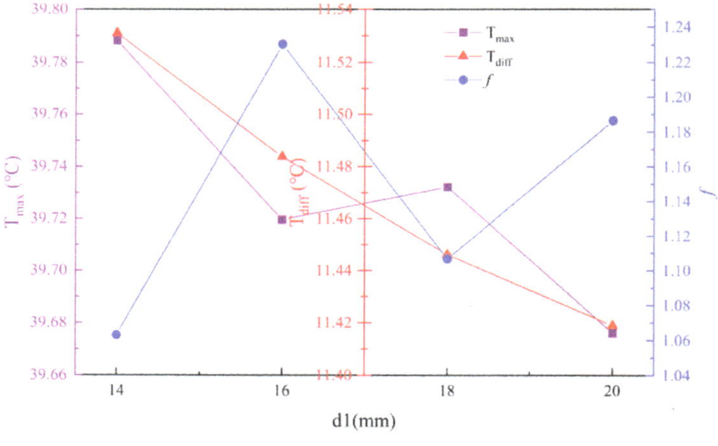

Figure 16. Influence of d_1 on evaluation indexes.

With d_2 as the horizontal coordinate and k as the vertical coordinate, the trend of the influence of the disturbance structure length on each evaluation index can be obtained as shown in Figure 17. As d_2 increases, the maximum temperature difference gradually increases, which indicates that increasing the length is not conducive to temperature uniformity. Actually, increasing d_2 means the distance between adjacent disturbance structures decreases, which is not conducive to the uniform distribution of the fluid, and the heat dissipation is better in the area where the disturbance structure exists. In addition, the maximum temperature and the friction factor show an opposite trend. On the one hand, increasing the length will cause the area of convective heat transfer inside the mini-channel to increase, thus improving the heat dissipation performance; on the other hand, increasing the length will disturb the original flow traces of the coolant, the velocity and pressure fields inside the mini-channel will change, and the coolant will diverge when it is close to the disturbance structure, so the frictional resistance inside the flow channel will also increase. However, when the length reaches 5 mm, increasing the length will cause the maximum temperature to rise, which indicates that when the disturbance structure is too long, the spacing will become too short, and the coolant will only flow from both sides of the disturbance structure. There will be a dead zone between the adjacent disturbance structures, and the friction factor will become smaller, but at the same time, the heat dissipation performance will be affected, and when the length is 5 mm, the enhanced heat transfer performance of the mini-channel liquid cooling plate is the best.

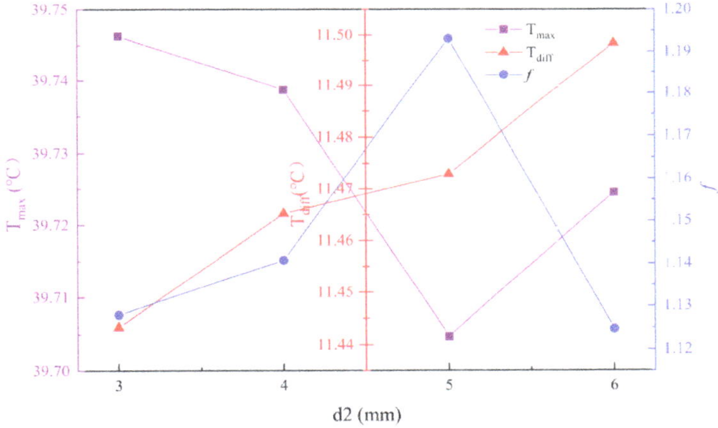

Figure 17. Influence of d_2 on evaluation indexes.

With d_3 as the horizontal coordinate and k as the vertical coordinate, the trend of the influence of the disturbance structure width on each evaluation index can be obtained, as shown in Figure 18. As d_3 increases, the maximum temperature decreases, but the friction factor gradually increases. On the one hand, increasing the width of the disturbance structure causes an increase in the convective heat transfer area inside the mini-channel; on the other hand, the disturbing effect on the fluid is enhanced, the stable flow field is abruptly changed, the boundary layer is generated in the middle region of the fluid, and the turbulence is enhanced. In addition, the overall trend of the maximum temperature difference is increasing, which indicates that increasing the width will cause the coolant to divert to both sides of the disturbance structure, and the coolant flow rate is lower in the area between adjacent disturbance structures, resulting in uneven distribution of coolant flow and preventing the battery surface from being uniformly dissipated, thus causing poor temperature uniformity.

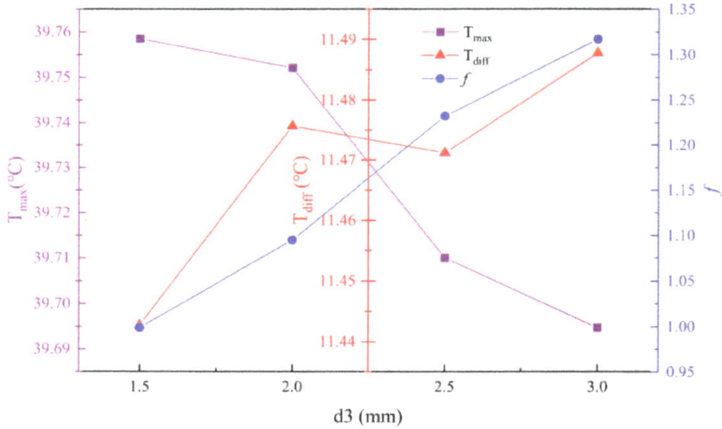

Figure 18. Influence of d_3 on evaluation indexes.

With β as the horizontal coordinate and k as the vertical coordinate, the trend of the influence of the disturbance structure width on each evaluation index can be obtained, as shown in Figure 19. The friction factor gradually decreases as β increases, which indicates that the coolant in the cavity is subject to less frictional resistance when flowing out of the cavity, which is more conducive to the natural flow of the fluid. However, the maximum temperature and the maximum temperature difference increase slightly and then decrease significantly when β increases to 30°, which indicates that the heat dissipation is the worst at this value. At this value, the coolant flow velocity is low at the opening angle of the cavity, and there is even a dead zone, so the coolant cannot flow evenly to dissipate heat. In addition, as shown in Table 5, the sensitivity of the friction factor to each evaluation index is the lowest among all factors, which indicates that it has the least influence on the enhanced heat transfer performance, and the thermal characteristics of the battery module and the performance of the flow channel are the best when the cavity opening angle is 60°.

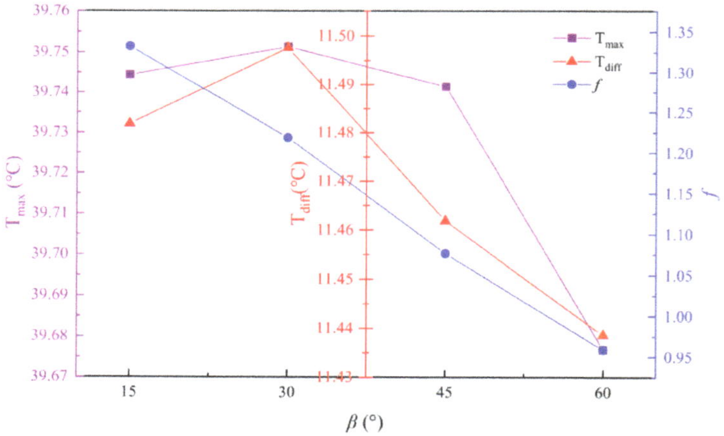

Figure 19. Influence of β on evaluation indexes.

Based on the determined parameters, a comparison between the BTMS in this paper and a BTMS with passive air cooling is performed. In the air-cooling system, the module is the same as that in this paper without cooling plates and thermal conductive sheets. All batteries are in contact with each other. The environmental conditions and discharging rate are the same. In addition, the computation domain is 20 mm larger than each side of the module, considering that the battery pack is compact in real-world applications. The results are shown in Table 6. It can be seen that the maximum average temperature cannot be controlled with a compact module structure under passive air cooling, and it reaches 49.06 °C. The maximum average temperature difference reaches 6.41 °C, 5.48 °C higher than before.

Table 6. Comparison with passive air cooling.

Cooling Type	Mini-Channel	Passive Air Cooling
Max average temperature (°C)	35.83	49.06
Max average temperature difference (°C)	0.93	6.41
Discharging rate	3 C	3 C

5. Conclusions

An efficient battery thermal management system plays an important role in electric vehicle operation. In this paper, a novel liquid-cooling system based on mini-channel plates with disturbance structures is proposed. In order to design the cooling system, the battery model is established first. Verified by heat generation experiments, the model achieves high accuracy with an error of less than 4%. Then, the battery module model consisting of 12 batteries and 6 cooling plates is established. Five plate designs are proposed first with zero, one, three, five, and seven disturbance structures, respectively. Plan 3 (five disturbance structures) is determined by considering the heat dissipation performance and flow channel performance. Then, four layout plans of the disturbance structures are proposed. Results show that plan 5 (disturbance structures distributed evenly) achieves the best performance both in the heat dissipation performance and flow channel performance. Under a 3 C discharging rate, the highest average temperature is 36.33 °C and the maximum average temperature difference is 0.16 °C. Based on plan 5, the orthogonal experiment and range analysis are adopted especially for the optimization of the disturbance structures. The factors include the space between adjacent disturbance structures d1, length d2, width d3, and tilt angle β. The evaluation indexes include the maximum temperature, maximum temperature difference, and friction factor. Results of the range analysis show that the best combination of the four parameters is d1 = 20 mm, d2 = 5 mm, d3 = 1.5 mm, and $\beta = 60°$.

Author Contributions: Methodology, study design, experiment, model, R.L.; software, data analysis, data collection, Y.Y.; writing, literature search, figures, F.L.; data collection, data analysis, writing modification, J.L.; study design, results discussion, methodology modification, X.C.; funding acquisition, J.L. and X.C. All authors have read and agreed to the published version of the manuscript.

Funding: The project is supported in part by the Prospective Study Funding of Nanchang Automotive Innovation Institute, Tongji University (No. QZKT2020-01), National Natural Science Foundation of China (No. 62103415).

Data Availability Statement: All the research data has been provided in this paper.

Conflicts of Interest: The authors declare no conflict of interest.

References

1. Farahani, G. DC–DC series-resonant converter with multi-stage current-driven for balance charger of series-connected lithium-ion batteries. *IET Power Electron.* **2021**, *14*, 992–1007. [CrossRef]
2. Li, R.; Wang, H.; Dai, H.; Hong, J.; Tong, G.; Chen, X. Accurate state of charge prediction for real-world battery systems using a novel dual-dropout-based neural network. *Energy* **2022**, *250*, 123853. [CrossRef]
3. Ashok, B.; Kannan, C.; Mason, B.; Ashok, S.; Indragandhi, V.; Patel, D.; Wagh, A.; Jain, A.; Kavitha, C. Towards safer and smarter design for lithium-ion-battery-powered electric vehicles: A comprehensive review on control strategy architecture of battery management system. *Energies* **2022**, *15*, 4227. [CrossRef]
4. Ouyang, Q.; Chen, J.; Zheng, J. State-of-charge observer design for batteries with online model parameter identification: A robust approach. *IEEE Trans. Power Electron.* **2019**, *35*, 5820–5831. [CrossRef]

5. Naguib, M.; Kollmeyer, P.; Emadi, A. Lithium-ion battery pack robust state of charge estimation, cell inconsistency, and balancing. *IEEE Access* **2021**, *9*, 50570–50582. [CrossRef]
6. Ghaeminezhad, N.; Ouyang, Q.; Hu, X.; Xu, G.; Wang, Z. Active cell equalization topologies analysis for battery packs: A systematic review. *IEEE Trans. Power Electron.* **2021**, *36*, 9119–9135. [CrossRef]
7. Liu, H.; Wei, Z.; He, W.; Zhao, J. Thermal issues about li-ion batteries and recent progress in battery thermal management systems: A review. *Energy Convers. Manag.* **2017**, *150*, 304–330. [CrossRef]
8. Kim, J.; Oh, J.; Lee, H. Review on battery thermal management system for electric vehicles. *Appl. Therm. Eng.* **2019**, *149*, 192–212. [CrossRef]
9. Samadi, M.; Saif, M. Integrated battery management system. In *Integrated Systems: Innovations and Applications*; Springer: Berlin/Heidelberg, Germany, 2015; pp. 173–193.
10. Khan, M.; Swierczynski, M.; Kær, S. Towards an ultimate battery thermal management system: A review. *Batteries* **2017**, *3*, 9. [CrossRef]
11. Zhao, J.; Rao, Z.; Li, Y. Thermal performance of mini-channel liquid cooled cylinder based battery thermal management for cylindrical lithium-ion power battery. *Energy Convers. Manag.* **2015**, *103*, 157–165. [CrossRef]
12. Mohammadian, S.; Zhang, Y. Temperature uniformity improvement of an aircooled high-power lithium-ion battery using metal and nonmetal foams. *J. Heat Transf.* **2016**, *138*, 114502. [CrossRef]
13. Wu, W.; Yang, X.; Zhang, G.; Chen, K.; Wang, S. Experimental investigation on the thermal performance of heat pipe-assisted phase change material based battery thermal management system. *Energy Convers. Manag.* **2017**, *138*, 486–492. [CrossRef]
14. Fathabadi, H. High thermal performance lithium-ion battery pack including hybrid active–passive thermal management system for using in hybrid/electric vehicles. *Energy* **2014**, *70*, 529–538. [CrossRef]
15. Chen, F.; Huang, R.; Wang, C.; Yu, X.; Liu, H.; Wu, Q.; Qian, K.; Bhagat, R. Air and PCM cooling for battery thermal management considering battery cycle life. *Appl. Therm. Eng.* **2020**, *173*, 115154. [CrossRef]
16. Saw, L.; Ye, Y.; Tay, A.; Chong, W.; Kuan, S.; Yew, M. Computational fluid dynamic and thermal analysis of lithium-ion battery pack with air cooling. *Appl. Energy* **2016**, *177*, 783–792. [CrossRef]
17. Chen, K.; Chen, Y.; Li, Z.; Yuan, F.; Wang, S. Design of the cell spacings of battery pack in parallel air-cooled battery thermal management system. *Int. J. Heat Mass Transf.* **2018**, *127*, 393–401. [CrossRef]
18. Ling, Z.; Cao, J.; Zhang, W.; Zhang, Z.; Fang, X.; Gao, X. Compact liquid cooling strategy with phase change materials for Li-ion batteries optimized using response surface methodology. *Appl. Energy* **2018**, *228*, 777–788. [CrossRef]
19. Malik, M.; Dinsor, I.; Rosen, M. Review on use of phase change materials in battery thermal management for electric and hybrid electric vehicles. *Int. J. Energy Res.* **2016**, *40*, 1011–1031. [CrossRef]
20. Chen, Y.; Nguyen, D.; Shen, M.; Yip, M.; Tai, N. Thermal characterizations of the graphite nanosheets reinforced paraffin phase-change composites. *Compos. A Appl. Sci. Manuf.* **2013**, *44*, 40–46. [CrossRef]
21. Huang, Y.; Cheng, W.; Zhao, R. Thermal management of Li-ion battery pack with the application of flexible formstable composite phase change materials. *Energy Convers. Manag.* **2019**, *182*, 9–20. [CrossRef]
22. Luo, X.; Guo, Q.; Li, X.; Tao, Z.; Lei, S.; Liu, J.; Kang, L.; Zheng, D.; Liu, Z. Experimental investigation on a novel phase change material composites coupled with graphite film used for thermal management of lithium-ion batteries. *Renew. Energy* **2020**, *145*, 2046–2055. [CrossRef]
23. Ali, H. Applications of combined/hybrid use of heat pipe and phase change materials in energy storage and cooling systems: A recent review. *J. Energy Storage* **2019**, *26*, 100986. [CrossRef]
24. Jiang, Z.; Qu, Z. Lithium–ion battery thermal management using heat pipe and phase change material during discharge–charge cycle: A comprehensive numerical study. *Appl. Energy* **2019**, *242*, 378–392. [CrossRef]
25. Lei, S.; Shi, Y.; Chen, G. A lithium-ion battery-thermal-management design based on phase-change-material thermal storage and spray cooling. *Appl. Therm. Eng.* **2020**, *168*, 114792. [CrossRef]
26. Lai, Y.; Wu, W.; Chen, K.; Wang, S.; Xin, C. A compact and lightweight liquid-cooled thermal management solution for cylindrical lithium-ion power battery pack. *Int. J. Heat Mass Transf.* **2019**, *144*, 118581. [CrossRef]
27. Shen, J.; Wang, Y.; Yu, G.; Li, H. Thermal management of prismatic lithium-ion battery with minichannel cold plate. *J. Energy Eng.* **2020**, *146*, 04019033. [CrossRef]
28. Madani, S.; Schaltz, E.; Kær, S. Thermal analysis of cold plate with different configurations for thermal management of a lithium-ion battery. *Batteries* **2020**, *6*, 17. [CrossRef]
29. Tan, X.; Lyu, P.; Fan, Y.; Rao, J.; Ouyang, K. Numerical investigation of the direct liquid cooling of a fast-charging lithium-ion battery pack in hydrofluoroether. *Appl. Therm. Eng.* **2021**, *196*, 117279. [CrossRef]
30. Wu, S.; Lao, L.; Wu, L.; Liu, L.; Lin, C.; Zhang, Q. Effect analysis on integration efficiency and safety performance of a battery thermal management system based on direct contact liquid cooling. *Appl. Therm. Eng.* **2022**, *201*, 117788. [CrossRef]
31. Zhou, H.; Zhou, F.; Zhang, Q.; Wang, Q.; Song, Z. Thermal management of cylindrical lithium-ion battery based on a liquid cooling method with half-helical duct. *Appl. Therm. Eng.* **2019**, *162*, 114257. [CrossRef]
32. Rao, Z.; Qian, Z.; Kuang, Y.; Li, Y. Thermal performance of liquid cooling based thermal management system for cylindrical lithium-ion battery module with variable contact surface. *Appl. Therm. Eng.* **2017**, *123*, 1514–1522. [CrossRef]
33. Shang, Z.; Qi, H.; Liu, X.; Ouyang, C.; Wang, Y. Structural optimization of lithium-ion battery for improving thermal performance based on a liquid cooling system. *Int. J. Heat Mass Transf.* **2019**, *130*, 33–41. [CrossRef]

34. An, Z.; Jia, L.; Li, X.; Ding, Y. Experimental investigation on lithium-ion battery thermal management based on flow boiling in mini-channel. *Appl. Therm. Eng.* **2017**, *117*, 534–543. [CrossRef]
35. Yates, M.; Akrami, M.; Javadi, A. Analysing the performance of liquid cooling designs in cylindrical lithium-ion batteries. *J. Storage Mater.* **2019**, *33*, 100913. [CrossRef]
36. Panchal, S.; Khasow, R.; Dincer, I.; Agelin-Chaab, M.; Fowler, M. Thermal design and simulation of mini-channel cold plate for water cooled large sized prismatic lithium-ion battery. *Appl. Therm. Eng.* **2017**, *122*, 80–90. [CrossRef]
37. Panchal, S.; Khasow, R.; Dincer, I.; Agelin-Chaab, M.; Fraser, R.; Fowler, M. Numerical modeling and experimental investigation of a prismatic battery subjected to water cooling. *Numer. Heat Transf. Part A Appl.* **2017**, *71*, 626–637. [CrossRef]
38. Huang, Y.; Mei, P.; Lu, Y.; Huang, R.; Yu, X.; Chen, Z.; Roskilly, A.P. A novel approach for lithium-ion battery thermal management with streamline shape mini channel cooling plates. *Appl. Therm. Eng.* **2019**, *157*, 113623. [CrossRef]
39. Liu, H.; Eze, C.; Zhao, J. Investigation into the effectiveness of nanofluids on the mini-channel thermal management for high power lithium ion battery. *Appl. Therm. Eng.* **2018**, *142*, 511–523. [CrossRef]

electronics

MDPI

Article

Analysis and Optimization Based on Factors Affecting the Spiral Climbing Locomotion of Snake-like Robot

Peng Zhang [1], Yong Zang [1,2], Ben Guan [1,2,*], Zhaolin Wu [1] and Zhiying Gao [1,2]

1 School of Mechanical Engineering, University of Science and Technology Beijing, Beijing 100083, China
2 Shunde Innovation School, University of Science and Technology Beijing, Foshan 528399, China
* Correspondence: guanben@ustb.edu.cn; Tel.: +86-10-62334197

Abstract: The snake-like robot is a limbless bionic robot widely used in unstructured environments to perform tasks with substantial functional flexibility and environmental adaptability in complex environments. In this paper, the spiral climbing motion of a snake-like robot on the outer surface of a cylindrical object was studied based on the three-dimensional motion of a biological snake, and we carried out the analysis and optimization of the motion-influencing factors. First, the spiral climbing motion of the snake-like robot was implemented by the angle control method, and the target motion was studied and analyzed by combining numerical and environmental simulations. We integrated the influence of kinematics and dynamics factors on the spiral climbing motion. Based on this, we established a multi-objective optimization function that utilized the influence factors to optimize the joint module. In addition, through dynamics simulation analysis, the change of the general clamping force of the snake-like robot's spiral climbing motion was transformed into the analysis of the contact force between the joint module and the cylinder. On the basis of the results, the effect of the control strategy adopted in this paper on the motion and change rule of the spiral climbing motion was analyzed. This paper presents the analysis of the spiral climbing motion, which is of great theoretical significance and engineering value for the realization of the three-dimensional motion of the snake-like robot.

Keywords: snake-like robot; spiral climbing; influencing factors; optimization design; locomotion analysis

Citation: Zhang, P.; Zang, Y.; Guan, B.; Wu, Z.; Gao, Z. Analysis and Optimization Based on Factors Affecting the Spiral Climbing Locomotion of Snake-like Robot. *Electronics* **2022**, *11*, 4002. https://doi.org/10.3390/electronics11234002

Academic Editor: Gianpaolo Vitale

Received: 30 October 2022
Accepted: 30 November 2022
Published: 2 December 2022

1. Introduction

The snake-like robot is bionic, with two free ends and multiple joints in tandem. Compared with the traditional legged mobile robots with a single or a few degrees of freedom, the snake-like robot has unique advantages and functional characteristics of multiple redundant degrees. It is, accordingly, extensively used in many fields, such as disaster rescue, work-state inspection, aerospace exploration, med-science research, and military reconnaissance [1–6].

Depending on the natural environment, snakes are capable of achieving a variety of locomotor gaits [7]. In particular, serpentine, rectilinear, and concertina locomotion are the most typical planar gaits. Moreover, these can accomplish many forward locomotor goals in planar environments. Three-dimensional gaits are developed to adapt to atypical environments and consist of sidewinding movements, intimidating movements when facing predators, and spiral climbing movements on tree trunks or the outer walls of pipes. Three-dimensional gaits are a complex gait of the snake that combines environmental factors and its characteristics with extreme functionality and adaptability. Among these, the spiral climbing motion extends the forward movement capabilities of the snake in planar motion to three-dimensional space, enriching the survival space of the snake in nature. Thus, studying the spiral climbing motion of snake-like robots is conducive to improving biomimicry and environmental interaction.

Generally, researchers have conducted studies related to the 3D motions of snake-like robots from different perspectives.

Hatton proposed a tread-based model for sidewinding. The new interpretation of the gait further admits a symmetry-based model reduction, and the behavior of an eclipse between the snake-like robot and the sloped surface in rolling contact was comprehensively analyzed [8]. Gong proposed a new and geometrically intuitive method to study the snake-like robot's steering strategy and turn rates of sidewinding gait by tapering the core cylinder into a cone [9]. Qi combined the hyperbolic curve with the helical curve in space to propose a new motion of helical wave propagation. By changing the hyperbolic function's parameters, the composite curve's stable propagation was achieved [10]. Yaqub developed a spiral curve gait whose joint angles are calculated by a Bellows model based on the curvature and torsion of the backbone curve, in which the rolling motion of the snake-like robot adapting to the variable diameter of the climbing pole was studied [11]. Rollinson found that the periodic gait of the snake-like robot was characterized by symmetry in form. They used the virtual chassis method to separate the different gaits of the snake-like robot into internal motion and external motion [12]. Zhou proposed a spring-like type of robot climbing pipe gait to solve the problem that the existing climbing pipe has a high demand of continuity of the pipe, employing variable curvature and variable torque discretization [13].

We divide the method that researchers in the current study on the 3D gait of snake-like robots into two main categories: curve approximation method [14] and curve discretization [13]. The former achieves the target curve's fitting by controlling the form of the snake-like robot. At the same time, the latter discretizes the ideal curve into nodes and controls the nodes' higher-order physical parameters to realize the snake-like robot's control. Although researchers can adopt both curve analysis and higher-order parameters to achieve 3D motion gait control, the correlation and influence between motion gait and the spatial shapes of snake-like robots still need to be improved in the current research.

Therefore, we studied the spiral climbing motion of the snake-like robot in this paper. The main contents are as follows: in Section 2, we realize the control of the spiral climbing motion of the snake-like robot, and the influencing factors affecting the motion state are obtained by combining kinematic and dynamic analysis methods. In Section 3, the cost function of each factor is established. Based on this, we create a multi-objective optimization function towards the robot's joint module, and thus, we complete the optimization analysis of the joint module design parameters. In Section 4, this paper shows a prototype of the snake-like robot, and the spiral climbing effect of the robot is verified and analyzed by simulation experiments. In the last chapter, we present the conclusion and discussion.

2. Analysis of Locomotion Patterns

There are two main types of spiral climbing motions of snakes [7,15], which are referred to as "folds and stretches" and "spirals and laterals" according to the different motion features. As shown in Figure 1A, the former type of movement is "head extension—head wrapping—tail contraction—tail wrapping". The snake achieves its entire spiral climbing through periodic changes of partial movements. In the latter case, as shown in Figure 1B, only part of the snake's body forms a spiral, while the other part forms a bend in the cylindrical surface and moves upward laterally, and this bend can alternate from side to side frequently.

The spiral climbing motion in nature is undoubtedly the most logical and efficient way to move [16], but it is too difficult to achieve for snake robots. The fundamental reason is that the joint module of the robots is less miniature and quantitative than a snake bone. Thus, researchers have modified this technically limited motion into a spiral rolling system.

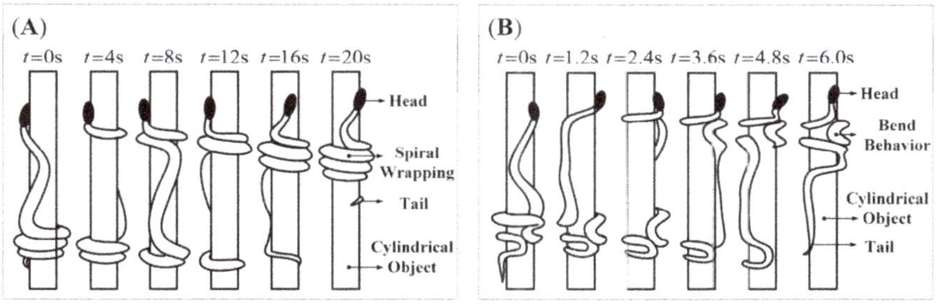

Figure 1. Different types of spiral climbing movement by a snake: (**A**) "Folds and Stretches"—wrapping helically and moving forward; and (**B**) "Spirals and Laterals"—Gripping during lateral undulation.

2.1. Method for Locomotion

In the current research, there are three main types of motion control methods for snake-like robots: the continuum method, the central pattern generator (CPG) method, and the function control method.

The Frenet–Serret expansion equation is the crucial point to the continuum model. Yamada first proposed the continuous model of active cord mechanism (ACM) and specifically classified it into the planar, Frenet–Serret, Bellows, and complete models according to the characteristics, and provided a method to deal with ACM's 3D shape [17]. Kamegawa realized cylinder locomotion with helical form by a snake-like robot using Yamada's ACM theory, and then the results were applied to a mechanical discrete snake-like robot model [18]. Qi developed a discrete control model for the joints of a snake-like robot using a continuous curve model and proposed a novel obstacle avoidance strategy for the robot wraps around the outside of a pipe [10,18]. Zhou's general parameter-based method discretized the spatial curve into the model with variable curvature and torque and applied the model of motion for gait by dividing functional areas to complete climbing over a stepped shaft and a discontinuous pipe [13]. Yaqub solved and calculated the joint angle of the snake-like robot based on the Bellows model utilizing a curvature integration algorithm [11]. Manzoor proposed a new algorithm for generating different rhythmic motions based on CPG, such as serpentine, sidewinding, two-step concertina, and four-step concertina [19]. Chirikjian presented an efficient kinematic modeling method for a snake-like robot based on the backbone curve, which reduces the inverse kinematics to the time-varying relative to the reference coordinate system of the backbone curve to describe the macroscopic geometric properties of the snake-like robot [20]. Lipkin described two categories of differential gaits: differentiable gaits; and segmented differentiable gaits. Based on the high redundancy characteristics of the snake-like robot, they established the function of the differentiable gaits and verified the feasibility of the gaits through experiments [21]. Choset proposed the CMU control model, which numbers the orthogonally connected joint modules of snake-like robots and outputs sinusoidal function control waves to the odd and even joint modules, respectively. Therefore the CMU control function is also known as a compound serpenoid curve [8,9,12,22]. Through the study of the spiral climbing gait of the snake-like robot, Sun proposed the angle control function model based on isometric spiral trajectory, which established the functional relationship between the form spiral trajectory angle and the joint angle. They analyzed the mechanical equilibrium performance of the snake-like robot's climbing gait [23]. Based on the serpentine curve proposed by Hirose, Wei compounded the motion model with the cylindrical helix equation to present a new simplified control function model, which established the relationship between the joint angle and time for the robot, and verified that it could accomplish a variety of motion gaits through experiments [24].

This paper focuses on the implementation and analysis of the spiral climbing motion developed by the snake-like robot, which requires both accurate output and control of the joint modules. In addition, it is significant that a robust motion control method is the key to the spiral climbing state of the snake robot. Consequently, this paper is unique due to the comprehensive analysis and optimization of the influencing factors on the robot's spiral climbing motion.

2.2. Control of Spiral Climbing Locomotion

The flexibility of the snake is derived from its multi-joint structure of physiological characteristics. Hence in this paper, we simplified the snake to a multi-link mechanism, as shown in Figure 2, assuming that no lateral sliding occurs during the spiral climbing motion [18,24–26].

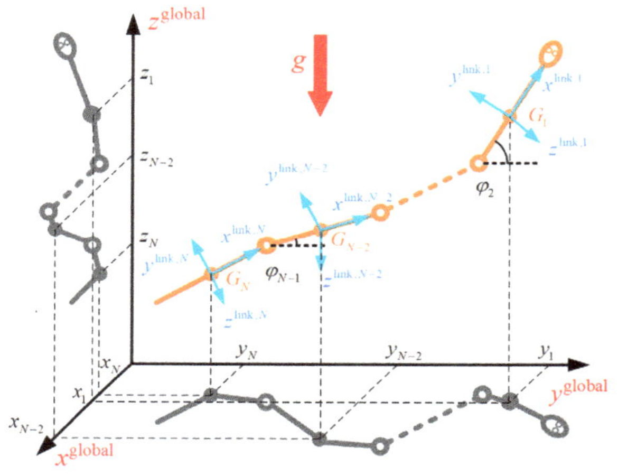

Figure 2. The Multi-link model of the snake-like robot.

Hirose first introduced the serpentine curve, which allows the serpentine gait of the snake-like robot by controlling the change of joint angle, and its control equation is described as:

$$\varphi_i = A\sin(\omega t + (i-1)\beta) + \lambda \tag{1}$$

The serpentine motion is a planar gait, but the spiral climbing motion is a 3D gait, for which a simplified cylindrical helix equation is introduced. As a result, we obtain the expression by compounding it with the angle control equation of the serpentine motion as follows:

$$\begin{cases} x = A\sin(\omega t) \\ y = A\cos(\omega t) \\ z = t \end{cases} \tag{2}$$

$$\varphi(i,t) = \begin{cases} A_{\text{even}}\sin(\omega t + (i-1)\beta_{\text{even}}) + \lambda_{\text{even}} \\ A_{\text{odd}}\sin(\omega t + (i-1)\beta_{\text{odd}}) + \lambda_{\text{odd}} \end{cases} \tag{3}$$

According to the connection order, we sort the joints of the snake-like robot into two categories, odd and even. A is the amplitude, which controls the rotational direction of the snake-like robot in the spiral climbing motion. ω is the frequency controlling the execution rate of the actuator (time part). λ is the compensation angle relative to the spiral centerline, and it controls the motion direction of the snake-like robot (space part). And β is the phase difference. According to the experiment, it is learned that the snake-like robot moves in rectilinear or inward climbing gait when $\beta = 0$, while spiral climbing gait when $\beta \neq 0$.

This paper presents the control parameters $A_{odd} = A_{even} = A$, $\beta_{odd} = \beta_{even} = \beta \neq [0 \quad \pi/2]$, and $\lambda_{odd} = \lambda_{even} = \lambda$. The joint angle control function for the spiral climbing motion of the snake robot is as follows:

$$\varphi(i,t) = A\sin(\omega t + \beta i) + \lambda \tag{4}$$

Figure 3 shows that we can obtain the change in the snake-like robot's spiral climbing gait by adjusting the control parameters A, β, and λ, respectively. We can learn that the parameter A affects the pitch of the spiral climbing gait, and the more significant A is, the more extensive the output range of the joint angle becomes. Accordingly, the complete spatial pattern of the robot shows a more twisted state at this time. Parameter β determines the radius of the robot's spiral, and the radius decreases with increasing β. In this case, the variation of the adjacent joint angle becomes more prominent, which in turn decreases the pitch of the spiral climbing gait. Thus, the robot shows a compressed state as a result. Parameter λ has no significant effect on the motion itself, except that it changes the direction of the robot's spiral at the global level. Thereby the effect of λ is not considered in the subsequent studies.

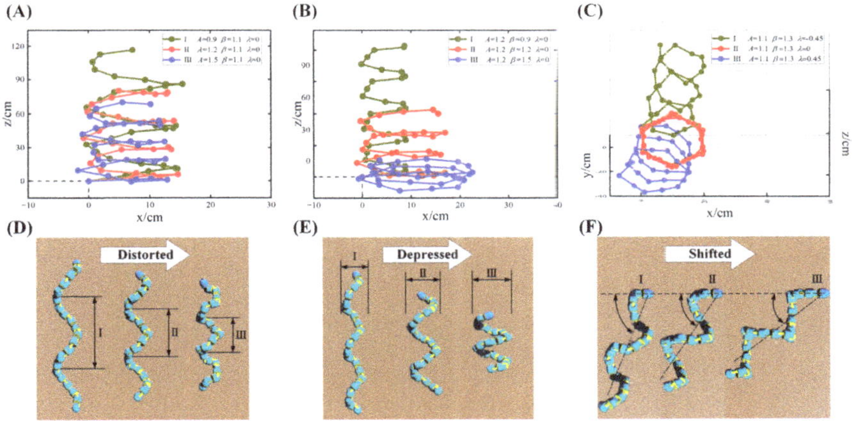

Figure 3. Different spiral climbing gaits when changing the parameters: (**A**,**D**) the effect of A on spiral climbing gait; (**B**,**E**) the effect of β on spiral climbing gait; (**C**,**F**) the effect of λ on spiral climbing gait.

2.3. *Influencing Factors*

2.3.1. Radius of Spiral Climbing Gaits

From the analysis of Section 2.2, we find that both A and β have a determinative effect on the spiral climbing gait, as reflected in that by changing a single parameter, the radius and pitch of the form spiral of the snake-like robot will change. As shown in Figure 4, there exists any one joint module in contact with the surface of the cylindrical object when the snake robot wraps its body around the surface of the cylinder. We describe the two ends of the module are described as $\mathbf{p}_i = \begin{bmatrix} x_i & y_i & z_i \end{bmatrix}^T$ and $\mathbf{p}_{i+1} = \begin{bmatrix} x_{i+1} & y_{i+1} & z_{i+1} \end{bmatrix}^T$, respectively, using the improved D-H parameter method, at which the distance between the joint module and the centerline of the cylindrical object can be expressed as:

$$d_i^{i+1} = (y_{i+1}x_i - y_i x_{i+1})\left((x_{i+1} - x_i)^2 + (y_{i+1} - y_i)^2\right)^{-1/2} \tag{5}$$

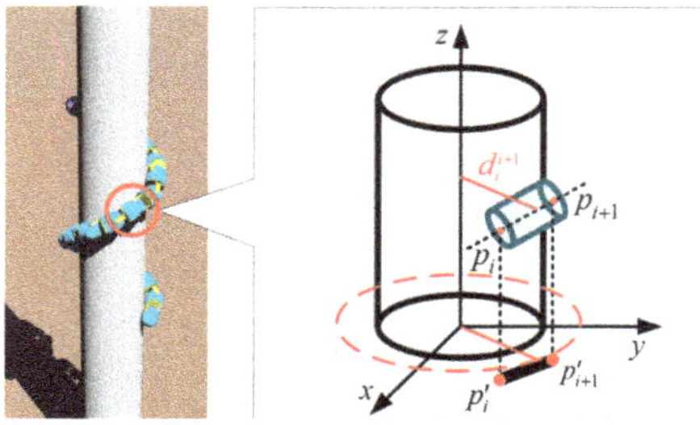

Figure 4. The single joint module in contact with the surface of a cylinder.

Ideally, the midpoint of the joint module should coincide with the generatrix of the cylinder, as described by the following equation:

$$\frac{x - x_0}{q_1} = \frac{y - y_0}{q_2} = \frac{z - z_0}{q_3} \tag{6}$$

Consequently, we equate the distance between the joint and the centerline of the cylinder to the radius of the form spiral of the robot's gait:

$$R_d = d_i^{i+1} = \frac{\begin{vmatrix} x_i - x_0 & y_i - y_0 & z_i - z_0 \\ x_{i+1} - x_i & y_{i+1} - y_i & z_{i+1} - z_i \\ q_1 & q_2 & q_3 \end{vmatrix}}{\left(\begin{vmatrix} y_{i+1} - y_i & z_{i+1} - z_i \\ q_2 & q_3 \end{vmatrix}^2 + \begin{vmatrix} z_i - z_0 & x_i - x_0 \\ q_3 & q_1 \end{vmatrix}^2 + \begin{vmatrix} x_i - x_0 & y_i - y_0 \\ q_1 & q_2 \end{vmatrix}^2 \right)} \tag{7}$$

The spatial spiral radius of the snake-like robot is associated with six parameters, including x_0, y_0, z_0, q_1, q_2, and q_3. In order to establish the relationship among these parameters, this paper uses the particle swarm optimization algorithm (PSO) to optimize it. We use 31 sets of data from 30 joint modules as samples, with the minimum mean square deviation of R_d as the optimization objective, set the number of examples of optimization $n = 31$, the maximum number of iterations of optimization $t_{ger} = 5000$, and the learning factor $c_1 = c_2 = 2.05$. Since there is uniqueness in the parameters obtained after optimization, this paper randomly selects one set of solutions, and Table 1 shows the data obtained by optimization:

Table 1. The radius of spiral climbing gaits by PSO.

	1	2	3	4	5	6	7	8	9	10
A	0.40	0.40	0.50	0.50	0.60	0.66	0.80	0.80	1.00	1.00
λ	1.00	1.30	0.90	1.35	1.30	0.70	1.00	1.20	1.20	1.35
R_d/cm	5.19	17.27	4.53	21.86	9.78	3.34	7.40	12.49	6.46	8.78

The fitting equation between the control parameters and the spatial spiral radius obtained based on the PSO establishment is as follows:

$$R_d = \frac{c_1 + c_2 A + c_3 A^2 + c_4 \lambda}{1 + c_5 A + c_6 A^2 + c_7 \lambda + c_8 \lambda^2} \tag{8}$$

Table 2 shows the 8 optimal values of Equation (8).

Table 2. The optimal values of the fitting equation.

	c_1	c_2	c_3	c_4	c_5	c_6	c_7	c_8
Optimal value	−0.0812	2.2446	−0.5158	0.0149	0.0187	0.0756	−1.2878	0.4128

2.3.2. Contact Point

During the ideal spiral climbing motion, the contact mode between the joint module of a snake-like robot and the surface of a cylindrical object can be classified into two categories, central point contact and non-central point contact [25], as shown in Figure 5. Since the snake robot is a tandem multi-joint robot, the contact mode of the first joint determines the general contact mode. Hence, the contact between the joint modules and the cylindrical surface is usually non-center point contact in the practical motion which is highly random.

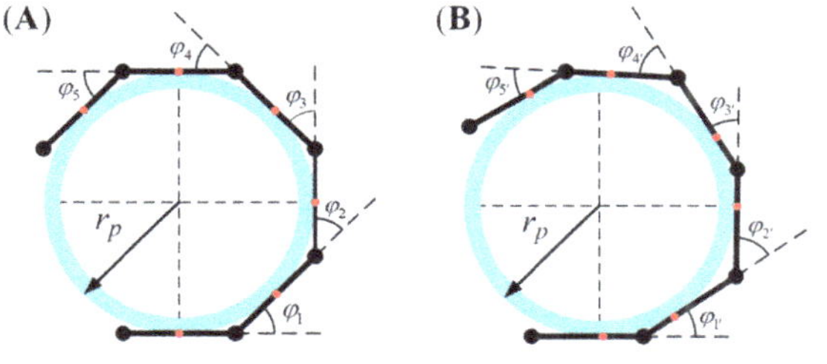

Figure 5. The contact models of a snake-like robot with the cylinder: (**A**) Central point contact model. (**B**) Non-central point contact model.

For further analysis of the contact circumstances, we simplify the adjacent joint modules into a shuttle-like structure with connected heads and tails, as shown in Figure 6.

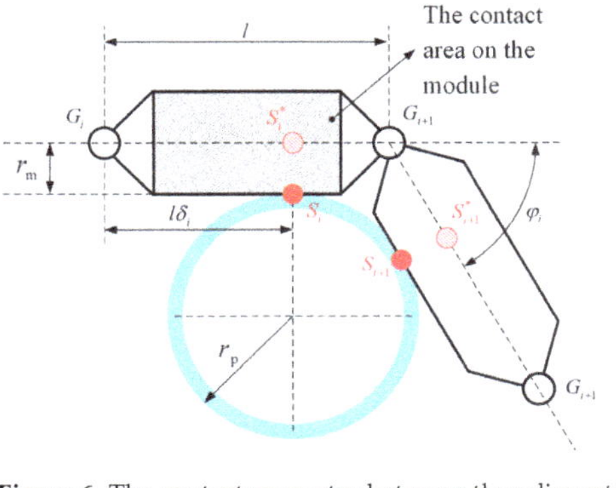

Figure 6. The contact geometry between the adjacent joint module and cylinder.

Taking $\delta_i = \overline{S^*G_i}/l \in [0,1]$ represents the coefficient of contact point position, and $\delta_i = 0.5$ means the central point contact. For non-central point contact, the larger δ_i is, the closer the contact point is to the backward joint. For t calculation, we determine that the coefficient of contact point position of the odd-joint module and the even-joint are to satisfy the requirement:

$$\delta_{\text{odd}} = 1 - \delta_{\text{even}} \tag{9}$$

The cylindrical coordinate system is established based on the cylinder, and then the contact point position matrix of the snake robot joint module is as follows:

$$\mathbf{S}_i = \left(r_p, \vartheta, z_{S_i}\right) \tag{10}$$

$$\vartheta = \sum_{i=0}^{n} \phi_i \tag{11}$$

$$\varphi_i = 2\arctan\left(\frac{(1-\delta_i l\cos\varepsilon_i)}{r_\mathrm{p}+r_\mathrm{m}}\right) \tag{12}$$

$$z_{S_i} = (i-\delta_0+\delta_i)l\sin\varepsilon_i \tag{13}$$

The end position of the joint module G_i is represented in the cylindrical coordinate system as:

$$G_i = \left(r_{G_i}, \phi_{G_i}, z_{G_i}\right) \tag{14}$$

$$\phi_{G_i} = \sum_{i=0}^{n} \phi_i + \phi_i/2 \tag{15}$$

$$r_{G_i} = \sqrt{(r_\mathrm{m}+r_\mathrm{p})^2+((1-\delta_i)l\cos\varepsilon_i)^2} \tag{16}$$

$$z_{G_i} = (1-c_0)l\cos\varepsilon_i \tag{17}$$

Figure 7 shows the projection of the two adjacent joint modules in contact with the cylinder in the $z = 0$ plane.

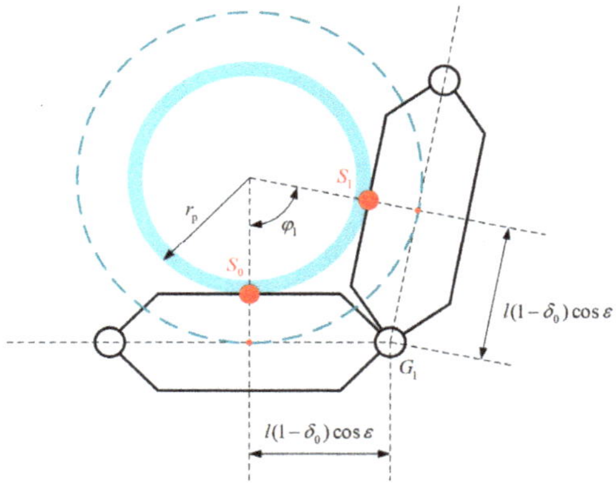

Figure 7. The contact between two adjacent modules and the cylinder.

The projected location of the contact point position on the joint centerline is expressed as follows:

$$\mathbf{S}_i^* = \mathbf{S}_i + \mathbf{r}_m \tag{18}$$

$$\mathbf{r}_\mathrm{m} = (r_\mathrm{m}, 0, 0) \tag{19}$$

$$\varphi_i = \pi - \arccos((\mathbf{a}_i \cdot \mathbf{b}_i)/\|\mathbf{a}_i\| \cdot \|\mathbf{b}_i\|) \tag{20}$$

$$\mathbf{a}_i = \mathbf{S}_i^* - \mathbf{G}_{i+1} \tag{21}$$

$$\mathbf{b}_i = \mathbf{G}_{i+1} - \mathbf{S}_{i+1}^* \tag{22}$$

$$\mathbf{a}_i \cdot \mathbf{b}_i = \|\mathbf{a}_i\| \cdot \|\mathbf{b}_i\| \cos(\pi - \varphi_i) \tag{23}$$

In the theoretical analysis, the spiral inclination angle of the robot's joint is $0 \le \varepsilon < \pi/2$ Simultaneously, in this paper, we consider that l and r_p, the length of the joint module of the snake-like robot and the radius of the cylinder, are satisfied to exist in a designing proportion. Due to the limitation of the mechanical structure, $\varphi_i \le \varphi_{\max} < \pi/2$. We assume that $l \ge 2r_p$ ($\varepsilon = 0$), in this case there will be a problem that the joint angle of the snake-like robot is all equal to $\pi/2$. Apparently, it does not meet the design requirements. Thus, we establish the design portion of the joints in this paper as follows:

$$l_{\max} < 2(r_{\mathrm{p}} + r_{\mathrm{m}}) \cos \varepsilon \tag{24}$$

Figure 8 shows the parameters of the joint module. The angle of the joint module's shuttle part is the same as the range of the rotation angle of the joint, which is $[0, \pi/2]$. Therefore, we consider that the maximum rotation angle of the joint and the inclination angle of the shuttle part are identical, both of which are $\varphi_{\max}$. The parameters of the joint module should meet the requirements as follows:

$$\max(r_{\mathrm{m}}) = \tan \varphi_{\max} l(1 - \delta_{\max}) \tag{25}$$

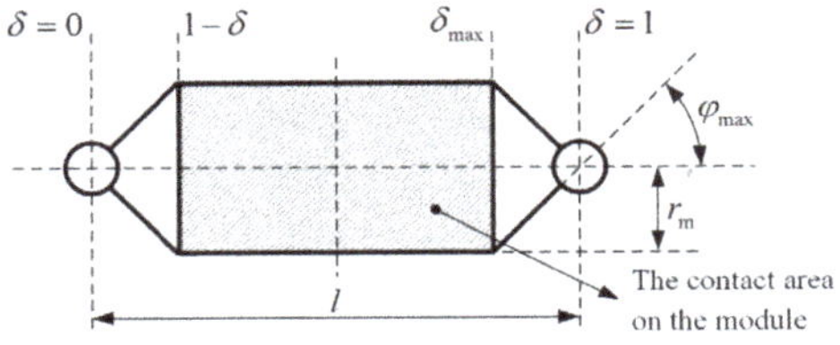

Figure 8. The parameters of a single joint module.

3. Analysis of Optimization

In the spiral climbing motion of the snake-like robot, influencing factors from kinematics and mechanics can disturb the motion itself. However, after the motion becomes stable, three critical factors are identified, including the number of joint modules of the snake-like robot, the forward velocity of the spiral climbing motion, and the output torque of the joints. As a result, this paper analyzes and optimizes the spiral climbing motion of the snake-like robot based on these three influencing factors.

3.1. Cost Function Based on Factors

3.1.1. The Number of Joint Modules

We learn that the number of joint modules impacts the state of the snake-like robot wrapping around the surface of the cylindrical object. Theoretically, the more joint modules connected by a snake-like robot, the larger the radius of the cylindrical object for the snake-like robot to wrap around, but the control cost will also increase. We can calculate the number of joint modules required to wrap spirally around the cylinder surface for one cycle by deriving the maximum angle between adjacent joint modules, and the cost of the spiral wrap based on the central point contact is expressed as follows:

$$cost_n = \frac{2\pi}{\max(\varphi_{\mathrm{odd}})} = \frac{2\pi}{\max(\varphi_{\mathrm{even}})} \tag{26}$$

In the case of non-central point contact, φ_{odd} and φ_{even} change with c_{odd} and c_{even}. Therefore, the cost of spiral wrapping is as follows:

$$cost_n = 2 \times \frac{2\pi}{\varphi_{\mathrm{odd}} + \varphi_{\mathrm{even}}} \tag{27}$$

3.1.2. The Forward Velocity of Spiral Climbing

The forward velocity of the snake-like robot's spiral climbing motion is related to the radius r_{m} of the joint module and the spiral inclination angle ε. As shown in Figure 9, the joint module with a larger radius has a faster forward velocity within the same spiral inclination angle ε.

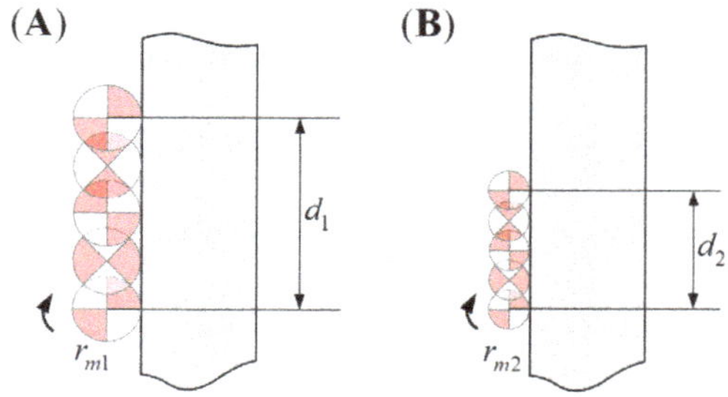

Figure 9. Comparison of modules with different radii: (**A**) joint module with radius of r_{m1}; and (**B**) joint module with radius of r_{m2}.

The forward velocity of the snake-like robot on the surface of the cylindrical object is the vertical component of the global velocity. Since the horizontal component of the global velocity causes the interaction between the robot itself and the cylinder, the cost function of the climbing velocity is as follows:

$$cost_v = \frac{\cos \varepsilon}{r_{\mathrm{m}}} \tag{28}$$

3.1.3. The Output Torque of Joints

The output torque of the joint is the critical parameter to ensure the motion of snake-like robots. As shown in Figure 10, we simplify the joint module of the snake-like robot in contact with the surface of the cylindrical object to a single cantilever module for analysis. We also neglect the anterior module reaction force to the analyzed joint. Moreover, the reaction force to the joint module cancels off with the normal force and the horizontal component of the friction force. Meanwhile, considering that the analyzed joint module is an ideal linkage without thickness, the vertical component of the friction force cancels off with the gravitational force of the linkage, expressed as follows:

$$f_f = W = \mu f_N \tag{29}$$

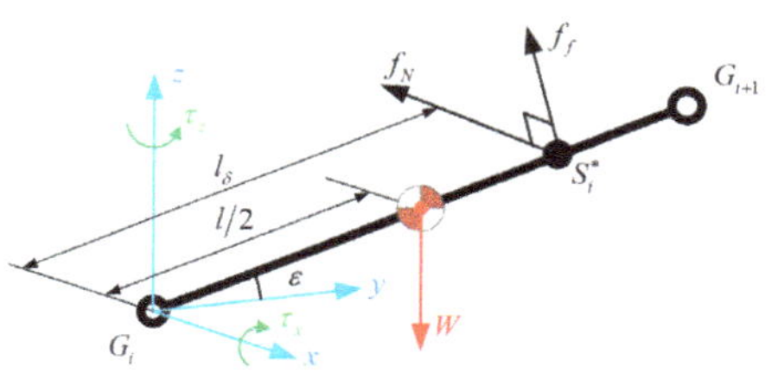

Figure 10. The forces of a single joint.

Since it is an ideal linkage without thickness, we can calculate the gravity of the joint module approximately as the volume of the module:

$$W = l r_{\mathrm{m}}^2 \tag{30}$$

$$\tau_x = 0.5 l \cos \varepsilon W - f_f l \delta_{\max} \tag{31}$$

$$\tau_z = f_N l_{\delta_{\max}} \tag{32}$$

Thus, the cost of the torque to the joint module is as follows:

$$cost_\tau = \sqrt{\tau_x^2 + \tau_y^2} \tag{33}$$

3.2. Optimization Design

From Section 3.1, we analyze that different factors affect the spiral climbing motion from different perspectives. Therefore, this paper adopts a linear combination to consider the cost of these factors comprehensively. In order to take the weights and sensitivities of different factors into account, the weight coefficients $w = [w_1 \quad w_2 \quad w_3]$ and sensitivity factors $\sigma = [\sigma_1 \quad \sigma_2 \quad \sigma_3]$ are introduced, respectively. Hence the multi-objective optimization function of the spiral climbing motion is created by the cost function as follows:

$$cost = \sigma_1 w_1 cost_n + \sigma_2 w_2 cost_v + \sigma_3 w_3 cost_\tau \tag{34}$$

where Equation (34) uses the sensitivity factor to normalize the three influencing factors and make it possible to calculate them in the same order of magnitude, and it uses the sensitivity factor by sensitivity analysis.

$$\sigma_j = \frac{1}{Max_j} \tag{35}$$

There are restrictions on the parameters of the joints in the individual cost functions. $\delta_{\max} = 0.5$ means that we choose the central contact model for calculation. Moreover, we define $l \in [0.2, 2]$ and $r_m \in [0.2, 2]$, both of which limit the joint module's size of the design to no larger than the cylinder's radius. Furthermore, $\varepsilon \in [0, \pi/2]$ requires that the snake-like robot can't be vertical relative to the ground. We calculate the sensitivity factor as $\sigma = [0.0819 \quad 0.3788 \quad 1.01]$. Each weight in the multi-objective optimization function has to be positive, representing the importance of the corresponding influencing factor compared to the other two factors. Also, the three need to be content with $w_1 + w_2 + w_3 = 1$. Consequently, we can obtain the optimization parameters of a single joint module according to different cylinder and spiral climbing. Table 3 shows the optimization input parameters and the optimized outputs.

Table 3. The optimal design parameters of joint module when different cases.

Symbol	Meaning	Case 1	Case 2	Case 3	Case 4
r_p	Radius of Cylinder	20	20	20	20
μ	Friction Coefficient	0.4	0.4	0.4	0.4
ε	Helical Pitch	10	10	10	10
$\varphi_{\max}/°$	Maximum Rotation Angle	75	75	75	75
δ_i	Coefficient of Contact Point	0.5	0.5	0.5	0.5
$l_{\min}/\text{cm}$	Minimum Length of Module	5	5	5	5
$r_m\|_{\max}/\text{cm}$	Maximum Length of Module	5	5	5	5
w_1	Weight of $cost_n$	0.15	0.15	0.70	0.33
w_2	Weight of $cost_v$	0.15	0.70	0.15	0.33
w_3	Weight of $cost_\tau$	0.70	0.15	0.15	0.33
l/cm	Length of Module	9.2326	11.2376	8.0152	12.1638
r_m/cm	Radius of Module	4.2781	4.5013	3.0201	4.1032

According to Equation (34) and the constraints mentioned above, we get four cases based on weights for the snake-like robot's joint module parameters. The data in Table 3 clearly show the optimized length and radius of the joint module. Moreover, Due to the linear combination among the weights, the sensitivity factors, and the cost functions, different weights have a magnification effect on different influence factors. Therefore, these four cases form a comparison and reference to each other, and the optimization results are compelling. Figure 11 shows the snake-like robot whose joint module is designed based on the specific case.

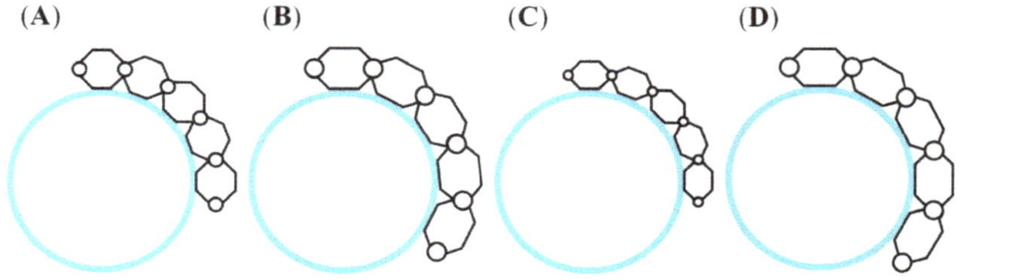

Figure 11. The snake-like robot whose joint module is designed based on the optimization cases: (**A**) case 1; (**B**) case 2; (**C**) case 3; and (**D**) case 4.

4. Simulation

In order to evaluate the performance of the snake-like robot with the optimized joint module. We design a snake-like robot with joint parameters $l = 12.16$ cm and $r_m = 4$ cm with a cylinder of $r_p = 20$ cm as the climbing target object, and the mechanical performance of this model is undertaken.

4.1. Modeling of the Snake-like Robot

Figure 12 illustrates the snake-like robot we designed, which has 20 joint modules connected orthogonally. A servo controls a single joint, and the rotation axes between two adjacent joints are perpendicular, making the snake-like robot have both ten pitch and ten yaw degrees of freedom, correspondingly.

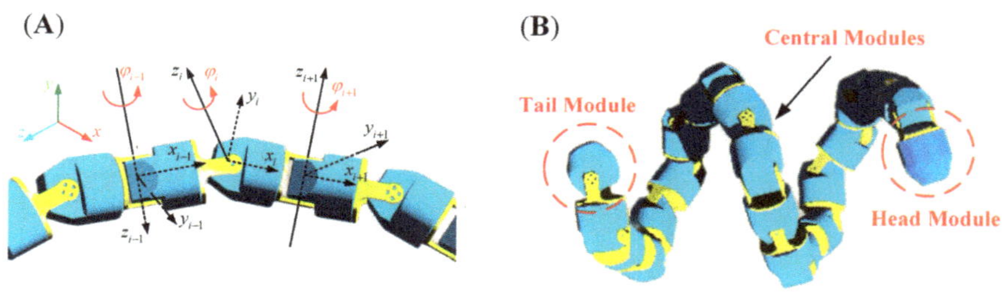

Figure 12. The snake-like robot with orthogonal joints: (**A**)the joint modules are connected orthogonally; and (**B**) the snake-like robot is modeled by 20 modules which can be divided into head, tail, and central modules.

We analyze the contact mode between the joint module and the cylinder in the process of climbing in Section 2.3. A reasonable contact state is an essential guarantee of the grip force required for the snake-like robot to wrap around the surface of the cylinder. Theoretically, the grip force is the sum of the frictional forces on all joint modules. However, the measurement and acquisition of the frictional forces are incredibly challenging to achieve in practice. Thus, this paper defines the contact force generated when the joint module is in contact with the cylinder as a collision. It can be obtained by dynamics simulation, which can transform the analysis of the grip force applied to the snake-like robot into that of contact force. Table 4 shows the setting parameters of the contact force constraint.

Table 4. The parameters of the contact force constraint.

Parameter	Value	Parameter	Value
Stiffness (N/mm)	2855.00	Dynamic Friction Coeff.	0.25
Damping $((N \cdot \text{s})/\text{mm})$	0.57	Static Friction Vel. (mm/s)	0.10
Exponent	1.10	Dynamic Friction Vel. (mm/s)	10.00
Penetration Depth (mm)	0.10	Coefficient of Restitution	0.80
Static Friction Coeff.	0.30	-	-

The spiral climbing motion of the snake-like robot is shown in Figure 13.

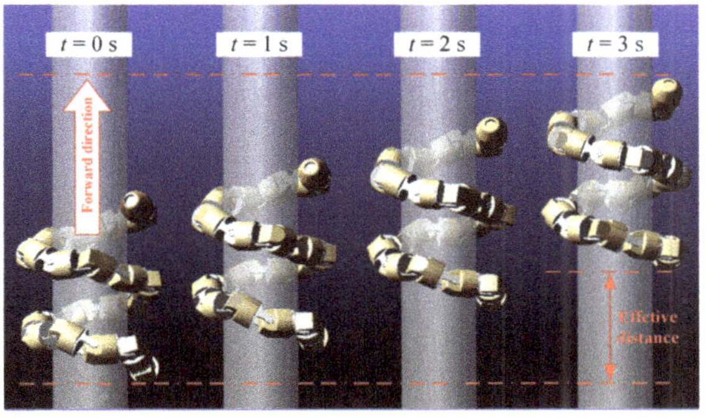

Figure 13. The Spiral climbing motion of the snake-like robot.

4.2. Simulations and Results

4.2.1. Vibration at Startup

The simulation indicates that in the moment of motion initiation, as the snake-like robot changes from the zero-moment state to the state of spiral climbing motion, the robot needs to break the original equilibrium state to another. Thereby, a shaking exists. Taking the head joint as an example, as shown in Figure 14, because of the contact between the head joint and the cylinder surface at the moment of the startup, the instantaneous contact force appears to change drastically, resulting in a vibration trend in the displacement of the head module. Furthermore, as the snake robot's posture gradually gets balanced, the joint module's displacement and contact force begin to change smoothly. Therefore, the first 0.5 s of the startup moment were ignored in the subsequent analysis to obtain a better analysis of the spiral climbing motion.

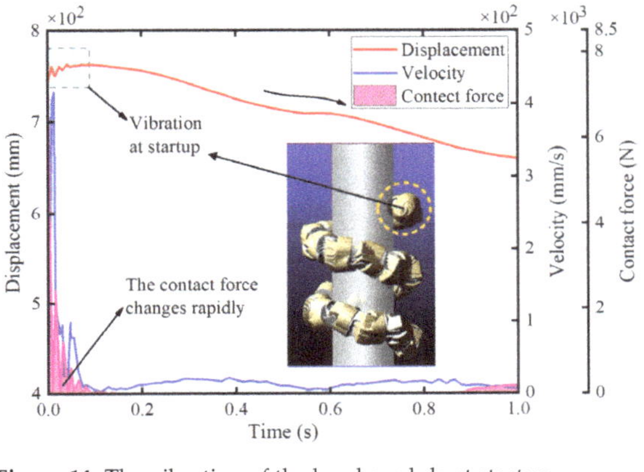

Figure 14. The vibration of the head module at startup.

4.2.2. Periodic Variation

Due to the tandem characteristics of the snake-like robot, there are interactions between adjacent joints. We select the central joint as the object of analysis in this paper.

The output torque and energy consumption of the 10th and 11th joints are analyzed as shown in Figure 15A. The output torque of the joints has the characteristic of periodic variation. However, there are multiple spikes in a single cycle, which is attributed to the gravity, friction, and inertia forces functioning simultaneously on the anterior and posterior joints at this time. This complicated force situation makes the combined force of the joint modules vary widely. Therefore, there are small fluctuations in the output within a single cycle, although the overall range of variation in output torque remains small. With the similarity between the joint module's instantaneous energy consumption and the variation of the output torque, that is, the cycle variation pattern and the energy consumption variation within a single cycle are not smooth curves.

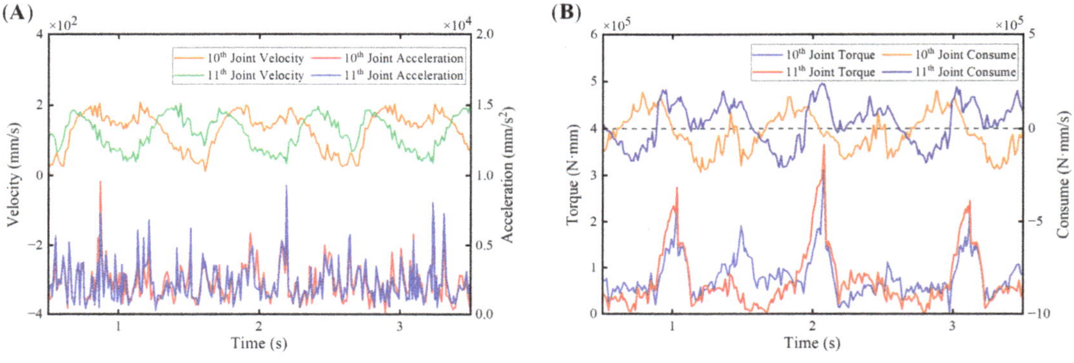

Figure 15. The mechanic characteristics of adjacent modules: (**A**) velocity and acceleration relative to time; and (**B**) joint torque and consumption relative to time.

Figure 15B shows the velocity and acceleration analysis of the 10th and 11th joints. The velocity variation of these joints maintains a good periodic pattern with a small scale. Due to the influence of the output torque fluctuation, acceleration is characteristic of periodic variation, but there are multiple spike bursts in a single cycle. However, the overall fluctuation of acceleration varies slightly, which verifies that the snake-like robot completes a smooth climbing motion on the cylinder after the changes from the starting vibration to the stable.

4.2.3. Contact Force

The snake-like robot continuously updates the state of contact with the cylinder surface through the change of joint output angle, as shown in Figure 16B, where the direction of the red arrow indicates the direction of the current joint contact force, and the length of the red line segment represents the magnitude. Figure 16A shows that the adjacent joints alternately have contact with the cylinder after being stable, and the contact force has a periodic characteristic. At the same time, multiple spikes of a sudden increase appear in the contact force within a single cycle but maintain a stable limit overall. The contact force in the two adjacent joints has a "delay" effect due to the phase difference between the output angle of the anterior and posterior joints.

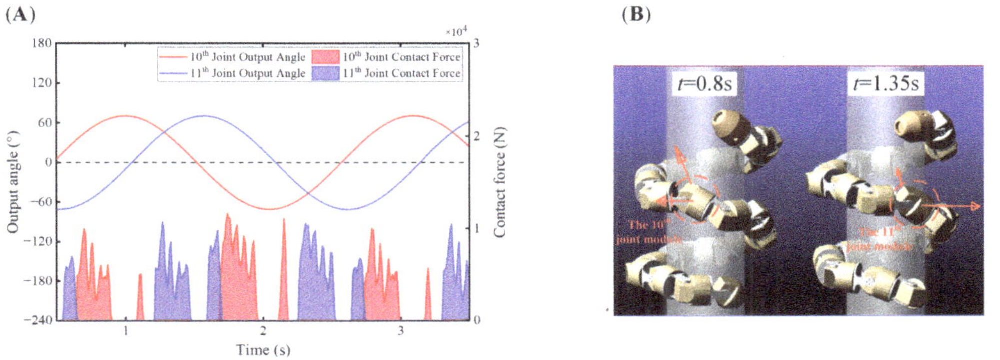

Figure 16. The output angle and contact force of two adjacent joints: (**A**) the output angle and contact force of joint modules; and (**B**) the contact force is applied to modules at different time points.

Figure 17 shows the pattern of output rotation angle and contact force variation of four successive adjacent central joints. In addition to the similar periodic regularity of contact force variation with two adjacent joints, not all the joints are in contact with the cylinder surface simultaneously. As illustrated in Figure 17B, only the 9th and 11th joints among the inspected joints are in contact with the cylinder surface when t = 1.35s. While t = 1.95s, only the 10th and 12th joints are in contact with the surface of the cylinder. Figure 17A clearly shows the delay effect between the contact cycles of the different joints. Different joints' contact forces will effectively cover the time axis to provide enough contact force for the snake-like robot to climb upwards stably within the same time frame.

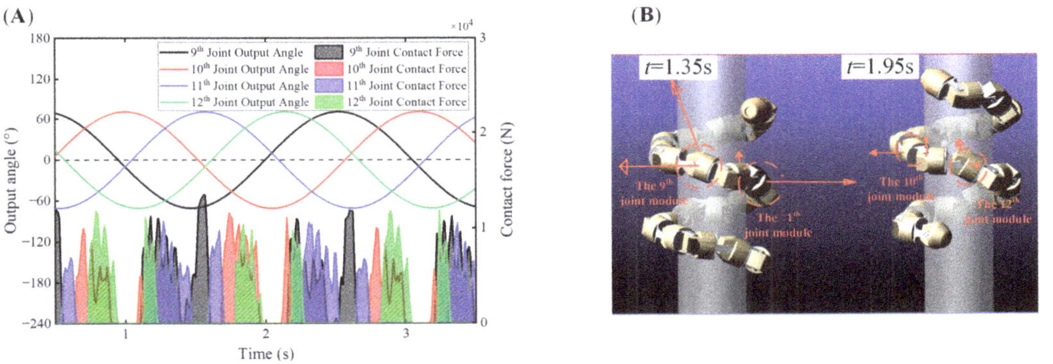

Figure 17. The output angle and contact force of four adjacent joints: (**A**) the output angle and contact force of joint modules; and (**B**) the contact force is applied to modules at different time points.

Figure 18 shows the joint output angles (JOA) and joint contact forces (JCF) of six successive adjacent joints. We consider the six successive adjacent joints of the snake-like robot we analyze as an integral unit, and that the output angles of different joints have a sinusoidal output pattern in one cycle. At any point under the cycle, the integral unit of the snake-like robot is in contact with the cylinder surface, which is necessary for generating sufficient grip force.

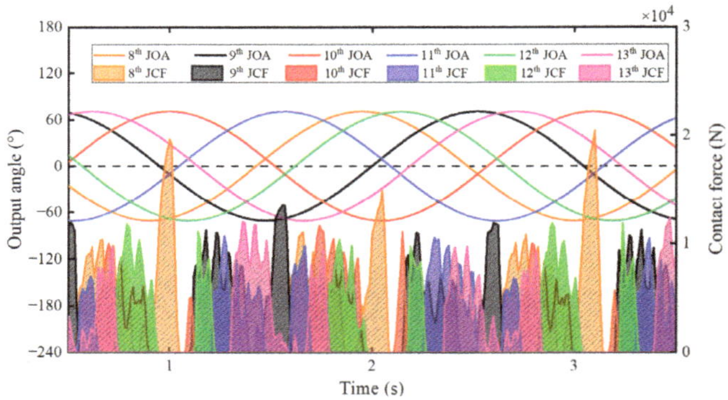

Figure 18. The output angle and contact force of six adjacent joints.

4.2.4. Non-Contact Zone

In the analysis mentioned above, we find that there are specific output angles with corresponding contact forces of joint modules at the same time. As shown in Figure 19, the joint module of the snake-like robot only makes contact with the surface of the cylinder in a particular range of angles during the spiral climbing motion, and the joint angles in the current state are not in contact with the surface of the cylinder in the range of $\pm 30°$.

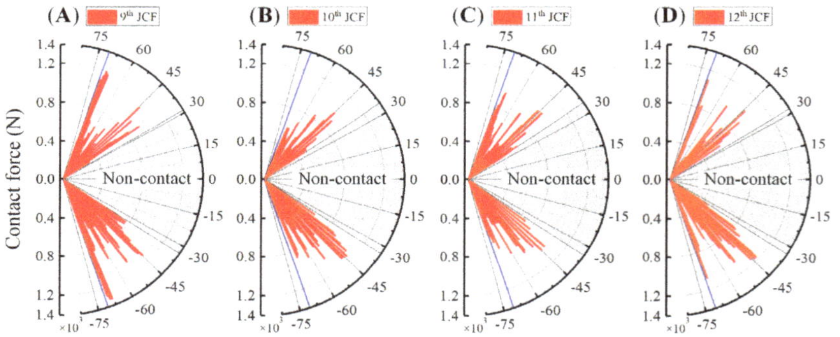

Figure 19. Non-contact zone: (**A**) JCF of the 9th module; (**B**) JCF of the 10th module; (**C**) JCF of the 11th module; and (**D**) JCF of the 12th module.

The non-contact zone is inevitable when the joint angle control method is applied to achieve spiral climbing motion. However, if the non-contact zone is too large, the contact between the joint modules and the cylindrical object will be reduced, which influences the climbing effect, while on the contrary, if the non-contact zone is too small, the joints will be in contact with the cylindrical object at all times, indirectly influencing the radius of the form spiral of the snake-like robot, which makes it incompatible for the robot to adapt to the cylinder of a larger diameter. Also, it still affects the climbing effect.

5. Conclusions

In this paper, we analyze the factors influencing the spiral climbing motion of a snake-like robot on the outer surface of a cylinder and its optimal design. Although many studies have been conducted on the spiral climbing motion, few have conducted analyses from the perspective of kinematics and dynamics-influencing factors.

First, the angle control method was used in this paper to realize the spiral climbing motion control of the snake-like robot. We analyzed the effect of control parameters A, β, and γ on the motion utilizing MATLAB R2022a and WEBOTS 2021a. Moreover, we establish a formulation of form spiral radius based on A and β to realize the directional control of the snake-like robot's spiral climbing motion. Then, a contact point location analysis determined the contact range of the joint module and the cylinder surface. A multi-objective optimization function with three principles was established based on the number of joint modules, the forward velocity of motion, and the output torque of the joint, to carry out a practical analysis and optimization of the design parameters of the joint module. In the final analysis, we transferred the grip force when the snake-like robot wrapped around the outer surface of the cylinder into the joint module's contact force. We clarified that the pattern of the grip force when taking the six consecutive adjacent joints as an integral unit shows that it is always in contact with the cylinder. Furthermore, the generation and influence of the non-contact zone were analyzed.

The work of the spiral climbing motion from kinematic and dynamic influences carried out in this paper is unique and instructive for analyzing the spatial motion of snake-like robots. In the subsequent research, we will focus on the spiral climbing motion of snake-like robots on the surface of cylinders with variable diameters.

Author Contributions: Methodology, formal analysis, investigation, writing—original draft preparation, P.Z.; supervision, validation, projection administration, Y.Z.; software, data curation, Z.W.; funding acquisition, writing—review and editing, B.G. and Z.G. All authors have read and agreed to the published version of the manuscript.

Funding: This research was funded by the Fundamental Research Funds for the Central Universities, grant number FRF-DF-20-13, and Foshan Science and Technology Innovation Fund, University of Science and Technology Beijing (USTB), China, grant number BK20BE007.

Acknowledgments: We would like to acknowledge the assistance of Wenhao Wang and Fan He, both of whom are the student at Shunde Innovation School, USTB (The former name is Shunde Graduate School, USTB). And we send our thanks to Xiaoya Zhao and Shanshan He for the software skills they have shared with us.

Conflicts of Interest: The authors declare no conflict of interest.

References

1. Seetohul, J.; Shafiee, M. Snake Robots for Surgical Applications: A Review. *Robotics* **2022**, *11*, 57. [CrossRef]
2. Zhu, Q.; Zhou, T.; Du, J. Upper-Body Haptic System for Snake Robot Teleoperation in Pipelines. *Adv. Eng. Inform.* **2022**, *51*, 101532. [CrossRef]
3. Li, C.; Zhang, T.; Goldman, D.I. A Terradynamics of Legged Locomotion on Granular Media. *Science* **2013**, *339*, 1408–1412. [CrossRef] [PubMed]
4. Rummel, J.D.; Beaty, D.W.; Jones, M.A.; Bakermans, C.; Barlow, N.G.; Boston, P.J.; Chevrier, V.F.; Clark, B.C.; de Vera, J.-P.P.; Gough, R.V.; et al. A New Analysis of Mars "Special Regions": Findings of the Second MEPAG Special Regions Science Analysis Group (SR-SAG2). *Astrobiology* **2014**, *14*, 887–968. [CrossRef] [PubMed]
5. Astley, H.C. Slithering across Worlds—Snake-Inspired Robots for Extraterrestrial Exploration. In *Biomimicry for Aerospace*; Elsevier: Amsterdam, The Netherlands, 2022; pp. 261–289. ISBN 978-0-12-821074-1.
6. Hou, X.; Shi, Y.; Li, L.; Tian, Y.; Su, Y.; Ding, T.; Deng, Z. Revealing the Mechanical Characteristics via Kinematic Wave Model for Snake-Like Robot Executing Exploration of Lunar Craters. *IEEE Access* **2020**, *8*, 38368–38379. [CrossRef]
7. Jayne, B.C. What Defines Different Modes of Snake Locomotion? *Integr. Comp. Biol.* **2020**, *60*, 156–170. [CrossRef] [PubMed]
8. Hatton, R.L.; Choset, H. Sidewinding on Slopes. In Proceedings of the 2010 IEEE International Conference on Robotics and Automation, Anchorage, AK, 3–7 May 2010; pp. 691–696.
9. Gong, C.; Hatton, R.L.; Choset, H. Conical Sidewinding. In Proceedings of the 2012 IEEE International Conference on Robotics and Automation, St Paul, MN, USA, 14–18 May 2012; pp. 4222–4227.
10. Qi, W.; Kamegawa, T.; Gofuku, A. Proposal of Helical Wave Propagate Motion for a Snake Robot to across a Branch on a Pipe. In Proceedings of the 2016 IEEE/SICE International Symposium on System Integration (SII), Sapporo, Japan, 13–15 December 2016; pp. 821–826.
11. Yaqub, S.; Ali, A.; Usman, M.; Zuhaib, K.M.; Khan, A.M.; An, B.; Moon, H.; Lee, J.-Y.; Han, C. A Spiral Curve Gait Design for a Modular Snake Robot Moving on a Pipe. *Int. J. Control Autom. Syst.* **2019**, *17*, 2565–2573. [CrossRef]
12. Rollinson, D.; Choset, H. Virtual Chassis for Snake Robots. In Proceedings of the 2011 IEEE/RSJ International Conference on Intelligent Robots and Systems, San Francisco, CA, USA, 25–30 September 2011; pp. 221–226.
13. Zhou, Y.; Zhang, Y.; Ni, F.; Liu, H. A Spring-like Pipe Climbing Gait for the Snake Robot. In Proceedings of the 2017 IEEE International Conference on Robotics and Biomimetics (ROBIO), Macau, Macao, 5–8 December 2017; pp. 1886–1891.

14. Yamada, H.; Hirose, S. Approximations to Continuous Curves of Active Cord Mechanism Made of Arc-Shaped Joints or Double Joints. In Proceedings of the 2010 IEEE International Conference on Robotics and Automation, Anchorage, AK, USA, 3–7 May 2010; pp. 703–708.
15. Jayne, B.C.; Newman, S.J.; Zentkovich, M.M.; Berns, H.M. Why Arboreal Snakes Should Not Be Cylindrical: Body Shape, Incline and Surface Roughness Have Interactive Effects on Locomotion. *J. Exp. Biol.* **2015**, *218*, 3978–3986. [CrossRef] [PubMed]
16. Byrnes, G.; Jayne, B.C. Gripping during Climbing of Arboreal Snakes May Be Safe but Not Economical. *Biol. Lett.* **2014**, *10*, 20140434. [CrossRef] [PubMed]
17. Yamada, H.; Hirose, S. Study on the 3D Shape of Active Cord Mechanism. In Proceedings of the Proceedings 2006 IEEE International Conference on Robotics and Automation, 2006. ICRA 2006, Orlando, FL, USA, 15–19 May 2006; pp. 2890–2895.
18. Kamegawa, T.; Harada, T.; Gofuku, A. Realization of Cylinder Climbing Locomotion with Helical Form by a Snake Robot with Passive Wheels. In Proceedings of the 2009 IEEE International Conference on Robotics and Automation, Kobe, Japan, 12–17 May 2009; pp. 3067–3072.
19. Manzoor, S. Neural Oscillator Based CPG for Various Rhythmic Motions of Modular Snake Robot with Active Joints. *J. Intell. Robot. Syst.* **2019**, *94*, 641–654. [CrossRef]
20. Chirikjian, G.S.; Burdick, J.W. A Modal Approach to Hyper-Redundant Manipulator Kinematics. *IEEE Trans. Robot. Automat.* **1994**, *10*, 343–354. [CrossRef]
21. Lipkin, K.; Brown, I.; Peck, A.; Choset, H.; Rembisz, J.; Gianfortoni, P.; Naaktgeboren, A. Differentiable and Piecewise Differentiable Gaits for Snake Robots. In Proceedings of the 2007 IEEE/RSJ International Conference on Intelligent Robots and Systems, San Diego, CA, USA, 29 October–2 November 2007; pp. 1864–1869.
22. Zhen, W.; Gong, C.; Choset, H. Modeling Rolling Gaits of a Snake Robot. In Proceedings of the 2015 IEEE International Conference on Robotics and Automation (ICRA), Seattle, WA, USA, 26–30 May 2015; pp. 3741–3746.
23. Sun, H.; Liu, X.; Ma, P. On the Tree-Climbing Staitic Mechanism of a Snake Robot Climbing Trees. *Robot* **2008**, *30*, 112–116. [CrossRef]
24. Wei, W.; Zhu, H. Research of Helical Rolling Gait of Snake-like Robot in Detection of Bridge Cables. *Comput. Eng. Des.* **2011**, *32*, 700–702. [CrossRef]
25. Goldman, G.; Hong, D. Considerations for Finding the Optimal Design Parameters for a Novel Pole Climbing Robot. In Proceedings of the Volume 2: 32nd Mechanisms and Robotics Conference, Parts A and B, New York, NY, USA, 3–6 August 2008; ASMEDC: Brooklyn, NY, USA, 2008; pp. 859–866.
26. Liu, X.; Gao, Z.; Zang, Y.; Zhang, L. Tribological Mechanism and Propulsion Conditions for Creeping Locomotion of the Snake-like Robot. *J. Mech. Eng.* **2021**, *57*, 189–201. [CrossRef]

electronics

MDPI

Article

A Strain Rate Dependent Damage Model for Evaluating the Dynamic Response of CFRTP Laminates with Different Stacking Sequence

Yiben Zhang [1] and Bo Liu [1,2,*]

[1] Beijing Key Laboratory of Lightweight Metal Forming, Beijing 100083, China
[2] School of Mechanical Engineering, University of Science and Technology Beijing, Beijing 100083, China
* Correspondence: liubo1@ustb.edu.cn

Abstract: Carbon fiber reinforced thermoplastic polymer (CFRTP) laminates can be used in packaging electronics components to reduce weight and shield external disturbance. The CFRTP structures in operation are inevitably to suffer dynamic loading conditions such as falling rocks, tools and impacts. In this study, a strain rate dependent material model for accurately evaluating the dynamic response of CFRTP laminates with different stacking sequence was proposed. The model was composed of three components: a strain rate dependent constitute model, a strain rate related damage initiation model and an energy-based damage evolution model. The strain rate effect of modulus and strength was described by a stacking sequence related matrix, and the damage initiation model could describe the matrix, fiber and delamination damage of CFRTP laminates without introducing cohesive elements. The material model was implemented into finite element software ABAQUS by user defines subroutine VUMAT. The low velocity impact tests of CFRTP laminates with quasi-isotropic and angle-ply stacking sequence were used to provide validation data. The dynamic response of CFRTP laminates from numerical results were highly consistent with the experimental results. The mechanical response of CFRTP laminates were affected by stacking sequence and impact energy, and the numerical error of proposed material model significantly decreased with the increasing impact energy especially for the laminae with damage occur.

Keywords: CFRTP; strain rate; stacking sequence; damage model; finite element simulation

Citation: Zhang, Y.; Liu, B. A Strain Rate Dependent Damage Model for Evaluating the Dynamic Response of CFRTP Laminates with Different Stacking Sequence. *Electronics* **2022**, *11*, 3728. https://doi.org/10.3390/electronics11223728

Academic Editor: Raffaele Giordano

Received: 22 October 2022
Accepted: 8 November 2022
Published: 14 November 2022

1. Introduction

Carbon fiber reinforced thermoplastic (CFRTP) laminates have considerable potential for lightweight use in electronic shields, aerospace, automotive, wind energy and marine due to their high specific stiffness and strength, corrosion resistance, fatigue performance and recyclability [1–3]. In structural applications, CFRTP laminates are inevitably exposed to the low velocity dynamic loading condition such as tool dropping, debris impact, and bird impact [4]. The dynamic loading issues cause barely visible impact damage (BVID), and should be crucially considered for strength assessment of composite structures.

To thoroughly investigate the dynamic response of composite laminates, many numerical investigations have been conducted in the past several decades, but few of them considered strain rate effects. Actually, due to the viscose-plastic of thermoplastic matrix, the strain rate sensitivity of CFRTP laminates cannot be neglected even if in the low strain rate loading condition. For example, Massaq et al. [5] claimed that the failure stress and failure energy of PA6/Glass showed obvious strain rate sensitivity in the strain rate range from 10^{-5} s^{-1} to 1 s^{-1} and 100 s^{-1} to 2500 s^{-1}, respectively. Chen et al. [6] showed the failure strain of PEEK composites increased apparently with the strain rate increasing from 0.001 s^{-1} to 1000 s^{-1}. Ou et al. [7] investigated the effect of strain rate on the mechanical properties and failure patterns of GFRP and reported that tensile strength, maximum strain and toughness increase with increasing strain rates from $1/600\ s^{-1}$ to $160\ s^{-1}$. The authors previous research also revealed that the strain rate sensitivity of CFRTP laminates was obvious from strain rate $2 \times 10^{-4}\ s^{-1}$ to $2200\ s^{-1}$ [8]. The strain rate sensitivity was one of the key factors an accurate assessment of composite tensile strength and fracture toughness under dynamic loadings [9,10].

The phase field initiated in 1990s, has received a significant development in the recent years [11–13]. It is widely used for composite laminates and the progressive failure [14,15]. The implementation of traditional damage model could be more easier as only the material properties on the integration points of the elements are required to be modified [16]. Thus, a material model involving strain rate effects based on continuum mechanics was proposed in this study. Normally, the constitute behaviour of composite laminate is assumed to be linear deformation and the non-linear mechanical response is mainly due to the damage formation and expansion. Thus, damage imitation and evolution model are the research focus in the past decades. Typical failure modes in composite laminates include matrix damage, fiber damage and delamination damage. Some widely used failure criteria such as Tai-Wu criterion [17], Hashin [18] and Hou [19,20] criteria can predict fiber and matrix damage, but none of them considered the effect of strain rate. Yen and Caiazoo [21,22] proposed a model to determine the stiffness and strength of composite materials at various strain rate levels [23,24]. Based on Y-C function, Wang et al. [25] proposed a three-dimensional strain-rate-dependent damage model which can predict the strain rate dependent contact force curve and damage modes. However, the aforementioned failure criteria cannot directly predict delamination damage of composite laminates. They additionally introduced cohesive zone elements to predict the mechanical behaviour of interface, which significantly increased the computation cost and may lead to the distortion of adjacent elements.

Additional, the dynamic behaviour of CFRTP laminates were affected by not only strain rate but also stacking sequence [26,27]. The strain rate sensitivity in matrix dominant direction was normally more obvious than fiber dominant direction. Hence, the effect of stacking sequence on the strain rate needed to be considered in the finite element modelling analysis of dynamic mechanical properties of CCFRT laminates.

The objective of this study is to present a strain rate related material model for accurately evaluating the dynamic response of CFRTP laminates with different stacking sequence. The established model included a strain rate related constitute model, a strain rate related damage initiation model, and an energy based damage evolution model. The strain rate related modulus and strength were evaluated by introducing a matrix for describing stacking sequence effect. The damage initiation model was established based on Hou criteria including fiber damage, matrix damage and delamination criteria. The material model was implemented in the ABAQUS/Explicit by user subroutine codes. Low velocity impact tests of CFRTP laminate plates with quasi-isotropic and angle-ply stacking sequence were used to validate the proposed model. Detailed strength and failure mode comparisons between the numerical predictions and experimental results were discussed.

2. Strain Rate Relate Dependent Material Model

To model the constitute behavior of CFRTP lamina, fiber direction, in-plane perpendicular to the fiber direction, out-of-plane perpendicular to the fiber direction were respectively defined as Direction 1, Direction 2 and Direction 3, as is shown in Figure 1.

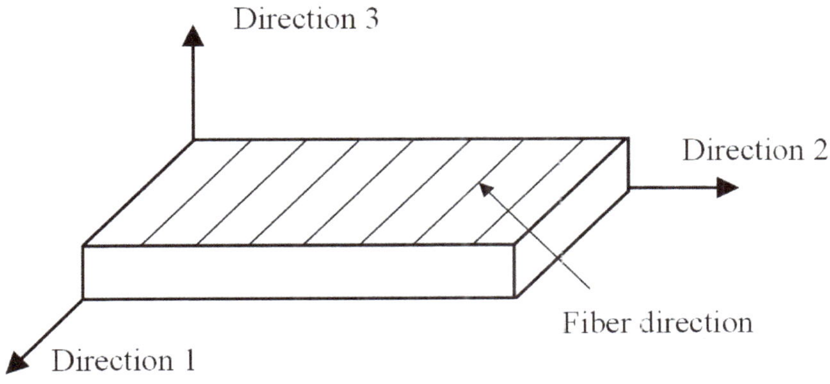

Figure 1. Material direction of composite lamina.

2.1. Constitute Model

A three dimensional constitute model for orthotropic composite can be expressed as:

$$\boldsymbol{\sigma} = \mathbf{C} \times \boldsymbol{\varepsilon} \tag{1}$$

where $\boldsymbol{\sigma}, \mathbf{C}$ and $\boldsymbol{\varepsilon}$ are stress matrix, stiffness matrix and strain matrix, respectively. $\boldsymbol{\sigma} = [\sigma_{11}\ \ \sigma_{22}\ \ \sigma_{33}\ \ \tau_{12}\ \ \tau_{23}\ \ \tau_{31}]^{T}$, $\boldsymbol{\varepsilon} = [\varepsilon_{11}\ \ \varepsilon_{22}\ \ \varepsilon_{33}\ \ \gamma_{12}\ \ \gamma_{23}\ \ \gamma_{31}]^{T}$.

In the static condition, the stiffness matrix of CFRTP laminate can be expressed as:

$$\mathbf{C} = \begin{bmatrix} C_{11} & C_{12} & C_{13} & & & \\ & C_{22} & C_{23} & & & \\ & & C_{33} & & & \\ & & & C_{44} & & \\ & & & & C_{55} & \\ & & & & & C_{66} \end{bmatrix} \tag{2}$$

where $C_{11} = \frac{1-\upsilon_{23}\upsilon_{32}}{E_{22}E_{33}\Delta}$, $C_{12} = \frac{\upsilon_{21}+\upsilon_{23}\upsilon_{31}}{E_{11}E_{33}\Delta}$, $C_{22} = \frac{1-\upsilon_{13}\upsilon_{31}}{E_{11}E_{33}\Delta}$, $C_{13} = \frac{\upsilon_{31}+\upsilon_{21}\upsilon_{32}}{E_{11}E_{22}\Delta}$, $C_{23} = \frac{\upsilon_{32}+\upsilon_{12}\upsilon_{31}}{E_{11}E_{22}\Delta}$, $C_{33} = \frac{1-\upsilon_{12}\upsilon_{21}}{E_{11}E_{22}\Delta}$, $C_{44} = E_{12}$, $C_{55} = E_{13}$, $C_{66} = E_{23}$, $\frac{\upsilon_{ij}}{E_{ii}} = \frac{\upsilon_{ji}}{E_{jj}}$, $\Delta = \frac{1-\upsilon_{13}\upsilon_{31}-\upsilon_{12}\upsilon_{21}-\upsilon_{23}\upsilon_{32}-2\upsilon_{12}\upsilon_{32}\upsilon_{31}}{E_{11}E_{22}E_{33}}$; E_{ij} and υ_{ij} $(i, j = 1, 2, 3)$ are the elastic modulus and poisson's ratio, respectively.

According to the logarithmic function established by Yen and Caiazzo [22], a matrix was introduced to describe the strain rate related modulus as:

$$\mathrm{DIF}_{ij}^{e} = 1 + m_{ij}^{e} \times ln(\dot{\varepsilon}/\dot{\varepsilon_0}), \quad E_{ij}^{*} = E_{ij} \times \mathrm{DIF}_{ij}^{e}(i, j = 1, 2, 3) \tag{3}$$

where $m_{ij}^{e}(i, j = 1, 2, 3)$ was the component in the matrix. $\dot{\varepsilon}$ was the loading strain rate. $\dot{\varepsilon}_0$ was the reference strain rate, which was 2×10^{-4} s^{-1} in this study. E_{ij}^{*} and E_{ij} were the strain rate related modulus and reference modulus, respectively.

2.2. Damage Initiation Model

Hou failure criteria including fiber damage and matrix damage were used to predict material damage initiation, and a traction separation model was introduced to predict delamination damage as [19,28,29]:

Fiber damage:

$$f_1^2 = \begin{cases} \left(\frac{\sigma_{11}}{\sigma_{11}^{f,t}}\right)^2 & \sigma_{11} \geq 0 \\ \left(\frac{\sigma_{11}}{\sigma_{11}^{f,c}}\right)^2 & \sigma_{11} < 0 \end{cases} \tag{4}$$

Matrix damage:

$$f_2^2 = \begin{cases} \left(\frac{\sigma_{22}}{\sigma_{22}^{f,t}}\right)^2 + \left(\frac{\sigma_{12}}{\sigma_{12}^{f}}\right)^2 + \left(\frac{\sigma_{23}}{\sigma_{23}^{f}}\right)^2 & \sigma_{22} \geq 0 \\ \left(\frac{\sigma_{22}+\sigma_{33}}{2\sigma_{23}^{f}}\right)^2 + \frac{\sigma_{22}^{f,c}\sigma_{22}}{\left(2\sigma_{12}^{f}\right)^2} - \frac{\sigma_{12}}{\sigma_{12}^{f}} + \left(\frac{\sigma_{12}}{\sigma_{12}^{f}}\right)^2 & \sigma_{22} < 0 \end{cases} \tag{5}$$

Delamination damage:

$$f_3^2 = \begin{cases} \left(\frac{\sigma_{33}}{\sigma_{33}^{f,t}}\right)^2 + \left(\frac{\sigma_{13}}{\sigma_{13}^{f}}\right)^2 + \left(\frac{\sigma_{23}}{\sigma_{23}^{f}}\right)^2 & \sigma_{33} \geq 0 \\ \left(\frac{\sigma_{13}}{\sigma_{13}^{f}}\right)^2 + \left(\frac{\sigma_{23}}{\sigma_{23}^{f}}\right)^2 & \sigma_{33} \geq 0 \end{cases} \tag{6}$$

where $\sigma_{ii}^{f,t}$ and $\sigma_{ii}^{f,c}(i = 1,2,3)$ represented the tensile and compression strength in direction i. σ_{12}^{f} is the in plane shear strength. σ_{13}^{f} and σ_{23}^{f} are the out of plane shear strength. f_i $(i = 1,2,3)$ represents the damage state: $f_i < 1$ represents undamaged state, and $f_i \geq 1$ represents damaged state.

A dynamic increased factor matrix is introduced to describe the strain rate related strength of CFRTP laminates as:

$$\mathrm{DIF}_{ij}^{s} = 1 + m_{ij}^{s} \times ln(\dot{\varepsilon}/\dot{\varepsilon_0}), \quad S_{ij}^{*} = S_{ij} \times \mathrm{DIF}_{ij}^{s}(i,j = 1,2,3) \tag{7}$$

where m_{ij}^{s} is the strain rate constants for describing the material strength strain rate sensitivity. $\dot{\varepsilon}$ is the loading strain rate. $\dot{\varepsilon_0}$ is the reference strain rate, which is $2 \times 10^{-4}\ s^{-1}$ in this study. S_{ij}^{*} and S_{ij} are the strain rate related strength and reference strength, respectively.

2.3. Damage Evolution Model

To described the damage evolution of CFRTP laminate, $d_i(i = 1,2,3)$ was defined to characterize the damage state of in material. d_1 represents the damage in direction 1, which can be quantified as the in-plane damage distribution density perpendicular to the fiber (Figure 2a). d_2 represents the damage in direction 2, which can be quantified as the in-plane damage distribution density along the fiber direction (Figure 2b). d_3 represents the damage in direction 3, which can be quantified as the out of plane damage distribution density (Figure 2c). The value of d_i is between 0 and 1. $d_i = 0$ represents the material is no damage in direction i; $d_i = 1$ represents the material fails in direction i [30].

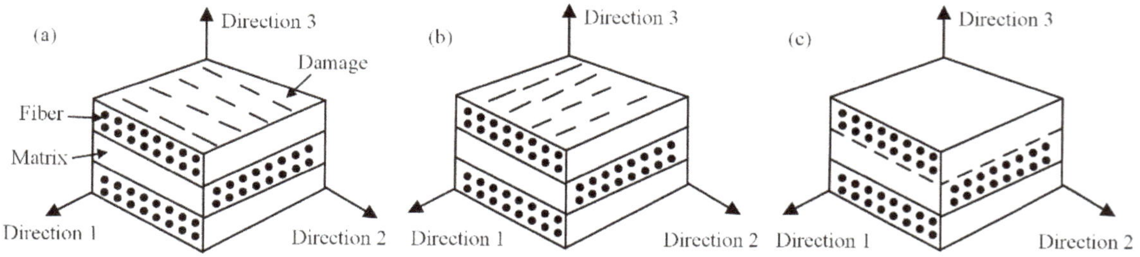

Figure 2. Damagedirection in CFRTP laminate (**a**) direction 1; (**b**) direction 2; (**c**) dirfection 3.

An energy-based criterion is used to describe the nonlinear damage evolution as [31]:

$$d_i = 1 - \exp\left(-\sigma_{ii}^f \delta_{eq,ii}^f (f_i - 1)/G_i\right)/f_i \quad (8)$$

where G_i represents the fracture energy in direction i; $\delta_{eq,ii}^f$ represents the equivalent displacement in relative direction and can be expressed as [32]:

$$\delta_{eq,ii}^f = \begin{cases} \frac{\sigma_{ii}^{f,t} L^C}{C_{ii}} & \sigma_{ii} \geq 0,\ i = 1,2 \\ \frac{\sigma_{ii}^{f,c} L^C}{C_{ii}} & \sigma_{ii} < 0,\ i = 1,2 \end{cases} \quad (9)$$

in which L^C is the feature length of element.

Once the damage initiation criterion is satisfied, the constitute model of composite laminate is defined as:

$$\boldsymbol{\sigma} = \mathbf{C}^d \times \boldsymbol{\varepsilon} \quad (10)$$

the damaged stiffness matrix $\mathbf{C}^d$ can be expressed as:

$$\mathbf{C}^d = \begin{bmatrix} C_{11}^d & C_{12}^d & C_{13}^d & & & \\ & C_{22}^d & C_{23}^d & & & \\ & & C_{33}^d & & & \\ & & & C_{44}^d & & \\ & & & & C_{55}^d & \\ & & & & & C_{66}^d \end{bmatrix} \quad (11)$$

$C_{11}^d = (1 - d_1)C_{11}$, $C_{22}^d = (1 - d_2)C_{22}$, $C_{33}^d = (1 - d_3)C_{33}$, $C_{12}^d = (1 - d_1)(1 - d_2)C_{12}$, $C_{23}^d = (1 - d_2)(1 - d_3)C_{23}$, $C_{13}^d = (1 - d_1)(1 - d_3)C_{13}$, $C_{44}^d = (1 - d_1)(1 - d_2)C_{44}$, $C_{55}^d = (1 - d_1)(1 - d_3)C_{55}$, $C_{66}^d = (1 - d_2)(1 - d_3)C_{12}$

2.4. Model Implementation

The strain rate related constitute model, strain rate related damage initiation model and energy based damage evolution model were implemented in finite element software ABAQUS by user subroutine VUMAT. The simulation flowchart is shown in Figure 3. Firstly, the mechanical parameters including material modulus and strength at the reference strain rate and the state variables in the previous increment were imported in the established finite element model. Secondly, the strain rate, as well as the strain rate related modulus and strength of the material in the current increment, were calculated according to the strain increment and time increment. Thirdly, the damage state is examined according to the intra-laminar and inter-laminar damage model. If the damage occurs, the stiffness degradation or delamination is conducted. Otherwise, status variable is updated and the model goes to the next incremental step. Finally, the time increment is calculated and the model goes to the next incremental step.

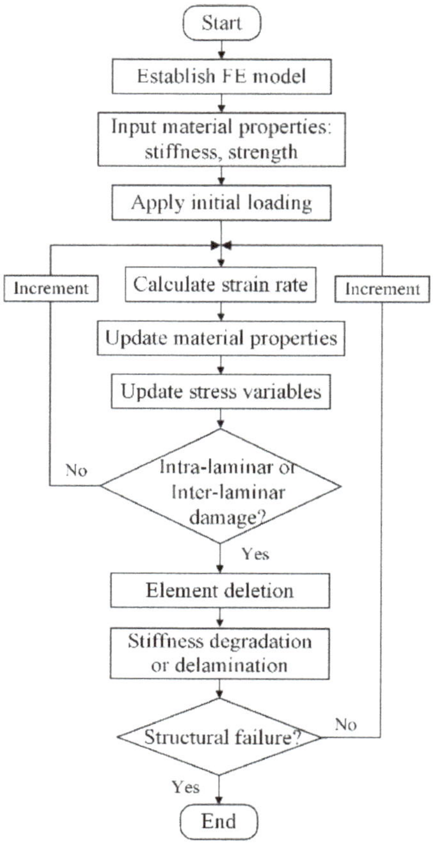

Figure 3. The simulation flowchart.

3. Experimental Method

The LVI experiments of CFRTP laminates was conducted with INSTRON CEAST 9350 impact test system in accordance with ASTM D7136 (Manufacture: American Society of Testing Materials; City: PA; Country: American) [33]. The rectangular specimens with length of 150 mm, width of 100 mm and thickness of 2.6 mm for quasi-isotropic and thickness 3.0 mm for angle-ply stacking sequence (Figure 4). The LVI specimens were fixed on the rigid support by four clamps to prevent the longitudinal vibration of the specimen. The fixture has 125 mm × 75 mm rectangular cut, and the impact point located at the center of rectangular cut. The steel hemispherical impactor was 5.5 kg weigh and 12.7 diameter. The LVI experiments were conducted at room temperature, and there is no obvious electromagnetic and vibration. For quasi-isotropic (QI) laminate, the matrix was PA and the reinforced component was carbon fiber; the impact energy was 10 J; For angle-ply (AP) laminate, the matrix was PC and the reinforced component was carbon fiber; the impact energy was 3 J and 6 J. The impact speed is determined by the impact energy and the initial height of the punch along the impact direction according to Formula (12)

$$E_{imp} = mgh = \frac{1}{2}mv_{imp}^2 \tag{12}$$

where E_{imp} is the impact energy; m is the impactor mass; *h* is the initial impactor height, and v_{imp} is the initial impactor velocity; The contact force history was recorded by a sensor in the test system.

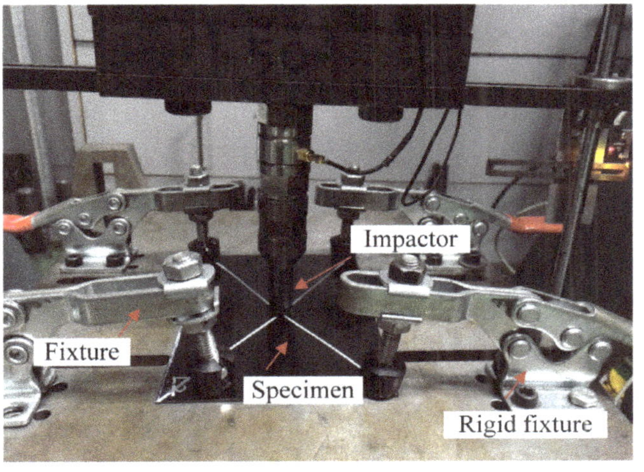

Figure 4. Low velocity impact experiment system.

4. Model Validation

Figure 5 showed the finite element model of LVI experiments of CFRTP laminates. The model included three components: a support, a sample and an impactor. The support and impactor were modeled by rigid bodies, and the composite laminate sample were modeled by C3D8R solid elements with a minimum element size of 0.8 mm × 0.8 mm. One element was used for each layer in the thickness direction. Rigid supports was constrained all the freedom in the transition and rotation direction. Four pressure heads on the specimen were simplified as constraint in the impact direction. The impactor was applied an initial impact speed according to experiments. The surface contact between the impactor and the CFRTP specimen was adopted. The material parameter of QI and AP laminate were listed in Tables 1 and 2, respectively.

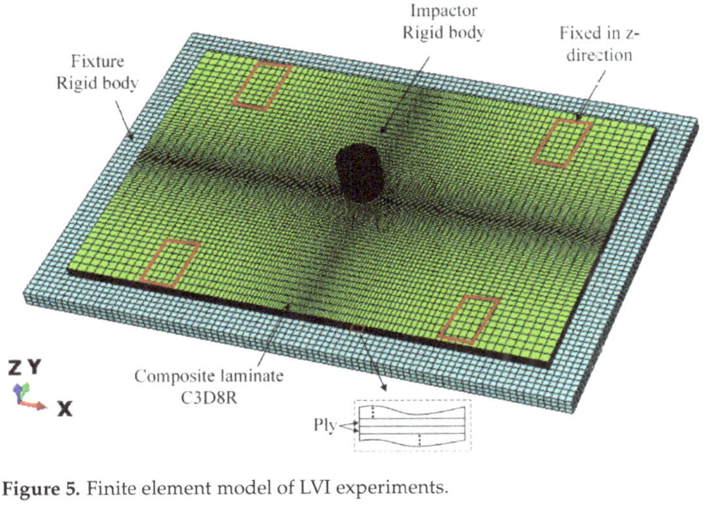

Figure 5. Finite element model of LVI experiments.

Table 1. Material parameter for QI laminate.

Elastic Parameter (GPa)		Strength Parameter (MPa)				Fracture Energy (N/mm)		Strain Rate Parameter	
E_{11}	42.9	$\sigma_{11}^{f,t}$	850	σ_{12}^{f}	105	G_1	0.3	m_1^e	0.005
E_{22}	4.5	$\sigma_{11}^{f,c}$	350	σ_{23}^{f}	105	G_2	0.2	m_2^e	0.064
E_{33}	4.5	$\sigma_{22}^{f,t}$	260	σ_{13}^{f}	105	G_3	0.2	m_3^e	0.032
E_{12}	1.2	$\sigma_{22}^{f,c}$	275					m_1^s	0.003
E_{23}	1.2	$\sigma_{33}^{f,t}$	260					m_2^s	0.042
E_{31}	1.2	$\sigma_{33}^{f,c}$	275					m_3^s	0.027

Table 2. Material parameter for AP laminate.

Elastic Parameter (GPa)		Strength Parameter (MPa)				Fracture Energy (N/mm)		Strain Rate Parameter	
E_{11}	115	$\sigma_{11}^{f,t}$	1524	σ_{12}^{f}	210	G_1	0.9	m_1^e	0.001
E_{22}	10.5	$\sigma_{11}^{f,c}$	945	σ_{23}^{f}	210	G_2	0.6	m_2^e	0.018
E_{33}	10.5	$\sigma_{22}^{f,t}$	615	σ_{13}^{f}	210	G_3	0.6	m_3^e	0.024
E_{12}	6.2	$\sigma_{22}^{f,c}$	425					m_1^s	0.001
E_{23}	6.2	$\sigma_{33}^{f,t}$	615					m_2^s	0.012
E_{31}	6.2	$\sigma_{33}^{f,c}$	425					m_3^s	0.014

Normally the modulus and strength of CFRTP in Direction 2 and Direction 3 were assumed to be equivalent as material properties in these two directions were affected by matrix. Thus, the strain rate parameter m_{22} and m_{33} was unified expressed by m_2, and m_{12}, m_{23} and m_{13} was unified expressed by m_3. The material parameters of QI and AP laminate were listed in Tables 1 and 2, respectively.

The contact force history of QI laminate under 10 J impact from experiment and simulation results were compared in Figure 6a. It can be seen that the contact force history curves from three simulation models and test have similar trends. Firstly, the contact force increased with impact propagation, and the growth rate decreased when the damage threshold was reached. With the damage propagation, the contact force continued to increase with a slow rate until reaching the peak value. Then, the impactor starts to rebound and the contact force gradually decreases to zero. The failure pattern of QI laminate under 10 J impact from experimental and simulation results were compared in Figure 6b–e. There were obvious cracks caused by fiber fracture on the laminated plate, and the cracks extend linearly along the 45° direction (the red circle). The damage area from SSD and SRD model was larger than that from the other two models as the failure evolution process were evidently accelerated according to the proposed criteria.

The contact force history of AP laminate under 3 J impact from experiment and simulation results were compared in Figure 7a. The contact force from experimental results, SSD and SRD, SSI and SRD, SSI and SRI simulation results were similar. The curve firstly increased with the increasing contact time, and then decreased after reaching the maximum value. The failure pattern of AP laminate under 3 J impact from experimental and simulation results were compared in Figure 6b–e. There were no obvious damage on the composite surface, which was reflected by all the three simulation results.

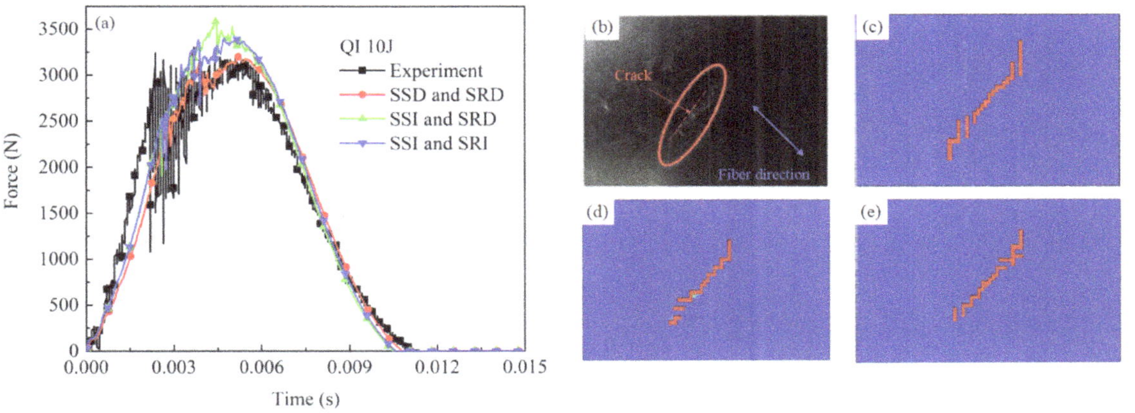

Figure 6. Comparison of experimental results and simulation results (**a**) contact force; (**b**) experimental failure pattern; (**c**) SSD and SRD (**d**) SSI and SRD (**e**) SSI and SRI simulation result.

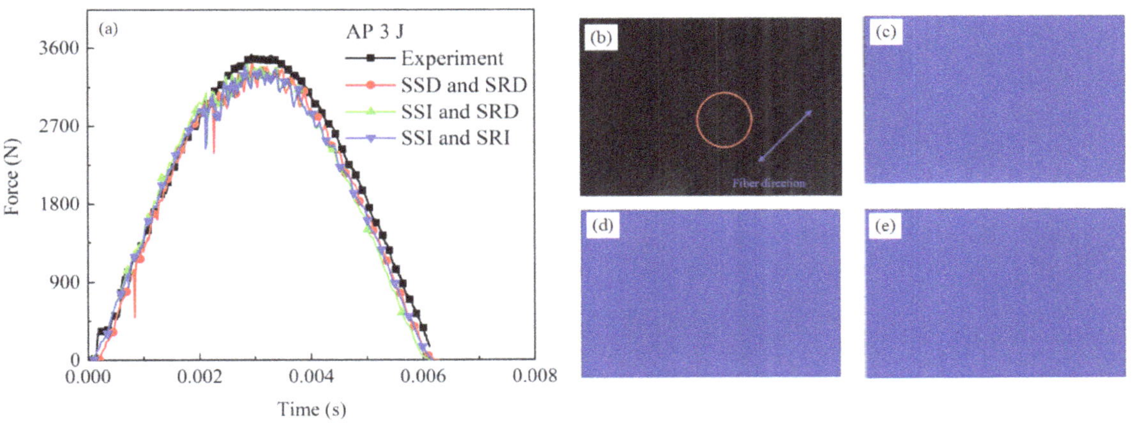

Figure 7. Comparison of experimental results and simulation results (**a**) contact force; (**b**) experimental failure pattern; (**c**) SSD and SRD (**d**) SSI and SRD (**e**) SSI and SRI simulation result.

The contact force history of AP laminate under 6 J impact from experiment and simulation results were compared in Figure 8a. Similarly, the contact force from experiments, SSD and SRD, SSI and SRD, SSI and SRI numerical models were quadratic function type. The failure pattern of AP laminate under 6 J impact from experimental and simulation results were compared in Figure 6b–e. The cracks from simulation results were on the impact point and along the fiber direction, which was similar to the experimental results. Moreover, the crack in SSD and SRD model was more continuity than other two models. It indicated that the failure evolution process were evidently in SSD and SRD model.

The detailed maximum contact force from experiment and simulation models were listed in Table 3. It can be seen that the simulation error decreased significantly in the SSD and SRD material model compared to the other two material model for both QI and AP laminate. Moreover, the simulation error for AP laminate under 3 J energy impact was decreased by 21.6% (from 2.9% to 3.7%), and the value was 40.5% (from 8.4% to 5%) for AP laminate under 6 J. The simulation error was decreased with the increasing strain rate especially for the laminae with damage occur.

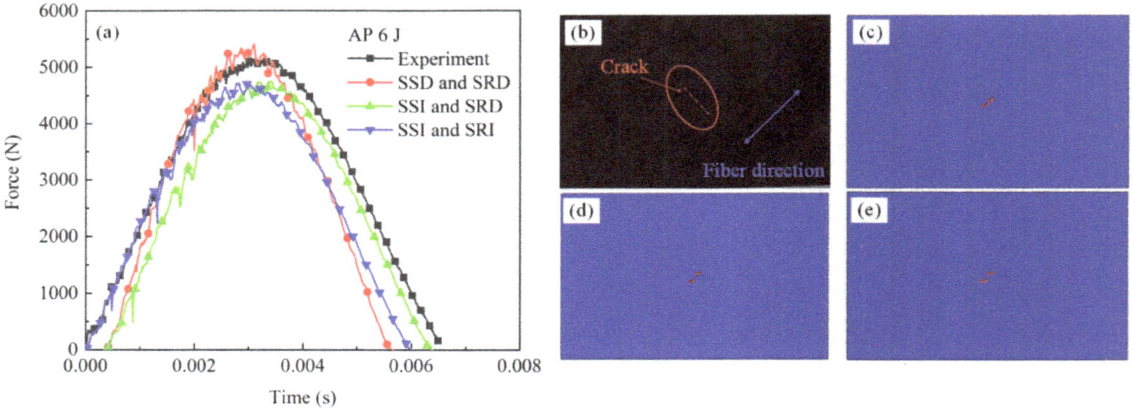

Figure 8. Comparison of experimental results and simulation results (**a**) contact force; (**b**) experimental failure pattern; (**c**) SSD and SRD (**d**) SSI and SRD (**e**) SSI and SRI simulation result.

Table 3. Maximum contact force comparison of the experimental and numerical results.

	Impact Energy (J)	Experiment (N)	SSD and SRD (N)	Error (%)	SSI and SRD (N)	Error (%)	SSI and SRI (N)	Error (%)
QI	10	3203	3200	0.1%	3581	11.8%	3408	6.4%
AP	3	3525	3423	2.9%	3412	3.2%	3396	3.7%
	6	5158	5416	5%	4759	7.7%	4727	8.4%

5. Conclusions

This paper proposed a strain rate dependent damage model for evaluating the dynamic response of CFRTP laminates with different stacking sequence. In this model, a strain-stress relationship as well as matrix damage, fiber damage and delamination damage criteria considering strain rate effect were established and implemented into finite element software ABAQUS by user defined subroutine. The low velocity impact experiments at different impact energies on quasi-isotropic laminate and angle-ply laminae were also used to validate the model. It was found that the contact force curves of LVI tests from numerical results were coincidence well with the experimental results. The simulation error was decreased with the increasing strain rate especially for the laminae with damage occur. The effect of stacking sequence on the strain rate sensitivity should be considered when analyzing the dynamic response of CFRTP laminates.

Author Contributions: Methodology, writing—original draft preparation Y.Z.; Project administration, writing-review and editing, supervision B.L. All authors have read and agreed to the published version of the manuscript.

Funding: This work was sponsored by Fundamental Research Funds for the Central Universities (No. FRF-BD-20-08A, No. FRF-BD-19-003A).

Institutional Review Board Statement: Not applicable.

Informed Consent Statement: Not applicable.

Data Availability Statement: We choose to exclude this statement.

Acknowledgments: This work was sponsored by Fundamental Research Funds for the Central Universities (No. FRF-BD-20-08A, No. FRF-BD-19-003A).

Conflicts of Interest: The authors declared no potential conflict of interest with respect to the research, authorship, and/or publication of this article.

References

1. Che, D.M.; Saxena, I.; Han, P.D.; Guo, P.; Ehmann, K.F. Machining of carbon fiber reinforced plastics/polymers: A literature review. *J. Manuf. Sci. Eng.* **2014**, *136*, 034001. [CrossRef]
2. Zhang, Y.B.; Sun, L.Y.; Li, L.J.; Xiao, H.Y.; Wang, Y.T. An efficient numerical method to analyze low-velocity impact response of carbon fiber reinforced thermoplastic laminates. *Polym. Compos.* **2020**, *41*, 2673–2686. [CrossRef]
3. Lu, Y.H.; Zhang, M.S.; Zheng, W.H. Preparation and properties of T300 carbon fiber-reinforced thermoplastic polyimide composites. *Appl. Polym. Sci.* **2006**, *102*, 646–654. [CrossRef]
4. Quaresimin, M.; Ricotta, M.; Martello, L.; Mian, S. Energy absorption in composite laminates under impact loading. *Compos. Part B* **2013**, *44*, 133–140. [CrossRef]
5. Massaq, A.; Rusinek, A.; Klosak, M.; Bahi, S.; Arias, S. Strain rate effect on the mechanical behavior of polyamide composites under compression loading. *Compos. Struct.* **2014**, *214*, 114–122. [CrossRef]
6. Chen, C.Y.; Zhang, C.; Liu, C.L.; Miao, Y.G.; Wong, S.C.; Li, Y.L. Rate-dependent tensile failure behavior of short fiber reinforced PEEK. *Compos. Part B* **2018**, *136*, 187–196. [CrossRef]
7. Ou, Y.F.; Zhu, D.J. Tensile behavior of glass fiber reinforced composite at different strain rates and temperatures. *Constr. Build. Mater.* **2015**, *96*, 648–656. [CrossRef]
8. Zhang, Y.B.; Sun, L.Y.; Li, L.J.; Wei, J.L. Effects of strain rate and high temperature environment on the mechanical performance of carbon fiber reinforced thermoplastic composites fabricated by hot press molding. *Compos. Part A* **2020**, *134*, 105905. [CrossRef]
9. Bie, B.X.; Han, J.H.; Lu, L.; Zhou, X.M.; Qi, M.L.; Zhang, Z.; Luo, S.N. Dynamic fracture of carbon nanotube/epoxy composites under high strain-rate loading. *Compos. Part A* **2015**, *68*, 282–288. [CrossRef]
10. Wang, S.Y.; Wen, L.H.; Xiao, J.Y.; Lei, M.; Hou, X.; Liang, J. The out-of-plane compression response of woven thermoplastic composites: Effects of strain rates and temperature. *Polymers* **2021**, *13*, 264. [CrossRef]
11. Zhang, P.; Tan, S.Y.; Hu, X.F.; Yao, W.A.; Zhuang, X.Y. A double-phase field model for multiple failures in composites. *Compos. Struct.* **2022**, *293*, 115730. [CrossRef]
12. Min, L.; Hu, X.F.; Yao, W.A.; Bui, T.Q.; Zhang, P. On realizing specific failure initiation criteria in the phase field model. *Comput. Methods Appl. Mech. Eng.* **2022**, *394*, 114881. [CrossRef]
13. Hu, X.F.; Xu, H.Q.; Xi, X.; Zhang, P.; Yang, S.T. Meso-scale phase field modelling of reinforced concrete structures subjected to corrosion of multiple reinforcements. *Constr. Build. Mater.* **2022**, *321*, 126376. [CrossRef]
14. Zhang, P.; Yao, W.A.; Hu, X.F.; Zhuang, X.Y. Phase field modelling of progressive failure in composites combined with cohesive element with an explicit scheme. *Compos. Struct.* **2021**, *262*, 113353. [CrossRef]
15. Zhang, P.; Hu, X.F.; Bui, T.Q.; Yao, W.A. Phase field modeling of fracture in fiber reinforced composite laminate. *Int. J. Mech. Sci.* **2019**, *161*, 105008. [CrossRef]
16. Bui, T.Q.; Hu, X.F. A review of phase-field models, fundamentals and their applications to composite laminates. *Eng. Fract. Mech.* **2021**, *248*, 107705. [CrossRef]
17. Tsai, S.; Wu, E.M. A general theory of strength for anisotropic materials. *J. Compos. Mater.* **1971**, *5*, 58–80. [CrossRef]
18. Hashin, Z.; Rotem, A. A fatigue failure criterion for fiber reinforced materials. *J. Compos. Mater.* **1973**, *7*, 448–464. [CrossRef]
19. Hou, J.P.; Petrinic, N.; Ruiz, C.; Hallett, S.R. Prediction of impact damage in composite plates Compos. *Sci. Technol.* **2000**, *60*, 273–281.
20. Hou, J.P.; Petrinic, N.; Ruiz, C. A delamination criterion for laminated composites under low-velocity impact. *Compos. Sci. Technol.* **2001**, *61*, 2069–2074. [CrossRef]
21. Yen, C.F. Ballistic impact modeling of composite materials. In Proceedings of the 7th International LS-DYNA Users Conference, Dearborn, MI, USA, 19–21 May 2002.
22. Yen, C.F.; Caiazzo, A. Innovative Processing of Multifunctional Composite Armor Forground Vehicles; ARL Technical Report ARL-CR-484; US Army Research Laboratory, Aberdeen Proving Ground, MD, USA. 2001. Available online: http://www.tjprc.org/publishpapers/2-67-1516433007-54.IJMPERDFEB201854.pdf (accessed on 7 November 2022).
23. Balkan, O.; Demirer, H.; Kayal, E.S. Effects of deformation rates on mechanical properties of PP/SEBS blends. *J. Achiev. Mater. Manuf. Eng.* **2011**, *47*, 26–33.
24. Brown, K.A.; Brooks, R.; Warrior, N.A. The static and high strain rate behaviour of a commingled E-glass/polypropylene woven fabric composite. *Compos. Sci. Technol.* **2010**, *70*, 272–283. [CrossRef]
25. Wang, K.K.; Zhao, L.B.; Hong, H.M.; Zhang, J.Y. A strain-rate-dependent damage model for evaluating the low velocity impact induced damage of composite laminates. *Compos. Struct.* **2018**, *201*, 995–1003. [CrossRef]
26. Taniguchi, N.; Nishiwaki, T.; Hirayama, N.; Nishida, H.; Kawada, H. Dynamic tensile properties of carbon fiber composite based on thermoplastic epoxy resin loaded in matrix-dominant directions. *Compos. Sci. Technol.* **2009**. *69*, 207–213. [CrossRef]
27. Lyu, Q.H.; Wang, B.; Zhao, Z.Q.; Guo, Z.Y. Damage and failure analysis of hybrid laminates with different ply-stacking sequences under low-velocity impact and post-impact compression. *Thin-Walled Struct.* **2022**, *180*, 109743. [CrossRef]
28. Rafiee, R.; Ghorbanhosseini, A. Investigating interaction between CNT and polymer using cohesive zone model. *Polym. Compos.* **2018**, *39*, 3903–3911. [CrossRef]
29. Li, X.; Ma, H.F.; Liu, H.F.; Tan, W.; Gong, X.J.; Zhang, C.; Li, T.L. Assessment of failure criteria and damage evolution methods for composite laminates under low-velocity impact. *Compos. Struct.* **2019**, *207*, 727–739. [CrossRef]

30. Robbins, D.H.; Chopra, I. Modelling of progressive damage in the adhesive bond layers of actuated plates. *J. Intell. Mater. Syst. Struct.* **2007**, *18*, 893–921. [CrossRef]
31. Linde, P.; Pleitner, J.; Boer, H.D.; Carmone, C. Modelling and simulation of fiber metal laminate. In *Proceedings of the ABAQUS Users' Conference*; Abaqus Inc.: Palo Alto, CA, USA, 2004; pp. 421–438.
32. Lapczyk, I.; Hurtado, A. Progressive damage modeling in fiber-reinforced materials. *Compos. Part A* **2007**, *38*, 2333–2341. [CrossRef]
33. *ASTM D7136–05*; Standard Test Method for Measuring the Damage Resistance of a Fibre-Reinforced Polymer Matrix Composite to a Drop-Weight Impact Event. ASTM International: West Conshohocken, PA, USA, 2005.

electronics

MDPI

Article

Collaborative Accurate Vehicle Positioning Based on Global Navigation Satellite System and Vehicle Network Communication

Haixu Yang [1,2,3], Jichao Hong [1,2,3,*], Lingjun Wei [4,*], Xun Gong [1] and Xiaoming Xu [2,3]

1 State Key Laboratory of Automotive Simulation and Control, Jilin University, Changchun 130025, China
2 School of Mechanical Engineering, University of Science and Technology Beijing, Beijing 100083, China
3 Shunde Innovation School, University of Science and Technology Beijing, Foshan 528399, China
4 Department of Waterway Transport and Maritime Management, Beijing Vocational College of Transportation, Beijing 102600, China
* Correspondence: qdbithong@163.com (J.H.); wljyal66@126.com (L.W.)

Abstract: Intelligence is a direction of development for vehicles and transportation. Accurate vehicle positioning plays a vital role in intelligent driving and transportation. In the case of obstruction or too few satellites, the positioning capability of the Global navigation satellite system (GNSS) will be significantly reduced. To eliminate the effect of unlocalization due to missing GNSS signals, a collaborative multi-vehicle localization scheme based on GNSS and vehicle networks is proposed. The vehicle first estimates the location based on GNSS positioning information and then shares this information with the environmental vehicles through vehicle network communication. The vehicle further integrates the relative position of the ambient vehicle observed by the radar with the ambient vehicle position information obtained by communication. A smaller error estimate of the position of self-vehicle and environmental vehicles is obtained by correcting the positioning of self-vehicle and environmental vehicles. The proposed method is validated by simulating multi-vehicle motion scenarios in both lane change and straight-ahead scenarios. The root-mean-square error of the co-location method is below 0.5 m. The results demonstrate that the combined vehicle network communication approach has higher accuracy than single GNSS positioning in both scenarios.

Keywords: intelligent vehicles; vehicle positioning; global navigation satellite system; vehicle network communication; multi-vehicle collaboration

Citation: Yang, H.; Hong, J.; Wei, L.; Gong, X.; Xu, X. Collaborative Accurate Vehicle Positioning Based on Global Navigation Satellite System and Vehicle Network Communication. *Electronics* **2022**, *11*, 3247. https://doi.org/10.3390/electronics11193247

Academic Editor: Taeshik Shon

Received: 14 September 2022
Accepted: 7 October 2022
Published: 9 October 2022

1. Introduction

In recent years, the degree of automation of intelligent vehicles has gradually increased [1,2]. Intelligent vehicles can primarily reduce the influence of human factors and decrease the occurrence of traffic accidents [3]. Highly accurate positioning is the basis for intelligent vehicles to achieve path planning and motion trajectory tracking. In the development of in-vehicle navigation, driver assistance, autonomous driving, intelligent transportation and other technologies [4], the location of the vehicle is a critical type of information. GNSS is currently an essential method in vehicle positioning [5,6]. However, the positioning can be inaccurate or even impossible if there are tunnels and other occlusions. This single positioning method cannot fully meet the needs of the growing automotive intelligence in terms of accuracy and reliability. Therefore, it is important to explore new vehicle localization methods to develop intelligent driving and intelligent transportation.

Researchers made many efforts in vehicle positioning, which contains methods of multi-sensor fusion, vehicle network communication and artificial intelligence. In fusing information from multiple sensors, Pankaj et al. [7] proposed an optical camera-based mobile vehicle localization scheme. Using the street light and camera as transmitter and receiver, respectively, it is able to achieve a positioning accuracy of less than 1 m. Ioannis et al. [8] used distance and velocity measurements to deal with localization and target tracking problems. This approach enabled localization despite the absence of GNSS signals.

Patrick et al. [9] performed position estimation by installing fixed points on the road and detecting body bumps and road surface imperfections. Hossain et al. [10] matched and fused the position information obtained from GNSS, vehicle network communication and radar. The simulation results showed a significant improvement in positioning accuracy. Wang et al. [11] proposed an auxiliary vehicle location system. The system consists of three base stations equipped with multiple inputs and multiple outputs. Accurate localization can be achieved based on the results of three cross-locations. Tao et al. [12] proposed a multi-sensor fusion localization strategy for intelligent vehicles using global pose map optimization and validated it on the ROS platform. Zhang et al. [13,14] proposed a collaborative positioning method based on 3D mapping-assisted GNSS. The technique utilized measurements from surrounding available GNSS receivers, eliminating systematic errors while also mitigating random errors. The multi-sensor fusion approach makes the vehicle more costly, and the sensors are more affected by the environment.

If only the vehicles are individually positioned by sensors, autonomous vehicle positioning may not be possible in the absence of signals. The development of Telematics technology allows the use of location data from nearby vehicles to improve or replace self-location. Nam et al. [15] proposed a cooperative adjacent vehicle localization system. The system quickly identifies the location of neighboring vehicles and communicates that information with neighboring vehicles. Zhu et al. [16] established a GNSS/dead reckoning/ultra-wide band fusion positioning algorithm with adaptive information allocation coefficients using the Kalman filter, which can improve positioning accuracy and reliability. Mahmoud et al. [17] used dedicated short-range communications to share data between vehicles, enabling the positioning of vehicles during GNSS signal outages. Hou et al. [18] proposed a displacement-based selection method that can reduce the effect of measurement errors in nearby vehicle information. The status information of nearby vehicles can be used to locate the target vehicle. Buehrer et al. [19] considered the application of IoT and 5G to co-location, demonstrating its superior coverage and accuracy. Ma et al. [20] proposed a BeiDou-based joint vehicle-road positioning method using the volumetric Kalman model to further improve the positioning accuracy. Ansari et al. [21] investigated the supplementation of dedicated short-range communication and V2V by terrestrial communication systems to examine the positioning performance of vehicles. Tong et al. [22] used information collection platforms to obtain driving status and roadside information to obtain the vehicle's location. The current method still has room for improvement in the integration of GNSS, vehicle network communication, and vehicle sensors. Positioning accuracy still needs to be improved.

With the advancement of computer technology, artificial intelligence methods are increasingly used in vehicle positioning. Kim et al. [23] developed an indoor vehicle location system using surveillance cameras. Such a system determines the location of a vehicle by extracting vehicle information from image information. Wan et al. [24] combined machine learning and edge computing and could obtain high accuracy in vehicle location estimation. Lee et al. [25] developed a machine learning method applicable to location prediction using the data generated by the localization algorithm, which improved the accuracy of location prediction by 10%. Kong et al. [26] proposed a vehicle localization system based on federated learning. The system makes full use of edge computing and can provide high-accuracy positioning correction. Gao et al. [27] proposed an error-weighting-based vehicle localization fusion algorithm. Distance and positioning compound errors are taken into account for single-vehicle positioning and trilateral positioning. Wang et al. [28] studied the effects of vehicle location potential features and vehicle association potential features. A routing algorithm based on vehicle location analysis is proposed to obtain more accurate vehicle location prediction. Watta et al. [29] proposed an intelligent system incorporating neural networks and geometric modeling. A neural network trained on V2V signals to obtain the location of remote vehicles. Figure 1 shows the different methods of vehicle positioning. Machine learning methods are considered to be black-box models. There is no way to know the direction of learning inside the model, and the application of such methods is still open to question. The collaborative localization method proposed in this study combines GNSS, vehicle network communication and in-vehicle sensors to achieve more accurate vehicle positioning.

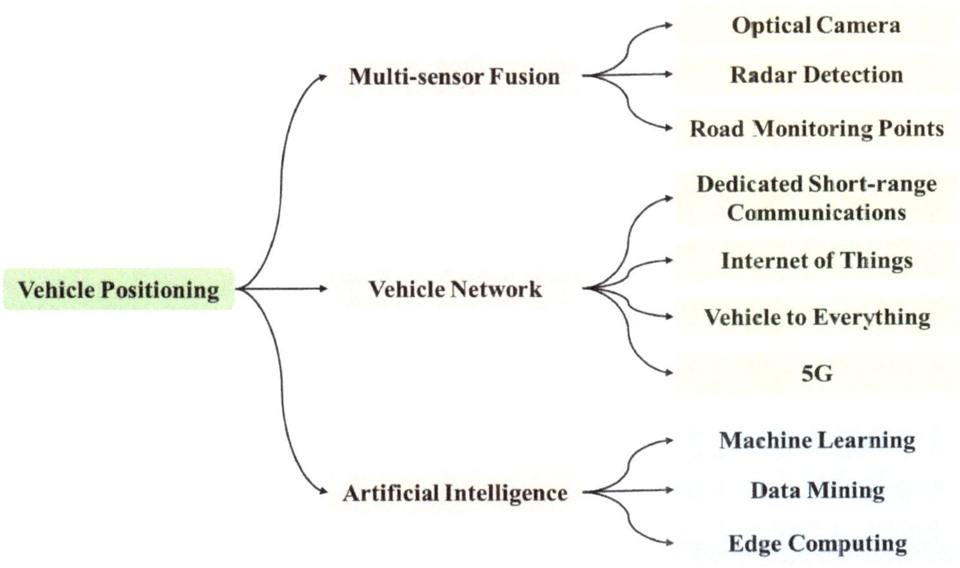

Figure 1. Different methods of vehicle positioning.

A collaborative vehicle positioning method using GNSS and vehicle network communication is proposed in order to overcome the shortcomings of GNSS positioning in terms of accuracy and reliability. Contributions of this paper are shown in the following aspects:

(1) GNSS and vehicle network communication: Combined positioning using GNSS and vehicle network communication to improve positioning accuracy and reliability.

(2) Multi-vehicle scenario validation: A multi-vehicle motion scenario was built, and the positioning was verified under straight-ahead and lane change conditions.

(3) Impact of communication interruption: Verification of positioning accuracy during communication interruption to check the effectiveness of vehicle network communication for positioning.

The remainder of this paper is structured as follows: Section 2 describes the vehicle model and the multi-vehicle motion scenario used for simulation. Section 3 compares the positioning accuracy in different positioning scenarios, and Section 4 gives the conclusion of the study.

2. Research Methodology

The proposed methodology uses GNSS and vehicle network communication. To verify the method's effectiveness, kinematic modeling of the vehicle is performed. A motion scene is created in the simulation environment and the positioning error is calculated.

2.1. Component Modules

The proposed multi-vehicle cooperative positioning system consists of three modules: GNSS, millimeter wave radar and vehicle network communication.

2.1.1. GNSS

GNSS is a navigation and positioning system that uses radio. It refers to all satellite navigation systems, such as GPS of the United States, Glonass of Russia, Galileo of Europe and BeiDou of China. It can provide 3D coordinates, velocity and time information at any of Earth's surfaces or near-Earth space. These three types of information are called PVT (Position Velocity and Time). In this study, absolute position information of the vehicle can be obtained using GNSS. The received coordinates are earth latitude and longitude coordinates, which can be converted to local area plane coordinates by calculation.

As shown in Figure 2, any location on the Earth's surface has a three-dimensional coordinate. GNSS satellites also have a coordinate. The distance between the satellite and the positioning target can be expressed in coordinates as

$$L = \sqrt{(x - x')^2 + (y - y')^2 + (z - z')^2} \tag{1}$$

where the coordinates of the satellite are known and the target point coordinates are unknown.

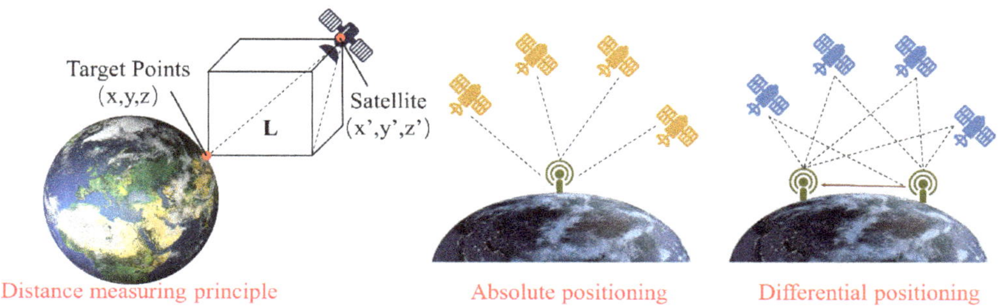

Figure 2. Satellite positioning principle.

The speed of signal transmission between the satellite and the target point can be regarded as the light speed. Then the distance between the satellite and the target point can be expressed as

$$L = (t - t') \cdot c \tag{2}$$

where t is the satellite time and t' is the time at the target point. Combining the two equations, we get

$$(t - t') \cdot c = \sqrt{(x - x')^2 + (y - y')^2 + (z - z')^2} \tag{3}$$

At this point, the coordinate values of the other three satellites are needed to solve the equation. It takes at least four satellites to determine the location of a target point on Earth. Depending on the number of signal receivers, positioning can also be divided into absolute positioning and differential positioning.

2.1.2. Millimeter Wave Radar

The use of millimeter wave radar can obtain the relative position information of nearby environmental vehicles to the present vehicle. This includes the distance and angle of the ambient vehicle relative to the vehicle, that is, the position of the ambient vehicle in the local coordinate system.

Sensors commonly used for environmental awareness in smart vehicles include cameras, Light Detection and Ranging (LIDAR), millimeter-wave radar and ultrasonic radar. The camera is a component based on the optical principle [30]. When the light passes through the lens, it is captured by the light sensor and then forms an image. LIDAR performs object detection and ranging by emitting laser pulses externally [31]. The laser will reflect when it reaches the surface of the target to be measured. LIDAR can acquire parameters such as the reflected signal's return time and signal strength. The information allows us to determine the target's distance, orientation, motion status and other characteristics. Millimeter wave radar is similar to LIDAR in principle. The difference is that the signal it emits is changed from laser to electromagnetic wave [32]. It calculates the distance from the time difference of electromagnetic wave returns and calculates the relative velocity of the measured target based on the Doppler effect. The principle of ultrasonic radar is also similar. The signal it generates is ultrasonic [33]. Due to the different principles of the above sensors, the application scenarios are different. Figure 3 illustrates the characteristics of several sensors and their application scenarios.

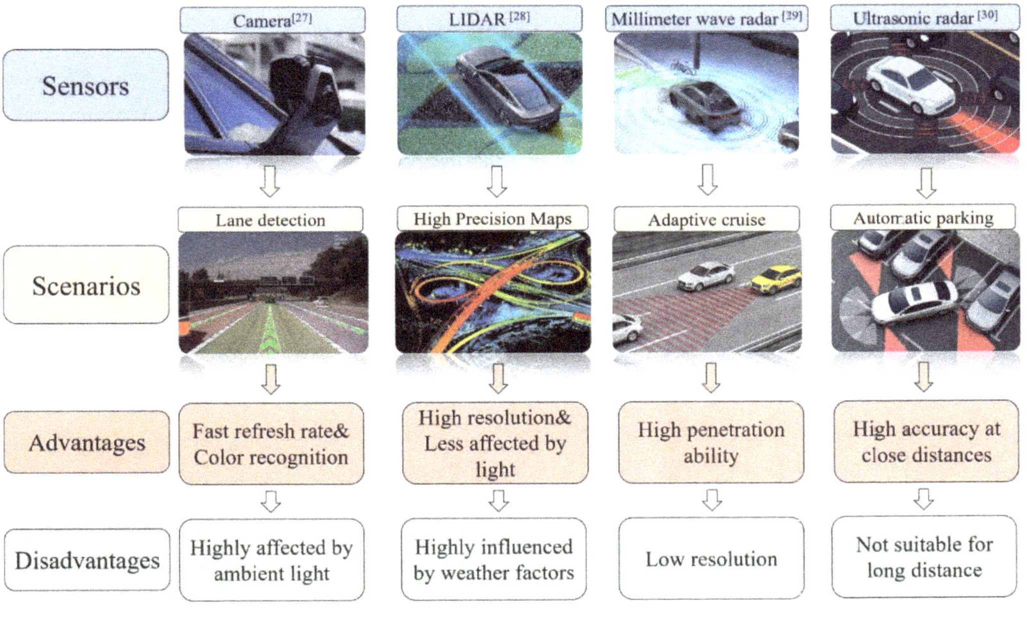

Figure 3. The characteristics of several sensors.

2.1.3. Vehicle Network Communication

Vehicle network refers to a system network for wireless communication and information exchange among vehicles, roads, pedestrians and the Internet based on intra-vehicle network, inter-vehicle network and in-vehicle mobile Internet, in accordance with agreed communication protocols and data interaction standards. Vehicle network is an intelligent three-dimensional architecture, including the data sensing layer, network transmission layer and platform application layer. The data sensing layer uses sensors to perceive information and obtain comprehensive information on road conditions. The network transmission layer connects the infrastructure to the platform application layer. It enables the transfer of information between various subjects. The platform application layer has the function of management and operation. It is capable of performing tasks such as traffic management and safety control.

The main units in the vehicle network communication are shown in Figure 4. Vehicle network communication can be based on dedicated short-range communication (DSRC) or 5G networks to enable the transfer of information between vehicles. This also represents the two main technical routes of the current vehicle network communication standards, namely DSRC and cellular vehicle to everything (C-V2X). DSRC has the technical characteristics of exclusive bandwidth and short-range communication. C-V2X highlights the advantages in capacity, latency, manageability and anti-interference algorithms. In collaborative positioning, when the GNSS signal is missing, the vehicle with the missing signal can establish communication with other vehicles with a normal signal through the vehicle network. Thus, the position of the signal-less vehicle can be determined by the position of the surrounding vehicles with normal signals.

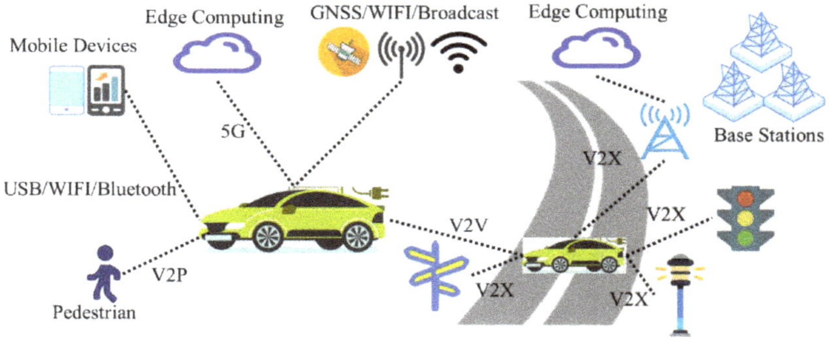

Figure 4. Vehicle network communication architecture.

2.2. *Overall Flow of the Method*

The overall workflow of multi-vehicle cooperative positioning is shown in Figure 5.

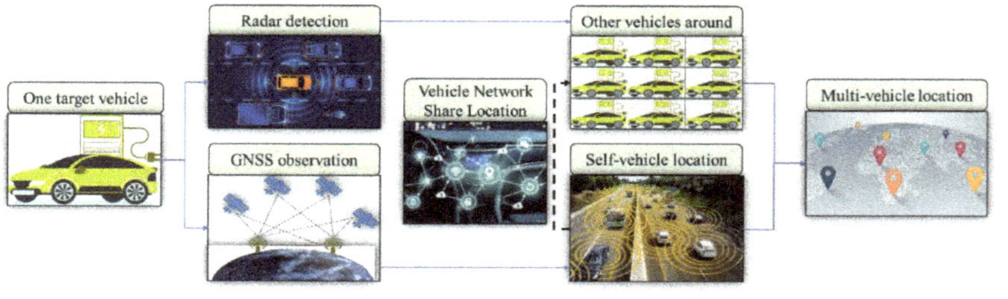

Figure 5. Multi-vehicle cooperative positioning process.

The GNSS module obtains absolute vehicle position information with noise [34]. By designing filters to process the signals, coarse position estimation based on self-vehicle observation information can be achieved. Vehicle network communication sends this coarse estimate of position to other vehicles in the vicinity. Combined with the relative position information detected by millimeter wave radar, multi-vehicle cooperative localization is enabled.

Four vehicles (A, B, C and D) are connected by a vehicle network. Each vehicle can position itself via GNSS. Vehicles can communicate with each other via vehicle networks to share location information with other vehicles. The vehicle can sense the relative position within a certain distance by millimeter wave radar. Combined with the information, the vehicle can estimate its own position as well as the position of other vehicles around it. The interrelationship between the vehicles is shown in Figure 6a.

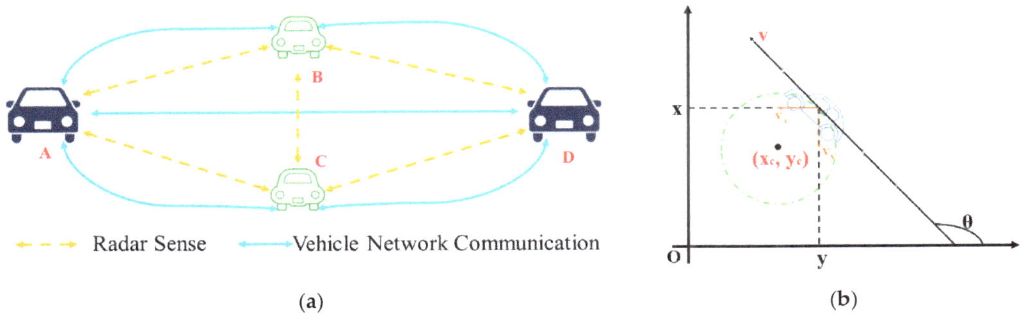

Figure 6. (**a**) Interrelationship between the vehicles; (**b**) Vehicle motion model.

2.3. Vehicle Modeling Based on Velocity Motion Model

The velocity motion model contains linear and angular velocities [35]. The control vector for vehicle motion is:

$$u_t = \begin{pmatrix} v_t \\ w_t \end{pmatrix} \tag{4}$$

where v_t represents the linear velocity of the vehicle and w_t represents the angular velocity at the moment t. When v_t is positive, the vehicle moves forward. When w_t is positive, the vehicle turns counterclockwise.

The state vector of the velocity motion model is:

$$x_t = \begin{pmatrix} x_t \\ y_t \\ \theta_t \end{pmatrix} \tag{5}$$

where x_t and y_t are the position coordinates in the right-angle coordinate system [36], θ_t is the velocity direction. The motion of the vehicle is shown in Figure 6b.

Noise can interfere with signal transmission and cause deviations in the state vector and control vector. The location information obtained will be inaccurate. In the ideal case (no noise influence), assume that the initial state vector is $(x,\ y, \theta)^T$, the control vector is $(v, w)^T$ and the state vector after Δt time is $(x', y', \theta')^T$. The state vector can be expressed as

$$\begin{pmatrix} x' \\ y' \\ \theta' \end{pmatrix} = \begin{pmatrix} x_c + \frac{v}{\omega}\sin(\theta + \omega\Delta t) \\ y_c - \frac{v}{\omega}\cos(\theta + \omega\Delta t) \\ \theta + \omega\Delta t \end{pmatrix} = \begin{pmatrix} x \\ y \\ \theta \end{pmatrix} + \begin{pmatrix} -\frac{v}{\omega}\sin\theta + \frac{v}{\omega}\sin(\theta + \omega\Delta t) \\ \frac{v}{\omega}\cos\theta - \frac{v}{\omega}\cos(\theta + \omega\Delta t) \\ \omega\Delta t \end{pmatrix} \tag{6}$$

where $x_c = x - \frac{v}{\omega}\sin\theta$, $y_c = y + \frac{v}{\omega}\cos\theta$. $(x_c,\ y_c)^T$ is the center of the circle for the motion in time Δt. Considering the error between the real motion and ideal motion of the vehicle and the effect of noise, the real motion model of the vehicle can be expressed as

$$\begin{pmatrix} x' \\ y' \\ \theta' \end{pmatrix} = \begin{pmatrix} x \\ y \\ \theta \end{pmatrix} + \begin{pmatrix} -\frac{\breve{v}}{\breve{\omega}}\sin\theta + \frac{\breve{v}}{\breve{\omega}}\sin(\theta + \breve{\omega}\Delta t) \\ \frac{\breve{v}}{\breve{\omega}}\cos\theta - \frac{\breve{v}}{\breve{\omega}}\cos(\theta + \breve{\omega}\Delta t) \\ \breve{\omega}\Delta t + \breve{\gamma}\Delta t \end{pmatrix} \tag{7}$$

$\breve{v}$, $\breve{\omega}$ should be the filtered value.

2.4. Estimation of Vehicle Location by Relative Positioning

When there are multiple vehicles in the environment for cooperative localization, one of the vehicles is selected as the target vehicle for ease of study and accuracy in presentation. The GNSS signal is filtered to obtain the position estimation result of this vehicle [37]. The results are sent to nearby vehicles via the vehicle network. At the same time, the vehicle can detect the position of nearby environmental vehicles relative to the vehicle through millimeter wave radar. Through the correspondence between the radar target and the ambient vehicle, the position of the ambient vehicle can be used to solve for the position of this vehicle.

The position of this vehicle is obtained from the position and relative distance of the environment vehicle as

$$\mu_{rt}^{i} = \left(\mu_{t,x}^{i} - d_t^i \cos\varphi_t^i,\ \mu_{t,y}^{i} - d_t^i \sin\varphi_t^i\right)^T \tag{8}$$

where $\mu_{t,x}^{i}$ and $\mu_{t,y}^{i}$ are the coordinates of the environment vehicle and d_t^i is the distance between this vehicle and the ambient vehicle. Assume that the error of the radar measurement at the moment t is

$$\varepsilon_{t,\mathrm{r}} = \left(\varepsilon_{t,\mathrm{rx}}, \varepsilon_{t,\mathrm{ry}}\right)^T \tag{9}$$

The two follow a normal distribution and are uncorrelated with each other. Then there are

$$\varepsilon_{t,\mathrm{rx}} \sim \mathrm{N}\left(0, \sigma_{\mathrm{rx}}^2\right),\ \varepsilon_{t,\mathrm{ry}} \sim \mathrm{N}\left(0, \sigma_{\mathrm{ry}}^2\right) \tag{10}$$

The covariance matrix of the radar range can be obtained as

$$\mathrm{P} = \begin{pmatrix} \sigma_{\mathrm{rx}}^2 & 0 \\ 0 & \sigma_{\mathrm{ry}}^2 \end{pmatrix} \tag{11}$$

2.5. Simulation Contents

To verify the effectiveness of the method, the simulation is performed by building a multi-vehicle motion scenario. The time duration of each simulation is set to 30 s, and the step size is 0.1 s.

2.5.1. Linear Motion Simulation

The motion scenario contains two lanes. The width of the lanes is 3.75 m, and there are five vehicles in each lane. Vehicles travel along the centerline of their respective lanes. Vehicles located in the fast lane have an initial speed of 50 km/h and a vehicle spacing of 80 m. Vehicles located in the slow lane have an initial speed of 30 km/h and a vehicle spacing of 50 m, as shown in Figure 7a. After the simulation starts, at $0 < t < 10$ s, the first car in both lanes moves according to the acceleration curve shown in Figure 7b. At $t > 10$ s, the first car in both lanes maintains uniform linear motion.

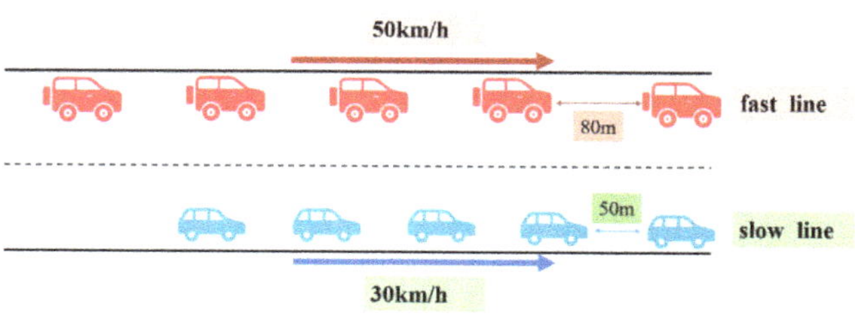

(a)

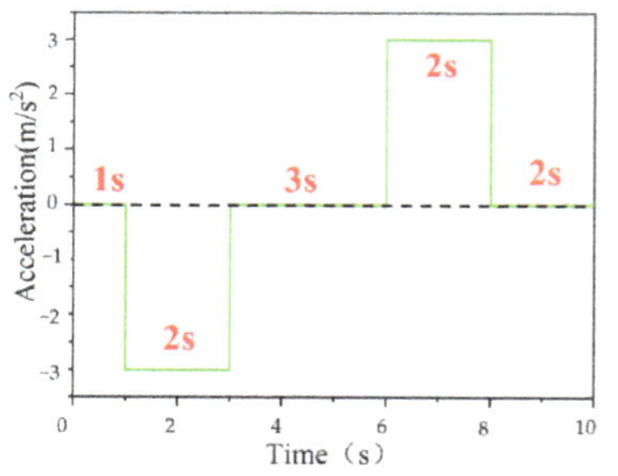

(b)

Figure 7. (**a**) Straight line simulation scenario; (**b**) Acceleration variation curve.

The eight cars after the first car respond to the speed fluctuations of the preceding car according to the linear-following model.

$$\ddot{x}_r(t+T) = \lambda\left(\dot{x}_f(t) - \dot{x}_r(t)\right) \tag{12}$$

where $\dot{x}_r(t)$ and $\dot{x}_f(t)$ are the distance between the rear car and the front car along the direction of the lane to the origin at time t. Take $\lambda = 0.3$, $T = 1$ s. The motion simulation takes into account the fluctuation of vehicle speed under linear motion and the change of surrounding environment vehicles with time.

2.5.2. Lane Change Motion Simulation

The motion scenario contains two lanes. The width of the lanes is 3.75 m, and there are five vehicles in each lane. Vehicles travel along the centerline of their respective lanes. The initial speed of both lanes is 30 km/h, and the distance between cars No.4 and No.5 and between cars No.6 and No.7 is 100 m. The distance between the front and rear of other vehicles is 50 m, as shown in Figure 8a.

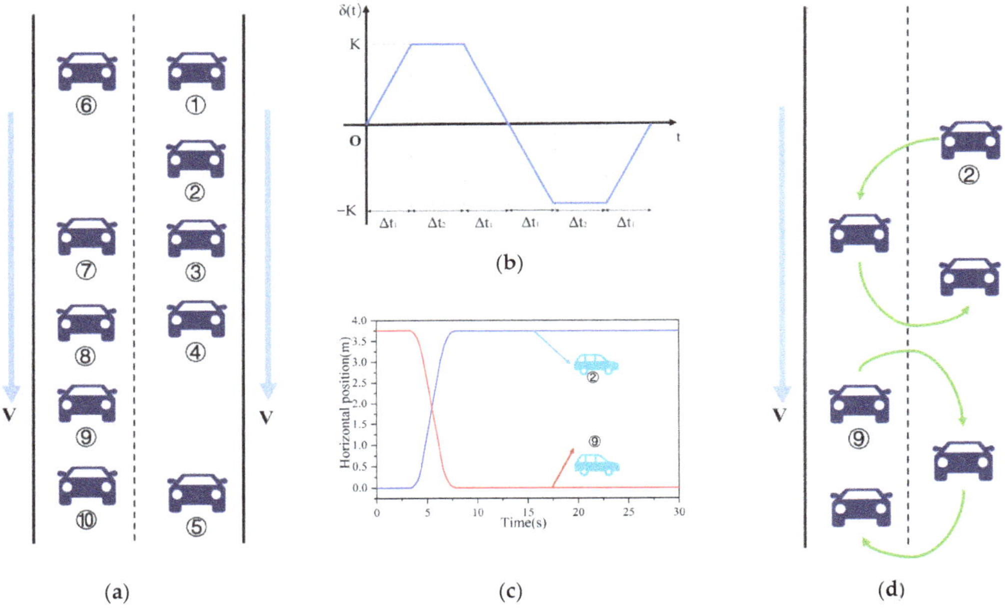

Figure 8. (**a**) Lane change simulation scenario; (**b**) Lane change steering angle model; (**c**) Lateral position curve for lane change motion; (**d**) Car 2 and Car 9 for lane change simulation.

After starting the simulation, cars 2 and 9 perform the lane change operation at 3 s < t < 8 s, while the other vehicles maintain uniform linear motion. The lane change steering angle is shown in Figure 8b. $\delta(t)$ is the steering angle at time t. Assuming a linear two-degree-of-freedom model for the vehicle, with constant longitudinal velocity and small angular offset in the heading. Then $\delta(t)$ function satisfies that the transverse pendulum angular velocity, lateral velocity and heading angular deflection are zero at the beginning and end of steering. Simplifying the velocity motion model assumes a zero lateral deflection angle. Therefore, the angular velocity of the vehicle transverse pendulum is proportional to the steering angle, and $\delta(t)$ can be replaced by $\omega(t)$.

To determine the parameters K, Δt_1, Δt_2 in this channel change model, the constraints are listed. The lateral displacement, maximum lateral acceleration and maximum lateral sharpness are constrained, respectively

$$\frac{y_l}{u} = \lim_{t\to\infty} \int_0^t \int_0^\tau \omega(\alpha) d\alpha \, dt \tag{13}$$

$$|a_{lateral}(t)| < 0.2\text{ g} \tag{14}$$

$$|\dot{a}_{lateral}(t)| < \frac{0.1\text{ g}}{s} \tag{15}$$

where y_l is the target lateral displacement, u is the vehicle longitudinal velocity,v is the lateral velocity of the vehicle, g is the acceleration of gravity and $a_{lateral}$ is the lateral acceleration

$$a_{lateral} = \dot{v} - u\omega \approx -u\omega \tag{16}$$

Substituting into the channel change model, we get

$$K\left(2\Delta t_1{}^2 + 3\,\Delta t_1 \Delta t_2 + \Delta t_2{}^2\right) = \frac{y_l}{u} \tag{17}$$

$$Ku < 0.2\text{ g} \tag{18}$$

$$\frac{Ku}{\Delta t_1} < \frac{0.1\text{ g}}{s} \tag{19}$$

where $y_l = 3.75$ m, $u = 30$ km/h, The value of the parameter satisfying the above constraint is solved as $K = 0.12$ rad/s, $\Delta t_1 = 1$ s, $\Delta t_2 = 0.5$ s. The corresponding lane change curve is shown in Figure 8c. Figure 8d displays Car 2 and Car 9 for lane change simulation.

At 9 s < t < 19 s, the first car in both lanes moves in the same way as Figure 7b: "uniform speed—uniform deceleration—uniform speed—uniform acceleration—uniform speed". The rear eight vehicles move according to a linear heel-chase model.

2.5.3. Module Parameter Setting

General millimeter-wave radar for longitudinal position detection accuracy than lateral position detection accuracy, the detection range of about 100 to 200 m. This simulation environment assumes that the radar can detect environmental vehicles within a radius of 100 m, with the vehicle's location as the center of the circle. The lateral and longitudinal errors of relative positions satisfy normal distribution with a standard deviation $\sigma_{rx} = 0.2$ m and $\sigma_{ry} = 0.1$ m. Data is updated every 0.1 s.

The accuracy of position information provided by satellite positioning is generally around 10 m, while speed detection accuracy is much higher. The simulation assumes that GNSS positioning can provide absolute position as well as velocity information of the vehicle. The lateral, longitudinal and heading angular errors all satisfy normal distribution. The standard deviations are $\sigma_{zx} = \sigma_{zy} = 3.33$ m, $\sigma_{z\theta} = \pi/180$, $\sigma_{zv} = 1$ m/s. Data is updated every 0.1 s.

The simulation assumes that the vehicle can receive location information sent from ambient vehicles within a radius of 100 m, with the vehicle's location as the center of the circle. To test the performance of the co-location method in the presence of unstable communication, it is assumed that the communication fails in the time 20 s < t < 25 s. Absolute location information of environmental vehicles is not available to all vehicles. At other times, all vehicles send and receive communication messages every 0.1 s.

3. Results and Discussion

The cooperative positioning and single vehicle positioning are simulated in the linear motion scenario and lane change motion scenario, respectively. Each scenario was simulated 100 times. The results of 100 simulations are averaged to eliminate the chance.

Figure 9a,b display the root-mean-square error of the co-location method. The meaning of the vertical coordinate of the curve is the root mean square of the positioning error of all vehicles in all simulations at a given time.

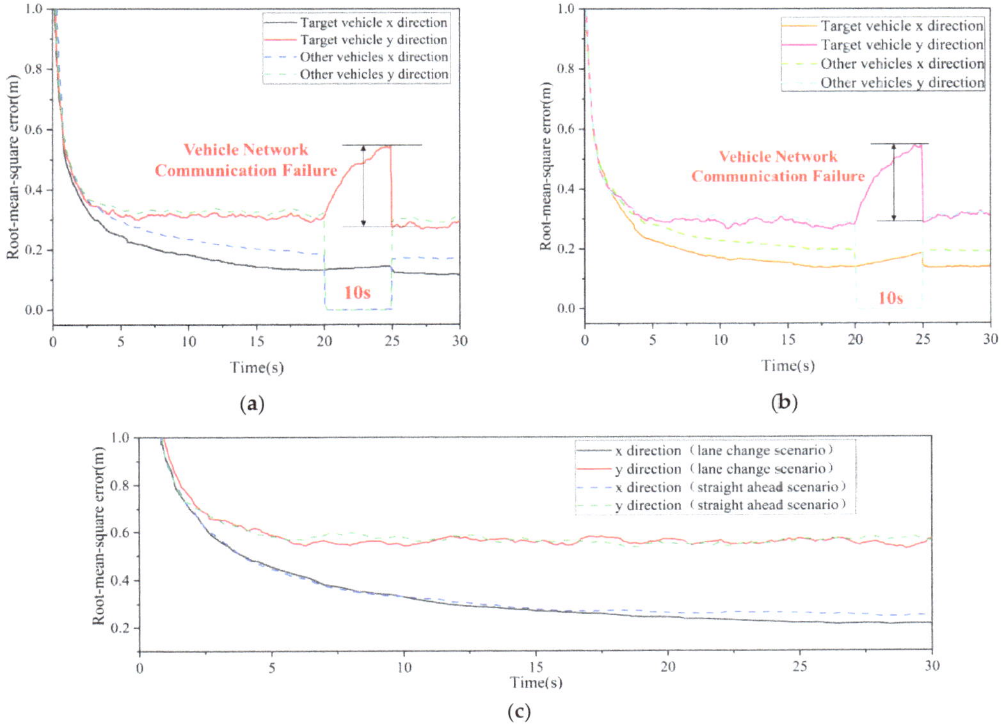

Figure 9. (**a**) Multi-vehicle cooperative positioning errors in lane change scenarios; (**b**) Multi-vehicle cooperative positioning error in straight-ahead scenario; (**c**) Individual vehicle positioning error.

Figure 9c demonstrates the root-mean-square error when a vehicle is positioned individually. Comparing the two methods shows that the error of cooperative positioning is significantly smaller than that of single-vehicle positioning. In addition, comparing the positioning error of the environmental vehicle with the positioning error of this vehicle, it is significantly larger in the x direction than in the y direction. This can be explained by the fact that radar has a more significant measurement error in the lateral direction than in the vertical direction.

Figure 10 shows the root-mean-square error of the two positioning methods during one lane change simulation (The simulation results for the straight-ahead scenario are very similar). The vertical coordinate of the curve means the root mean square of the positioning error of all vehicles at the same time in one simulation. It can be seen that the co-location is less error-prone and more stable than the individual positioning in the same lane change simulation.

Table 1 shows the comparison of the positioning error after averaging the two positioning methods in the time domain. It can be found that co-location has higher positioning accuracy in the case of communication failure, which is consistent with the results shown in Figure 9. The co-location error level is kept below 0.5 m overall. The current positioning accuracy of GNSS for civilian use is basically within 10 m. It can be seen that the collaborative positioning method can significantly improve the positioning accuracy. However, at the same time, the collaborative approach uses vehicle networks and in-vehicle sensors. In terms of future real-world applications, more comprehensive considerations are needed.

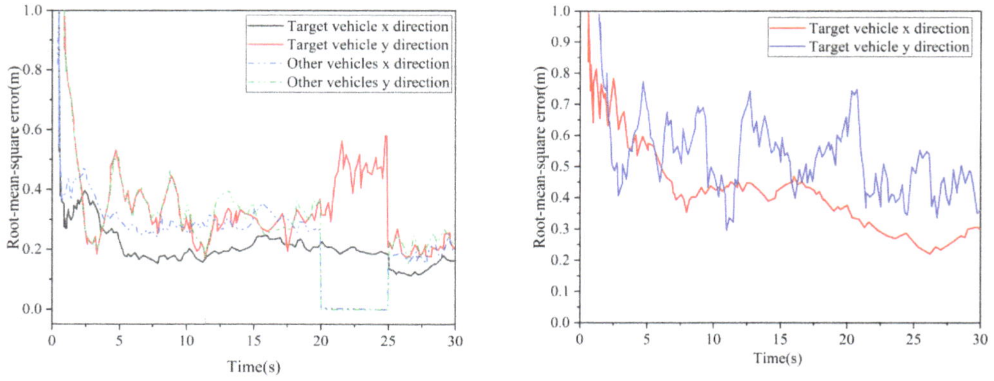

(a) (b)

Figure 10. Error variation in a particular channel change simulation: (**a**) Multi-vehicle cooperative positioning error; (**b**) Individual vehicle positioning error.

Table 1. The average value of the error in the time domain for the two positioning methods.

Method	**Lane Change Scenario**		**Straight Ahead Scenario**	
Error (m)	**Co-Positioning**	**Single Positioning**	**Co-Positioning**	**Single Positioning**
Root mean square error in the x-direction	0.20	0.36	0.27	0.38
Root mean square error in y-direction	0.36	0.61	0.34	0.62
Total root mean square error	0.42	0.73	0.44	0.74

4. Conclusions

This paper proposes a collaborative positioning method based on GNSS and vehicle network communication. The target vehicle first obtains its location information through GNSS and then transmits the information to the nearby environment vehicles through vehicle network communication. The target vehicle can correct its position with the environmental vehicle information. A multi-vehicle motion scenario was established to verify this method's effectiveness. The results show that the multi-vehicle cooperative localization method is more accurate than the single localization by GNSS. The root-mean-square error of positioning can be controlled to less than 0.5 m. This study was conducted in a simulation environment, and the effectiveness of the method was verified by the simulation results. The limitation of the study is that it still lacks the verification of real vehicles, which is the direction of our future efforts. With the continuous development of intelligent vehicles, vehicle position information is an important parameter in the vehicle's driving status. This paper explores the role of vehicle network communications in vehicle location. In the near future, this approach can be further applied to the precise positioning of actual vehicles.

Author Contributions: Conceptualization, H.Y. and L.W.; data curation, H.Y.; funding acquisition, J.H. and X.G.; investigation, X.G.; methodology, H.Y.; project administration, J.H.; resources, L.W.; software, H.Y.; supervision, J.H.; validation, J.H. and L.W.; visualization, X.X. All authors have read and agreed to the published version of the manuscript.

Funding: This research was funded by the Foundation of State Key Laboratory of Automotive Simulation and Control (No. 20210211), National Natural Science Foundation of China (No. 52107220), Postdoctoral Research Fund Project of China (No. 2021M690353), Scientific and Technological Innovation Foundation of Foshan (No. BK21BE012), Postdoctor Research Foundation of Shunde Innovation School of University of Science and Technology Beijing (No. 2021BH007) and Interdisciplinary Research Project for Young Teachers of USTB(Fundamental Research Funds for the Central Universities, NO. FRF-IDRY-21-013).

Data Availability Statement: Not applicable.

Acknowledgments: The authors are grateful to the editors and reviewers for their constructive suggestions.

Conflicts of Interest: The authors declare no conflict of interest.

References

1. Wang, Q.; Xu, N.; Huang, B.; Wang, G. Part-Aware Refinement Network for Occlusion Vehicle Detection. *Electronics* **2022**, *11*, 1375. [CrossRef]
2. Hong, J.; Wang, Z.; Yao, Y. Fault prognosis of battery system based on accurate voltage abnormity prognosis using long short-term memory neural networks. *Appl. Energy* **2019**, *251*, 113381. [CrossRef]
3. Schinkel, W.; van der Sande, T.; Nijmeijer, H. State estimation for cooperative lateral vehicle following using vehicle-to-vehicle communication. *Electronics* **2021**, *10*, 651. [CrossRef]
4. Hong, J.; Wang, Z.; Chen, W.; Yao, Y. Synchronous multi-parameter prediction of battery systems on electric vehicles using long short-term memory networks. *Appl. Energy* **2019**, *254*, 113648. [CrossRef]
5. Williams, T.; Alves, P.; Lachapelle, G.; Basnayake, C. Evaluation of GPS-based methods of relative positioning for automotive safety applications. *Transp. Res. Part C Emerg. Technol.* **2012**, *23*, 98–108. [CrossRef]
6. Adegoke, E.I.; Zidane, J.; Kampert, E.; Ford, C.R.; Birrell, S.A.; Higgins, M.D. Infrastructure Wi-Fi for connected autonomous vehicle positioning: A review of the state-of-the-art. *Veh. Commun.* **2019**, *20*, 100185. [CrossRef]
7. Singh, P.; Jeon, H.; Yun, S.; Kim, B.W.; Jung, S.Y. Vehicle Positioning Based on Optical Camera Communication in V2I Environments. *Comput. Mater. Contin.* **2022**, *72*, 2927–2945. [CrossRef]
8. Sarras, I.; Marzat, J.; Bertrand, S.; Piet-Lahanier, H. Collaborative multiple micro air vehicles' localization and target tracking in GPS-denied environment from range–velocity measurements. *Int. J. Micro Air Veh.* **2018**, *10*, 225–239. [CrossRef]
9. McLaughlin, P.; Vagg, C. A New Method of Vehicle Positioning Using Bumps and Road Surface Defects. *IEEE Trans. Intell. Transp. Syst.* **2021**, *23*, 13655–13665. [CrossRef]
10. Hossain, M.; Elshafiey, I.; Al-Sanie, A. Cooperative vehicle positioning with multi-sensor data fusion and vehicular communications. *Wirel. Netw.* **2019**, *25*, 1403–1413. [CrossRef]
11. Wang, H.; Wan, L.; Dong, M.; Ota, K.; Wang, X. Assistant vehicle localization based on three collaborative base stations via SBL-based robust DOA estimation. *IEEE Internet Things J.* **2019**, *6*, 5766–5777. [CrossRef]
12. Tao, X.; Zhu, B.; Xuan, S.; Zhao, J.; Jiang, H.; Du, J.; Deng, W. A Multi-Sensor Fusion Positioning Strategy for Intelligent Vehicles Using Global Pose Graph Optimization. *IEEE Trans. Veh. Technol.* **2021**, *71*, 2614–2627. [CrossRef]
13. Zhang, G.; Ng, H.F.; Wen, W.; Hsu, L.T. 3D mapping database aided GNSS based collaborative positioning using factor graph optimization. *IEEE Trans. Intell. Transp. Syst.* **2020**, *22*, 6175–6187. [CrossRef]
14. Zhang, G.; Wen, W.; Hsu, L.T. Rectification of GNSS-based collaborative positioning using 3D building models in urban areas. *GPS Solut.* **2019**, *23*, 1–12. [CrossRef]
15. Nam, S.; Lee, D.; Lee, J.; Park, S. CNVPS: Cooperative neighboring vehicle positioning system based on vehicle-to-vehicle communication. *IEEE Access* **2019**, *7*, 16847–16857. [CrossRef]
16. Zhu, B.; Tao, X.; Zhao, J.; Ke, M.; Wang, H.; Deng, W. An integrated GNSS/UWB/DR/VMM positioning strategy for intelligent vehicles. *IEEE Trans. Veh. Technol.* **2020**, *69*, 10842–10853. [CrossRef]
17. Mahmoud, A.; Noureldin, A.; Hassanein, H.S. Integrated positioning for connected vehicles. *IEEE Trans. Intell. Transp. Syst.* **2019**, *21*, 397–409. [CrossRef]
18. Hou, X.; Luo, L.; Cai, W.; Guo, B. E^2T-CVL: An Efficient and Error-Tolerant Approach for Collaborative Vehicle Localization. *IEEE Internet Things J.* **2021**, *9*, 3481–3494. [CrossRef]
19. Buehrer, R.M.; Wymeersch, H.; Vaghefi, R.M. Collaborative sensor network localization: Algorithms and practical issues. *Proc. IEEE* **2018**, *106*, 1089–1114. [CrossRef]
20. Ma, Z.; Sun, S. Research on Vehicle-Road Co-Location Method Oriented to Network Slicing Service and Traffic Video. *Sustainability* **2021**, *13*, 5334. [CrossRef]
21. Ansari, K. Cooperative position prediction: Beyond vehicle-to-vehicle relative positioning. *IEEE Trans. Intell. Transp. Syst.* **2019**, *21*, 1121–1130. [CrossRef]
22. Tong, Q.; Yang, Z.; Chai, G.; Wang, Y.; Qi, Z.; Wang, F.; Yin, K. Driving state evaluation of intelligent connected vehicles based on centralized multi-source vehicle road collaborative information fusion. *Environ. Dev. Sustain.* **2022**. [CrossRef]
23. Kim, S.T.; Fan, M.; Jung, S.W.; Ko, S.J. External Vehicle Positioning System Using Multiple Fish-Eye Surveillance Cameras for Indoor Parking Lots. *IEEE Syst. J.* **2020**, *15*, 5107–5118. [CrossRef]
24. Wan, L.; Zhang, M.; Sun, L.; Wang, X. Machine learning empowered IoT for intelligent vehicle location in smart cities. *ACM Trans. Internet Technol. (TOIT)* **2021**, *21*, 1–25. [CrossRef]

25. Lee, W.C.; Jeon, Y.B.; Han, S.S.; Jeong, C.S. Position Prediction in Space System for Vehicles Using Artificial Intelligence. *Symmetry* **2022**, *14*, 1151. [CrossRef]
26. Kong, X.; Gao, H.; Shen, G.; Duan, G.; Das, S.K. Fedvcp: A federated-learning-based cooperative positioning scheme for social internet of vehicles. *IEEE Trans. Comput. Soc. Syst.* **2021**, *9*, 197–206. [CrossRef]
27. Gao, C.; Wang, J.; Lu, X.; Chen, X. Urban Traffic Congestion State Recognition Supporting Algorithm Research on Vehicle Wireless Positioning in Vehicle–Road Cooperative Environment. *Appl. Sci.* **2022**, *12*, 770. [CrossRef]
28. Wang, L.L.; Gui, J.S.; Deng, X.H.; Zeng, F.; Kuang, Z.F. Routing algorithm based on vehicle position analysis for internet of vehicles. *IEEE Internet Things J.* **2020**, *7*, 11701–11712. [CrossRef]
29. Watta, P.; Zhang, X.; Murphey, Y.L. Vehicle position and context detection using V2V communication. *IEEE Trans. Intell. Veh.* **2020**, *6*, 634–648. [CrossRef]
30. Haider, A.; Hel-Or, H. What Can We Learn from Depth Camera Sensor Noise? *Sensors* **2022**, *22*, 5448. [CrossRef]
31. Li, N.; Ho, C.P.; Xue, J.; Lim, L.W.; Chen, G.; Fu, Y.H.; Lee, L.Y.T. A Progress Review on Solid-State LiDAR and Nanophotonics-Based LiDAR Sensors. *Laser Photonics Rev.* **2022**, 2100511. [CrossRef]
32. Zhu, B.; Sun, Y.; Zhao, J.; Zhang, S.; Zhang, P.; Song, D. Millimeter-Wave Radar in-the-Loop Testing for Intelligent Vehicles. *IEEE Trans. Intell. Transp. Syst.* **2021**, *23*, 11126–11136. [CrossRef]
33. Vargas, J.; Alsweiss, S.; Toker, O.; Razdan, R.; Santos, J. An overview of autonomous vehicles sensors and their vulnerability to weather conditions. *Sensors* **2021**, *21*, 5397. [CrossRef]
34. Park, K.W.; Park, J.I.; Park, C. Efficient methods of utilizing multi-SBAS corrections in multi-GNSS positioning. *Sensors* **2020**, *20*, 256. [CrossRef]
35. San Martín, J.; Cortés, A.; Zamora-Cadenas, L.; Svensson, B.J. Precise positioning of autonomous vehicles combining UWB ranging estimations with on-board sensors. *Electronics* **2020**, *9*, 1238. [CrossRef]
36. Yang, J.A.; Kuo, C.H. Integrating Vehicle Positioning and Path Tracking Practices for an Autonomous Vehicle Prototype in Campus Environment. *Electronics* **2021**, *10*, 2703. [CrossRef]
37. Li, W.; Shen, Y.Z. The consideration of formal errors in spatiotemporal filtering using principal component analysis for regional GNSS position time series. *Remote Sens.* **2018**, *10*, 534. [CrossRef]

electronics

Article

Experimental Study, Simulation and Analysis of the Fracture Failure of the Drum Shaft of a Casting Bridge Crane

Dong Xiang [1,*], Yan Li [2], Yuqing Zhang [2] and Feng Xu [1]

[1] The School of Mechanical Engineering, University of Science and Technology Beijing, Beijing 100083, China
[2] China United Engineering Corporation Limited, Hangzhou 310052, China
* Correspondence: xdustb@163.com; Tel.: +86-010-62332538

Abstract: This study investigated the fatigue fracture of bilateral drive drum shafts in casting bridge cranes including its fracture morphology and factors, such as materials, manufacturing processes, and loads. Seven conditions were designed to test the effects of changes in the speed and torque of the drum shafts during startup, commissioning and braking under different loads. A dynamic model was developed for the structure and control system of the hoisting mechanism. Changes in the speed and torque of the motor and drum shafts were simulated under common operating conditions such as speed and load changes of the motor, control asynchrony and single-motor towing. The results showed that asynchronous motor starting and braking, motor dragging and other behaviors led the left and right drum shafts undergo oscillated torque with a value reached 2×10^5 N·m in a period of approximately 13 s, and a residual torque about 3×10^4 N·m was retained after braking. The torques on the drum shafts changed suddenly during the processes of starting, shifting and braking. Dynamic loading was the root cause of fatigue fracture of the drum shafts.

Keywords: crane; drum shaft; fatigue; fracture

Citation: Xiang, D.; Li, Y.; Zhang, Y.; Xu, F. Experimental Study, Simulation and Analysis of the Fracture Failure of the Drum Shaft of a Casting Bridge Crane. *Electronics* **2022**, *11*, 3043. https://doi.org/10.3390/electronics11193043

Academic Editor: Domenico Mazzeo

Received: 11 August 2022
Accepted: 15 September 2022
Published: 24 September 2022

1. Introduction

Fatigue fracture, which accounts for more than 50% of the fractures of mechanical parts [1], has long been the focus of fracture failure research. Recent studies have considered many factors that contribute to fatigue fracture [2–11], such as materials, structure, manufacturing processes, service conditions, and dynamic behavior.

Fatigue fracture of the core components used for the transmission of rotation and torque in the drive shaft is very dangerous to machines and humans. The motor shaft was the object of the world's first fatigue testing machine [12]. Currently, research on fatigue fracture of drive shafts has been conducted for various machinery and equipment, such as automobiles, construction machinery, rail vehicles, and air compressors. [13–18] However, the causes of fatigue have not been related to the differences in materials, structure, and manufacturing processes, leading to continuous research effort focusing on different service conditions and equipment.

Casting bridge cranes, which are mainly used to lift molten iron and molten steel, have the characteristics of large lifting capacity, high operation rate and harsh working conditions. Once fatigue fracture occurs hereby, it will result extremely dangerous accidents. After an accident, in which a ladle of molten steel overturned at Qinghe Special Steel Co., Ltd. in 2007, casting bridge cranes were required to use a bilateral main hoisting mechanism for greater safety [19].

The bilateral propelled hoisting mechanism contains two load drums with shafts at their ends. One load drum shaft is connected to the output shaft of the reducer through a coupling, and the other load drum shaft is connected to the other load drum through a coupling too. The load drum, as a component in the hoisting mechanism that experiences very complex forces, carries the dynamic load produced not only by the lifting and dropping of the workpiece but also by the operation, control and behavior of the motor, transmission system, braking system and other components. Thus, failure of load drums is responsible for approximately 25% [20], the largest proportion, of crane accidents. The load drum shaft is the part of the load drum that is most likely to fail. More importantly, once the load drum shaft breaks, it causes downtime and even serious accidents due to falling objects. For example, at Hebei Tangshan Steel Co., Ltd., a serious accident occurred in which the load drum shaft broke when the casting crane was lifting a ladle of molten iron. The coupling that connected the load drum and the reducer tore, and the load drum fell off the frame; accordingly, the ladle was inclined and molten iron overflowed, and consequently, the production line burned [21]. Although the failure of load drum shafts has always been a concern of the crane industry, current research into load drum shaft failure is still mainly based on qualitative analysis [22–26].

Taking the failure of a 50/10 t casting bridge crane load drum shaft as an example, a testing and analysis of dynamic factors in the fatigue fracture of the load drum shaft in the hoisting mechanism are carried out after a failure factor investigation and qualitative analysis considering typical working conditions. The sources of the loads are clarified, and their dynamic responses are quantified. As a result, design accuracy and operational control must be improved.

2. Investigation of the Reasons for Fractures

Fracture failure usually involves a combination of multiple factors. To determine the causes of a fracture, the possible factors are investigated, such as fracture appearance, materials, manufacturing processes, loads, etc.

2.1. Fracture Appearance Analysis

A fracture of a load drum shaft is shown in Figure 1. Unfortunately, the fracture features were not completely preserved because of friction. However, the cracks nearer to the shaft surface are smoother, and the cross-section does not have obvious necking, and the cracks are inclined in one direction with a large depth inside. In addition, as shown in Figure 1b, the source of the crack in the drum shaft is near the surface of the shaft, the final crack area is located inside the shaft, and scallops are located between the two areas. Therefore, the fracture was caused by rotational bending fatigue; the load drum shaft experienced a large torsional force prior to fractured.

(a)

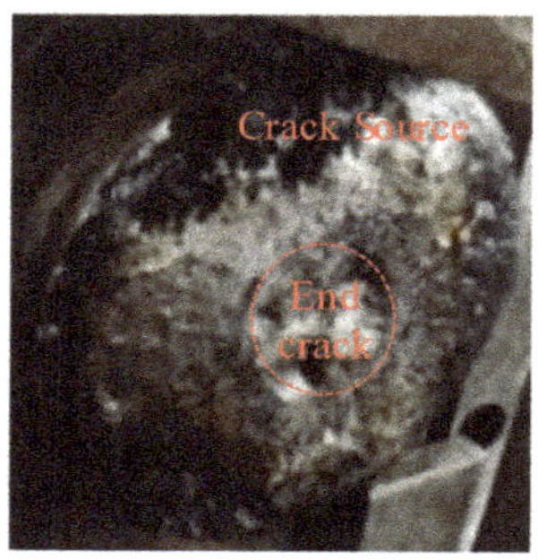

(b)

Figure 1. Appearance of the fracture of a load drum shaft: (**a**) cross-section of the left load drum shaft; (**b**) cross-section of the right load drum shaft.

The load drum shaft broke at the neck journal root with diameter Φ220 mm, and a magnified image of the transition fillet is shown in Figure 2. According to the standards in *Parts Rounding and Chamfering*, when the shaft shoulder diameter difference is more than 40 mm, the fillet radius should be 4 mm. However, the actual fillet radius of this shaft shoulder was approximately 1.5 mm, which may have led to a high stress concentration.

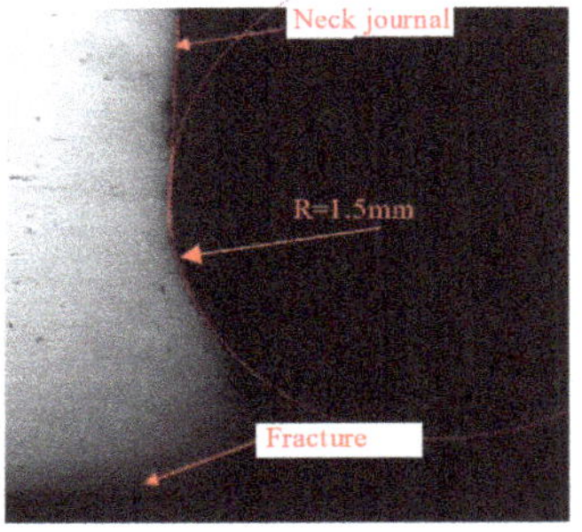

Figure 2. Transition fillets of the drum shaft, 20X.

2.2. Material Analysis

According to the design requirements, the material of the failed load drum shaft should be #35 steel, and it must be normalized and tempered, so that its hardness reaches 137–163 HBW. After testing, the hardness of this load drum shaft was distributed in the range of 153–161 HBW, and the core hardness was lower than the surface hardness, satisfying the technical requirements.

Samples were drilled respectively at the spaces of near the surface, 1/2 radius and core of the shaft for analysis of the chemical composition. As shown in Table 1, the material composition met the requirements for #35 steel in manufacturing standard *GB/T699*. From the metallographic structure shown in Figure 3, the structure of the shaft was normalized pearlite and ferrite with few inclusions, which also meeting the performance requirements of the spool shaft material.

Table 1. Chemical Composition of the Drum Shaft, w%.

Location / Elements	Surface	1/2R	Core	Required
C	0.37	0.34	0.35	0.32–0.39
Mn	0.59	0.58	0.59	0.50–0.80
Si	0.22	0.22	0.22	0.17–0.37
P	0.018	0.017	0.017	≤0.035
S	0.002	0.002	0.002	≤0.035
Cr	0.01	0.01	0.01	≤0.10
Ni	0.01	0.01	0.01	≤0.30

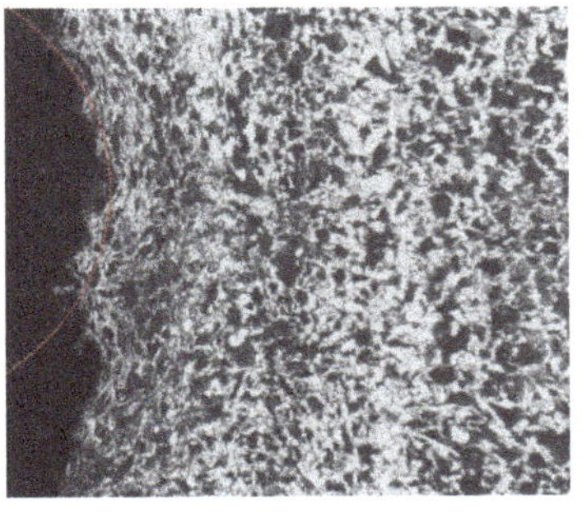

Figure 3. Microstructure of a cross-section, 100X.

2.3. Manufacturing Process Analysis

In the processing of the load drum and the load drum shaft, manufacturing and acceptance were carried out according to the manufacturing standard *DHQ. JS001*, circumferential and longitudinal welds were 100% UT inspected, and the quality reached BI grade and BII grade in manufacturing standard *GB/T11345*. Short shaft and plate welds were 100% MT inspected 48 h after welding, and the quality reached level 2 according to manufacturing standard *JB/T6061–2007*. In addition, a total annealing treatment after welding was carried out to improve the tissue defects and residual stress of the load drum. Therefore, the influence of the manufacturing process on the fracture failure can be ignored.

2.4. Service Condition Analysis

Table 2 shows the working level of the crane and its hoisting mechanism with a broken load drum shaft. The working level means that the crane frequently withstands cyclic loads when it is operated, which may lead to fatigue.

Table 2. Working rank of the crane and its lifting mechanism.

Items	Work Level	Use Level	Loads Level
Crane	A7	U5	Q4
Hoisting mechanism	M8	T7	L4

2.5. Static Analysis of the Load Drum Group

As shown in Figure 4, the load drum group of the bilateral propelled hoisting mechanism contained two load drums. Both ends of the two load drums were load drum shafts, one load drum shaft was connected to the output shaft of the reducer, and the other load drum shaft was connected to the other load drum through a GIICL type drum gear coupling. The load drum group had four supports. The left and right ends were supported by the reducer box, and the load drum shafts (in addition to the coupling) were supported by the bearing pedestal.

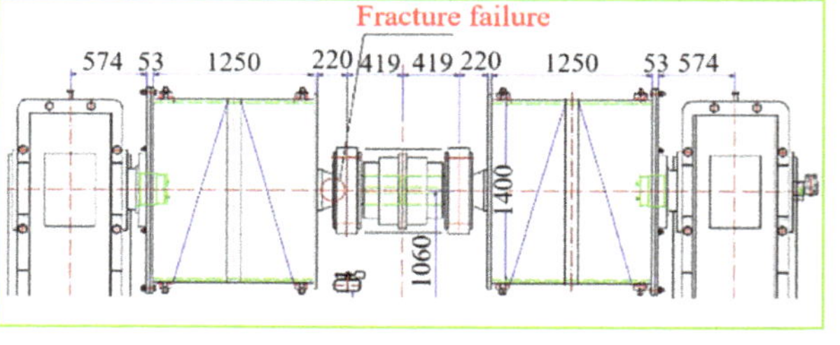

Figure 4. Bilateral drive drum.

The simplified static model of the bilateral drive load drum is shown in Figure 5. The values and meanings of the codes in Figure 5 are shown in Table 3. The following assumptions were made:

1. The coupling connecting the left and right load drum shafts can transmit torque without loss, and it was simplified to a hinged connection.
2. The axial movement of the load drum group was ignored, and the bearing pedestal of the two load drum shafts was converted into a hinge support.
3. The weight of the load drum and the coupling were uniform loads, the wire rope tension was the concentrated force acting on the center of the load drum, and the quality of the load drum shaft was ignored.

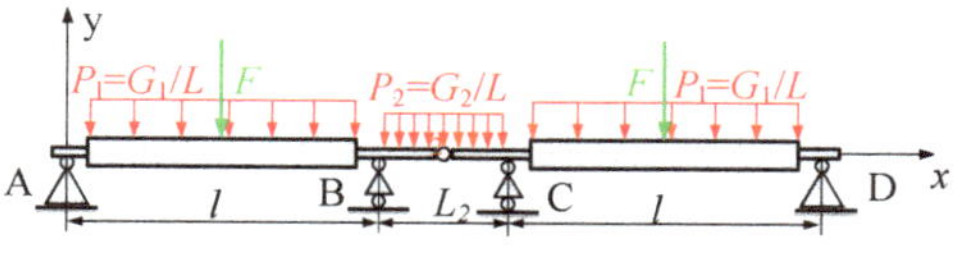

Figure 5. Simplified static model of the bilateral drive drum.

Table 3. Values and meanings of the codes in Figure 5.

Codes	Values
Load drum length L/mm	1331
Load drum shaft length L_1/mm	220
Coupling length L_2/mm	838
Load drum group quality G_1/kN	44
Coupling quality G_2/kN	8
Unilateral lifting weight G_3/kN	250
Pulley set magnification N	2
Single load drum load F/kN	125

Considering the above assumptions and the symmetrical structure of the couplings, the force and flexural moment of the equivalent hinge of the coupling were both 0, and the static model in Figure 5 was simplified to the beam structure shown in Figure 6a. The gear engagement of coupling generated a bending moment Ma in the axial direction due to friction. According to *JB/T 8854.2-2001*, $M_a = 0.1T_{max}$, where T_{max} is the maximum torque acting on the coupling. Obtained by calculations using the mechanics theory of materials science, the shearing force diagram and bending moment diagram are shown in Figure 6b,c, and the maximum absolute value of the shearing force and bending moment for the load drum and load drum shaft were calculated using Equations (1)–(3).

$$\begin{aligned} |F_q|_{max1} &= |F_q|_{max2} = \left|-F_{By} + \tfrac{G_2}{2}\right| \\ &= \left|-\tfrac{F+G_1}{2} - \tfrac{1}{l}\left(M_a + \tfrac{G_2L_2}{8}\right)\right| \end{aligned} \tag{1}$$

$$\begin{aligned} |M|_{max1} &= M^+_{max} = \tfrac{F_{Ay}l}{2} - \tfrac{G_1L_1}{8} \\ &= \tfrac{(F+G_1)l}{4} - \tfrac{1}{2}\left(M_a + \tfrac{G_1L_1}{4} + \tfrac{G_2L_2}{8}\right) \end{aligned} \tag{2}$$

$$|M|_{max2} = M^-_{max} = \left|-\left(M_a + \frac{G_2L_2}{8}\right)\right| \tag{3}$$

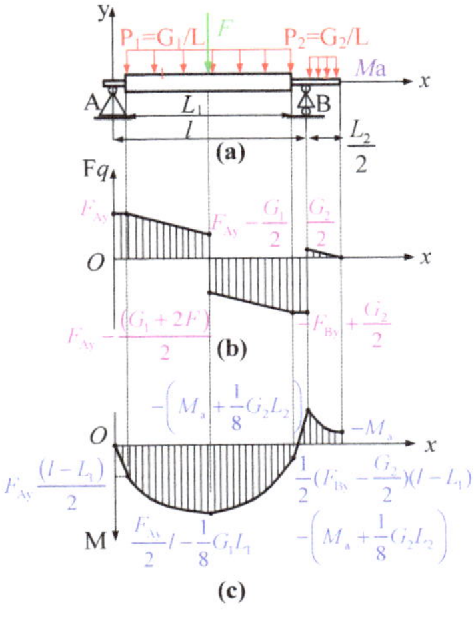

Figure 6. The force and flexural moment analysis of drum: (**a**) static model of the bilateral drive drum; (**b**) the shear force diagram; (**c**) bending moment diagram.

After calculations using Hooke's law and the moment deformation equation of the simply supported beams, the maximum normal bending stress of the load drum body section and load drum shaft section were written respectively as follows.

$$\sigma_{max1} = \frac{|M|_{max1}}{W_1} = \frac{M_{max1}D}{2I_1} \quad (4)$$

$$\sigma_{max2} = \frac{|M|_{max2}}{W_2} = \frac{M_{max2}d_2}{2I_2} \quad (5)$$

where σ_{max1} is the maximum normal bending stress of the load drum body section, σ_{max2} is the maximum normal bending stress of the load drum shaft section, I_1 is the cross-sectional moment of inertia of the load drum body, D is the outside diameter of the load drum body, d_1 is the inside diameter of the load drum body, d_2 is the diameter of the load drum shaft, and I_2 is the cross-sectional moment of inertia of the load drum shaft.

If the cross-sectional maximum shearing stress of the load drum body and load drum shaft are both distributed on the neutral axis, then the maximum bending shear stress of the load drum body and load drum shaft can be written as:

$$\tau_{max1} = 16\frac{|Fq|_{max1}}{\pi(D^2 - d_2{}^2)} \quad (6)$$

$$\tau_{max2} = \frac{16}{3}\frac{|Fq|_{max2}}{\pi d_1{}^2} \quad (7)$$

Because T_{max} is small in the load drum group of the bilateral propelled hoisting mechanism, M_a can be ignored. After calculation, the values of σ_{max1}, τ_{max1}, σ_{max2}, τ_{max2} are 1.7 MPa, 2.2 MPa, 5.2 MPa, 7.75 MPa, respectively. All of these values are far less than the ultimate strength of the load drum and load drum shaft, which are 235 MPa and 315 MPa, respectively.

2.6. Dynamic Load Analysis of the Hoisting Mechanism

As shown in Figure 7, when the hoisting mechanism operating, the motor drives the transmission components, such as the reducer, the load drum, and the wire rope, to lift heavy objects. The hoisting mechanism bears dynamic loads that include the dynamic electromagnetic torque generated by the motor under the speed control system, and the loads, which are the forces generated by the structure with friction, assembly tolerance, interstice and damping behavior, and the forces generated by weight while lifting, lowering and braking. Particularly for the bilateral propelled lifting mechanism, high dynamic loads are carried for the asymmetric structure, and the nonsynchronous control system changes the force distribution in its driving system.

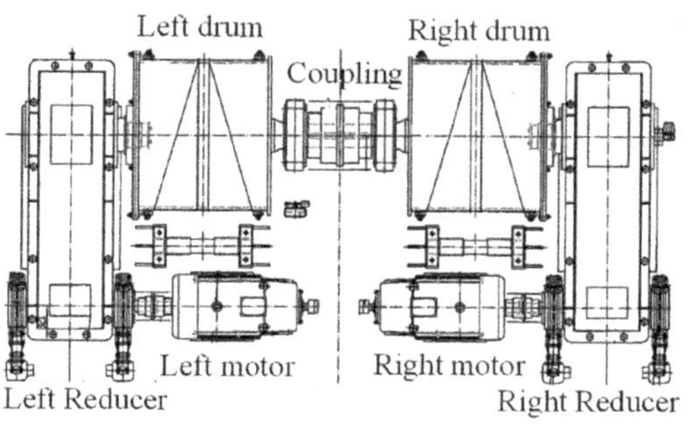

Figure 7. Structural layout diagram of the lifting mechanism.

3. Dynamic Test of the Lifting Mechanism

3.1. Test System Setup

To study the dynamic behavior of the hoisting mechanism, a test system was constructed on a cast bridge crane with a broken drum shaft, as shown in Figure 8. The test system included a strain-torque telemetry equipment, strain acquisition sensor, data acquisition equipment and other modules. Table 4 summarizes the performance parameters of the cast bridge crane, and Table 5 summarizes the types of main instruments for the test.

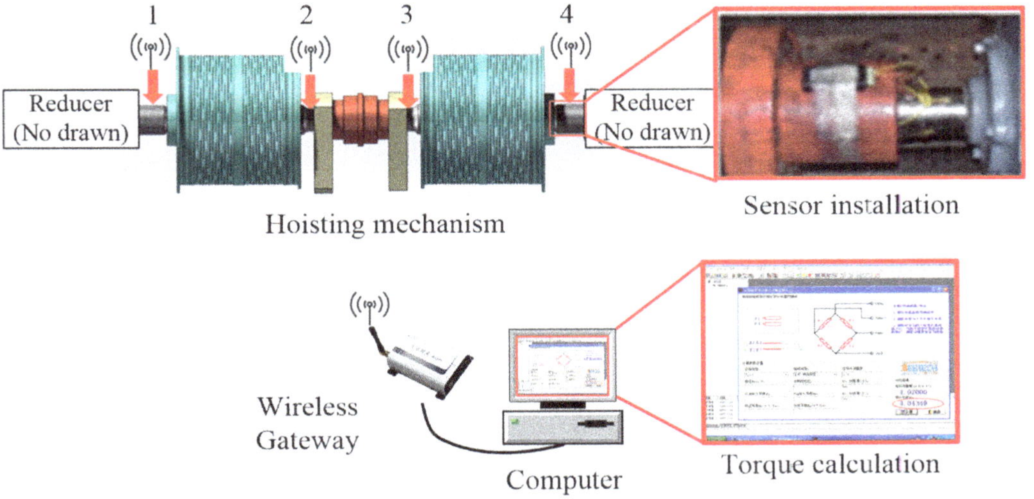

Figure 8. Composition of the dynamic test system for the lifting mechanism.

Table 4. Main parameters of a casting bridge crane.

Properties	Lifting Capacity	Length	Beam Weight	Trolley Weight	Motor Power	Rated Speed
Units	t	m	t	t	kW	rpm
Values	50	22.5	183	29	75	750

Table 5. Main instruments for testing.

No.	Instrument Model	Application
1	KFW-2-120-D16-11 L1M2S	Strain test
2	TQ201 No. 2126	Torque and speed test
3	BS903	Wireless receiving gateway
4	BeeData	Software for signal acquisition and processing

To compare the dynamic behavior under different working conditions, 7 kinds of working conditions, listed in Table 6, were tested. The speeds in four gears of the crane during test are listed in Table 7. In each working condition, the rotational speed and stator current of the motor and the torque of the load drum shaft were measured; the detection positions were named positions 1, 2, 3 and 4, as shown in Figure 8.

Table 6. Working conditions for the tests and simulations.

Type	Object Weight	Speed and Direction	Operation Method
Type 1	None	Static	The hook is static without weights, and the sensor is set to zero
Type 2	None	4th gear and dropping	After the hook falls a certain vertical distance, quickly decelerate from the 4th gear to the 1st gear and finally stop in the air
Type 3		4th gear and lifting	After the hook lifts a certain vertical distance, quickly decelerate from the 4th gear to the 1st gear and finally stop in the air
Type 4	40.44 t	4th gear and lifting	After the object is lifted a certain vertical distance, quickly decelerate from the 4th gear to the 1st gear and finally stop in the air
Type 5		4th gear and dropping	After the object falls a certain vertical distance, quickly decelerate from the 4th gear to the 1st gear and finally stop in the air
Type 6	≤40.44 t	4th gear and dropping or lifting	The heavy object is always placed on the ground, and the hoisting mechanism repeatedly lifts and descends in 1 block
Type 7	40.44 t	4th gear and dropping	After the object falls a certain vertical distance, quickly decelerate from the 4th gear to the 1st gear and finally stop on the ground

Table 7. Gears and speeds.

Gear	Percentage of Maximum Speed	Motor Speed	Load Drum Speed
Units	%	r/min	r/min
4	100	745	4.3
3	30	223.5	1.29
2	20	149	0.86
1	10	74.5	0.43

3.2. Test Results Analysis

Since working condition 6 is most frequently used, and the dynamic behavior of the crane in this condition is most representative for it involves operations such as lifting, dropping, braking and direction adjustment, working condition 6 was taken as an example for analysis. The motor speed, gear position and torque of the load drum shaft for 30 s are shown in Figure 9.

Figure 9a illustrates the motor speed and gear of the hoisting mechanism when it repeatedly lifts, drops and brakes with objects on the ground. Figure 9b illustrates the torques in positions 2 and 3. It can be concluded that the torque on the load drum shaft changed suddenly due to the change in the hoisting weight when the gear changed from the zero point and the motor accelerated the lifting object. Additionally, the torque on the load drum shaft changed suddenly during braking and then stabilized at a constant value, just like it was affected by the step signal. It is clear that frequent starting and braking brought periodic reciprocating vibrations and load changes for the transmission system. Otherwise, as shown in Figure 9a, the motor speed lagged behind the gear change, and the motor first reversed and then rotated forward due to insufficient lifting torque at low gear—that is, it was dragged by heavy objects and in a state of power generation.

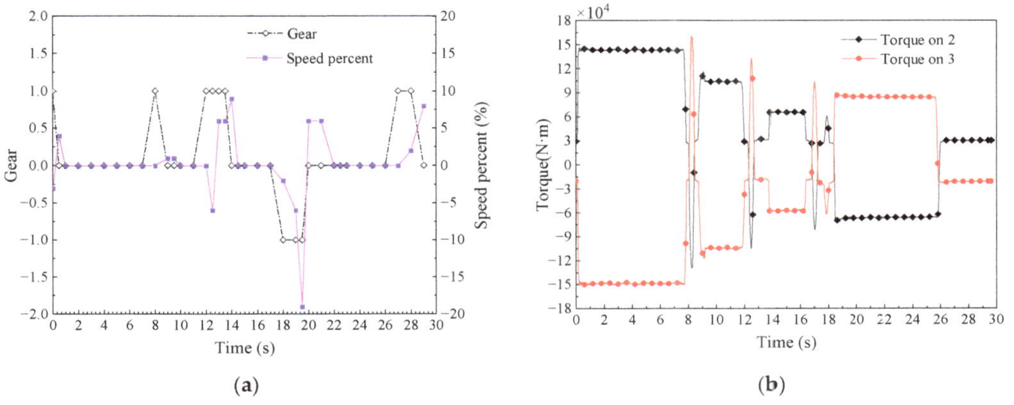

Figure 9. Crane lifting weight fast and frequent starting-braking test: (**a**) The motor speed and gear of the hoisting mechanism; (**b**) frequent starting-braking drum shaft torque change.

The dynamic behavior of the crane under other working conditions was similar to that under working condition 6. Once the gear position changed, lifting, falling or braking would consequently occur, and vibrations and impacts would occur in the hoisting mechanism. When objects rose or fell, there was a certain phase difference between the torques on the two load drum shafts, as shown in Figure 10.

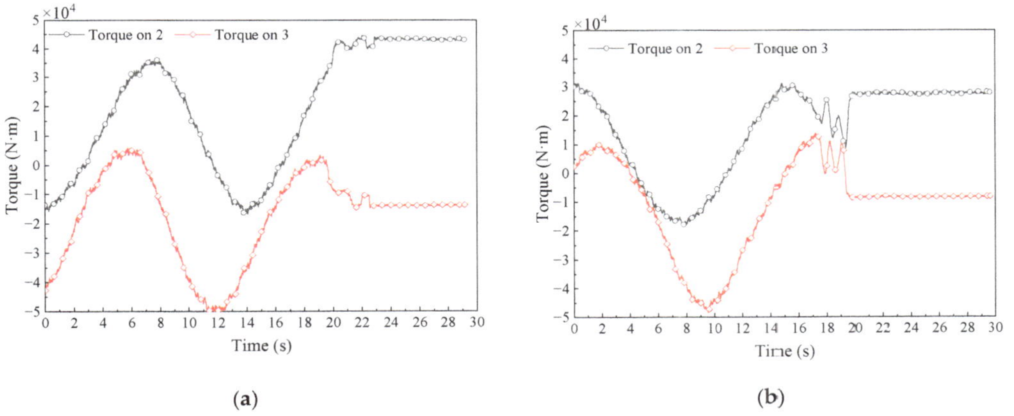

Figure 10. Torque change of the drum shaft before and after braking of the bilateral drive lifting mechanism: (**a**) lifting brake in 1st gear when lifting 40.44 t; (**b**) lowering brake in 1st gear when lifting 40.44 t.

Figure 10a,b show the torques measured at points 2 and 3 when the crane worked in the 1st gear, and the object weight was 40.44 t. All braking started at 20 s. Prior to braking, when the object was being lifted or dropped, the torques on the load drum shafts vibrated with a high amplitude and low frequency, and there was a phase difference between them. After braking, the torques on the load drum shafts remained stable after short-term vibrations at high frequency.

The test results showed that the service conditions strongly influenced the dynamic performance of the hoisting mechanism, and the torque on both load drum shafts of the hoisting mechanism changed periodically; the period was approximately 13 s. Otherwise, after braking, the torque values of the two load drum shafts changed abruptly, but the changes were in opposite directions. The analysis showed that this may be related to motor drag and asynchronous brake holding in the bilateral drive hoisting mechanism.

For an in-depth explanation of the behavior observed in the above-described test and improve the design of the spool shaft, a dynamic model was constructed and the hoisting mechanism was simulated.

4. Electromechanical Coupling Dynamics Model of the Bilateral Propelled Hoisting Mechanism

4.1. Dynamic Simulation Process

As shown in Figure 11, the hoisting mechanism is divided into three modules: the motor and its control, the mechanical transmission system and hoisting weight. Each modules contain its own structural parameters, control parameters, and working parameters, and there are dynamic loads, motion parameters, and control parameters to realize data transmission and feedback between them. Therefore, it is necessary to build an electromechanical coupling dynamics model of the motor, its control system and the mechanical transmission system, in order to analyze the dynamic response of the hoisting mechanism.

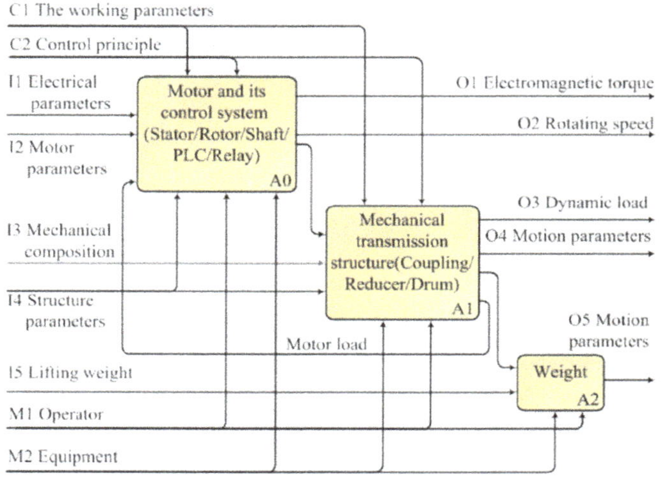

Figure 11. Structural layout diagram of the hoisting mechanism.

However, it is difficult to carry out a dynamic simulation if the electromechanical model is too complex. Therefore, the following assumptions are made to simplify the model.

1. The vibration of the system in the horizontal plane and the swing of the weight are ignored, and only the vertical motion and the torsion of the structure are considered.
2. The friction between the wire rope and the drum, and the rigid resistance of the wire rope as well are ignored.
3. The contact stiffness at the connection for the reducer and the drum, which are rigidly connected, is ignored. That is, only the torque stiffness and torsional damping of the connecting shaft are considered.
4. The torsional deformation of the reducer, drum and coupling are ignored.

Based on the above assumptions, considering the electromechanical coupling effect and simplifying the transmission components into a mechanical equivalent model composed of mass, stiffness, and damping, as shown in Figure 12, the motor, its control system and transmission mechanicals are also considered. The relevant parameters in the vertical translation-torsion dynamic equivalent model are shown in Table 8.

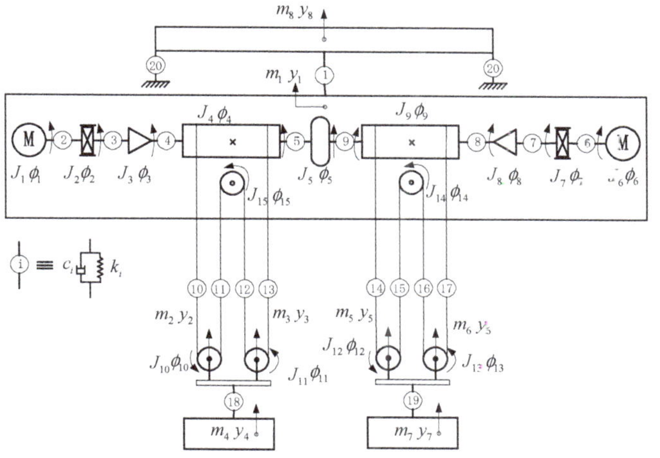

Figure 12. Vertical translational-torsional dynamic model of the bilaterally driven lifting mechanism.

Table 8. The parameters for the bilateral driven crane mechanisms and their meanings are shown in Figure 12.

Parameters	Object Weight	Parameters	Operation Method
m_1	Equivalent weight of the hoisting mechanism	$m_2 + m_3, m_5 + m_6$	Weight of the hooks
m_4, m_7	Weight of the objects	m_8	Weight of the crane girder
J_1, J_6	Moment of the motor	J_2, J_7	Moment of the brake
J_3, J_8	Moment of the reducer	J_4, J_9	Moment of the load drum
J_5	Moment of the coupling	$J_{10} \sim J_{15}$	Moment of the pulley block
k_i $i = 1,10,11,12 \ldots 20$	Translational stiffness coefficient	k_i $i = 2,3, \ldots 9$	Rotational stiffness coefficient
c_i $i = 1,10,11,12 \ldots 20$	Translation damping coefficient	c_i $i = 2,3, \ldots 9$	Rotational damping coefficient
r_4, r_9	Radius of the load drum	r_i $i = 10,11, \ldots 15$	Radius of the pulley
y_i $i = 1,2 \ldots 8$	Displacement of the crane masses	θ_i $i = 1,2,3 \ldots 15$	Rotation angle of the crane parts

Since this research mainly focuses on the influence of the bilateral driving arrangement on the dynamic load distribution in the hoisting mechanism, it is assumed that the structure is symmetrical. That is, the motors, brakes, reducers, drum sets, pulley sets and hook sets on both sides are considered to be identical, and the manufacturing errors and assembly errors of individual components are ignored. In addition, it is assumed that the stiffness and damping of each wire rope in the unilateral pulley block are the same. The relationship between the parameters in Figure 12 are given in Table 9.

Table 9. Relationship between the parameters in Table 7.

Number	Name	Relationship
1	Weight	$m_2 = m_3 = m_5 = m_6 = m_h$, $m_2 + m_3 = m_5 + m_6 = m_j$
2	Displacement	$y_2 = y_3 = y_1$, $y_5 = y_6 = y_r$
3	Radius	$r_{10} = r_{11} = r_{12} = r_{13} = r_{14} = r_{15} = r_h$
4	Moment	$J_1 = J_6, J_2 = J_7, J_3 = J_8, J_4 = J_9$
5	Stiffness	$J_{10} = J_{11} = J_{12} = J_{13} = J_{14} = J_{15} = J_h$
6	Damping	$k_2 = k_6, k_3 = k_7, k_4 = k_8, k_5 = k_9, k_{18} = k_{19}$

4.2. Motor and Control System

The hoisting mechanism use a YZR-type wound three-phase asynchronous motor. As to solve the voltage equation, flux equation, torque equation and motion equation of the motor, the motor model is usually converted from the three-phase stationary A, B, and C coordinate systems to the two-phase synchronous rotating coordinate system, as shown in Equations (8)–(10) [27], through Clarke coordinates and a Park transformation.

Voltage equation:

$$\begin{cases} u_{sd} = R_s i_{sd} + p\psi_{sd} - \psi_{sq}\omega_1 \\ u_{sq} = R_s i_{sq} + p\psi_{sq} + \psi_{sd}\omega_1 \\ u_{rd} = R_r i_{rd} + p\psi_{rd} - \psi_{rq}(\omega_1 - \omega_s) \\ u_{rq} = R_r i_{rq} + p\psi_{rq} - \psi_{rd}(\omega_1 - \omega_s) \end{cases} \tag{8}$$

Flux equation:

$$\begin{cases} \psi_{sd} = L_s i_{sd} + L_m i_{rd} \\ \psi_{sq} = L_s i_{sq} + L_m i_{rq} \\ \psi_{rd} = L_m i_{sd} + L_r i_{rd} \\ \psi_{q2} = L_m i_{sq} + L_r i_{rq} \end{cases} \tag{9}$$

Torque equation:

$$T_e = n_p L_m (i_{sq} i_{rd} - i_{rq} i_{sd}) \tag{10}$$

Motion equation:

$$T_e = T_m + \frac{J_e}{n_p} \cdot \frac{d\omega}{dt} \tag{11}$$

where u_{sd}, u_{sq}, u_{rd}, and u_{rq} are the components of the stator voltage and rotor voltage on the d and q coordinate axes, respectively. i_{sd}, i_{sq}, i_{rd} and i_{rq} are the components of the stator current and rotor current on the d and q coordinate axes, respectively. R_s and R_r are the winding resistances of the stator and rotor, respectively. ψ_{sd}, ψ_{sq}, ψ_{rd} and ψ_{rq} are the components of the stator and rotor flux linkages in the d and q coordinate axes, respectively. L_s, L_r and L_m are the inductance of stator and rotor and the mutual inductance between the stator and rotor. ω_1 is the rotational angular velocity of the d, q coordinate axis system, and, where f is the AC power frequency for the motor. ω_s is the slip velocity, and $\omega_s = (\omega_1 - \omega)$, where ω is the rotor speed. J_e is the rotational inertia of the unit. n_p is the polar number of the motor. T_m is the motor load resistance torque.

To realize the stable operation of the hoisting mechanism, it is necessary to control the motor by the AC speed control system. A crane is a special equipment, and its motor control system typically uses a stable and reliable thyristor stator regulating voltage to achieve speed regulation. The torque equation [28] of the rotor is:

$$T_e = \frac{3n_p U_s^2 R_r'/s}{\omega_1\left[(R_s + R_r'/s)^2 + \omega_1{}^2(L_{ls} + L_{lr}{}')^2\right]} \tag{12}$$

where n_p is the pole pair of the motor, U_s and ω_1 are the stator phase voltage and supply angular frequency of the motor, s is the slip ratio, R_s and $R_r{}'$ are the resistance of each phase of the stator and the resistance of each phase of the rotor is converted to the stator side, L_{ls} and $L_{lr}{}'$ are the leakage inductance of each phase of the stator and the leakage inductance of each phase of the rotor is converted to the stator side.

4.3. Mass-Stiffness-Damping a Model of Transmission Mechanisms

Usually, the dynamic equation of transmission mechanisms can be written as the differential equation in the form matrix shown in Equation (13).

$$\boldsymbol{M}\ddot{\boldsymbol{q}} + \boldsymbol{C}\dot{\boldsymbol{q}} + \boldsymbol{K}\boldsymbol{q} = \boldsymbol{Q}(\boldsymbol{q}, \dot{\boldsymbol{q}}, \boldsymbol{t}) \tag{13}$$

where $\boldsymbol{M}$, $\boldsymbol{C}$ and $\boldsymbol{K}$ are the mass matrix, damping matrix and stiffness matrix of the system, respectively; $\boldsymbol{q}$, $\dot{\boldsymbol{q}}$ and $\ddot{\boldsymbol{q}}$ are the displacement matrix, velocity matrix and acceleration matrix of the system in generalized coordinates, respectively; and $\boldsymbol{Q}(\boldsymbol{q}, \dot{\boldsymbol{q}}, \boldsymbol{t})$ is the generalized force matrix of the system, which is composed of the gravity and external moment of each equivalent mass unit of the crane.

According to the kinetic model and parameters in Figure 12, the components of Equation (13) can be expressed in matrix form as follows.

$\boldsymbol{q}$, the generalized coordinate displacement matrix, can be expressed as:

$$q_{21\times 1} = [y_1, y_l, y_r, y_4, y_7, y_8, \phi_1, \phi_2, \cdots, \phi_{15}]^T \tag{14}$$

$\boldsymbol{M}$, the mass matrix, is a 21×21 diagonal matrix that can be expressed as:

$$\boldsymbol{M} = \begin{bmatrix} m_{6\times 6} & \mathbf{0} \\ \mathbf{0} & J_{15\times 15} \end{bmatrix} \tag{15}$$

where $\boldsymbol{m_{6\times 6}}$ is a diagonal matrix of equivalent masses, as shown in Formula (16). $\boldsymbol{J_{15\times 15}}$ is the matrix of the rotational inertia, as shown in Formula (17), and $\boldsymbol{J_h}$ is a 6 × 6 rotational inertia matrix, as shown in Formula (18).

$$\boldsymbol{m_{6\times 6}} = \begin{bmatrix} m_1 & 0 & & \cdots & & 0 \\ 0 & m_j & & & & \\ & & m_j & \ddots & & \vdots \\ \vdots & & \ddots & m_4 & & \\ & & & & m_7 & 0 \\ 0 & & \cdots & & 0 & m_8 \end{bmatrix} \tag{16}$$

$$\boldsymbol{J_{15\times 15}} = \begin{bmatrix} J_1 & 0 & \cdots & & 0 \\ 0 & J_2 & & & \\ \vdots & & \ddots & & \vdots \\ & & & J_9 & 0 \\ 0 & \cdots & & 0 & \boldsymbol{J_h} \end{bmatrix} \tag{17}$$

$$\boldsymbol{J_h} = diag(J_h, J_h, J_h, J_h, J_h, J_h) \tag{18}$$

$\boldsymbol{C}$, the damping matrix is a 21×21 symmetric matrix that can be expressed by Equation (19), and $\boldsymbol{C} = \boldsymbol{C}^T$.

$$\boldsymbol{C} = \begin{bmatrix} \boldsymbol{C}_{11} & \boldsymbol{C}_{12} & \boldsymbol{C}_{13} & 0 \\ \boldsymbol{C}_{21} & \boldsymbol{C}_{22} & \boldsymbol{C}_{23} & \boldsymbol{C}_{24} \\ \boldsymbol{C}_{31} & \boldsymbol{C}_{32} & \boldsymbol{C}_{33} & \boldsymbol{C}_{34} \\ 0 & \boldsymbol{C}_{42} & \boldsymbol{C}_{43} & \boldsymbol{C}_{44} \end{bmatrix} \tag{19}$$

where $\boldsymbol{C}_{11}$, $\boldsymbol{C}_{21}$, $\boldsymbol{C}_{22}$, $\boldsymbol{C}_{31}$, $\boldsymbol{C}_{32}$, $\boldsymbol{C}_{33}$, $\boldsymbol{C}_{42}$, $\boldsymbol{C}_{43}$, and $\boldsymbol{C}_{44}$ can be expressed by Equations (20)–(28).

$$\boldsymbol{C}_{11} = \begin{bmatrix} (c_1 + 4c_l + 4c_r) & -4c_l & -4c_r & 0 & 0 & -c_1 \\ -4c_l & (4c_l + c_{18}) & 0 & -c_{18} & 0 & 0 \\ -4c_r & 0 & (4c_r + c_{19}) & 0 & -c_{19} & 0 \\ 0 & -c_{18} & 0 & c_{18} & 0 & 0 \\ 0 & 0 & -c_{19} & 0 & c_{19} & 0 \\ -c_1 & 0 & 0 & 0 & 0 & (c_{20} + c_1) \end{bmatrix}_{6\times 6} \tag{20}$$

$$C_{21} = \begin{bmatrix} 0 & & & \cdots & 0 \\ 0 & & & & \\ 0 & & & \ddots & \vdots \\ 2c_l r_4 & 2c_l r_4 & & & \\ 0 & 0 & 0 & \cdots & 0 \end{bmatrix}_{5\times 6} \tag{21}$$

$$C_{22} = \begin{bmatrix} c_2 & -c_2 & 0 & 0 & 0 \\ -c_2 & (c_2+c_3) & -c_3 & 0 & 0 \\ 0 & -c_3 & \left(c_3 + \frac{1}{n^2}c_4\right) & -\frac{c_4}{n} & 0 \\ 0 & 0 & -\frac{c_4}{n} & \left(c_4 + c_5 + 2c_l {r_4}^2\right) & -c_5 \\ 0 & 0 & 0 & -c_5 & (c_5 + c_9) \end{bmatrix}_{5\times 5} \tag{22}$$

$$C_{31} = \begin{bmatrix} 0 & 0 & \cdots & \cdots & & 0 \\ 0 & 0 & \vdots & & & \vdots \\ 0 & 0 & \cdots & \cdots & & 0 \\ 2c_r r_9 & 0 & -2c_r r_9 & 0 & 0 & 0 \end{bmatrix}_{4\times 6} \tag{23}$$

$$C_{32} = \begin{bmatrix} 0 & & \cdots & 0 & 0 \\ \vdots & \ddots & & & 0 \\ & & \ddots & \vdots & 0 \\ 0 & \cdots & & 0 & -c_9 \end{bmatrix}_{4\times 5} \tag{24}$$

$$C_{33} = \begin{bmatrix} c_6 & -c_6 & 0 & 0 \\ -c_6 & (c_6+c_7) & -c_7 & 0 \\ 0 & -c_7 & \left(c_7 + \frac{1}{n^2}c_8\right) & -\frac{c_8}{n} \\ 0 & 0 & -\frac{c_8}{n} & \left(c_8 + c_9 + 2c_r {r_9}^2\right) \end{bmatrix}_{4\times 4} \tag{25}$$

$$C_{42} = \begin{bmatrix} 0 & \cdots & 0 & c_l r_h r_4 & 0 \\ & \ddots & 0 & -c_l r_h r_4 & 0 \\ \vdots & & 0 & 0 & 0 \\ & & & \ddots & \vdots \\ 0 & & \cdots & & 0 \end{bmatrix}_{6\times 5} \tag{26}$$

$$C_{43} = \begin{bmatrix} 0 & \cdots & 0 & 0 \\ & \ddots & \vdots & 0 \\ \vdots & & 0 & c_r r_h r_9 \\ 0 & & 0 & -c_r r_h r_9 \\ & \ddots & \vdots & 0 \\ 0 & \cdots & 0 & 0 \end{bmatrix}_{6\times 4} \tag{27}$$

$$C_{44} = \begin{bmatrix} 2c_l {r_h}^2 & 0 & 0 & 0 & 0 & c_l {r_h}^2 \\ 0 & 2c_l {r_h}^2 & 0 & 0 & 0 & c_l {r_h}^2 \\ 0 & 0 & 2c_r {r_h}^2 & 0 & c_r {r_h}^2 & 0 \\ 0 & 0 & 0 & 2c_r {r_h}^2 & c_r {r_h}^2 & 0 \\ 0 & 0 & c_r {r_h}^2 & c_r {r_h}^2 & 2c_r {r_h}^2 & 0 \\ c_l {r_h}^2 & c_l {r_h}^2 & 0 & 0 & 0 & 2c_l {r_h}^2 \end{bmatrix}_{6\times 6} \tag{28}$$

Because, C, the damping matrix, is a symmetric matrix; thus, $C_{12} = {C_{21}}^T$, $C_{13} = {C_{31}}^T$, $C_{23} = {C_{32}}^T$, $C_{24} = {C_{42}}^T$, and $C_{34} = {C_{43}}^T$.

K, the stiffness matrix is a 21×21 symmetric matrix that can be expressed by Equation (29), and $\boldsymbol{K} = \boldsymbol{K}^T$.

$$\boldsymbol{CK} = \begin{bmatrix} \boldsymbol{K}_{11} & \boldsymbol{K}_{12} & \boldsymbol{K}_{13} & 0 \\ \boldsymbol{K}_{21} & \boldsymbol{K}_{22} & \boldsymbol{K}_{23} & \boldsymbol{K}_{24} \\ \boldsymbol{K}_{31} & \boldsymbol{K}_{32} & \boldsymbol{K}_{33} & \boldsymbol{K}_{34} \\ 0 & \boldsymbol{K}_{42} & \boldsymbol{K}_{43} & \boldsymbol{K}_{44} \end{bmatrix} \tag{29}$$

where $\boldsymbol{K}_{11}$, $\boldsymbol{K}_{21}$, $\boldsymbol{K}_{22}$, $\boldsymbol{K}_{31}$, $\boldsymbol{K}_{32}$, $\boldsymbol{K}_{33}$, $\boldsymbol{K}_{42}$, $\boldsymbol{K}_{43}$, and $\boldsymbol{K}_{44}$ can be expressed by Equations (30)–(38).

$$\boldsymbol{K}_{11} = \begin{bmatrix} (k_1 + 4k_l + 4k_r) & -4k_l & -4k_r & 0 & 0 & -k_1 \\ -4k_l & (4k_l + k_{18}) & 0 & -k_{18} & 0 & 0 \\ -4k_r & 0 & (4k_r + k_{19}) & 0 & -k_{19} & 0 \\ 0 & -k_{18} & 0 & k_{18} & 0 & 0 \\ 0 & 0 & -k_{19} & 0 & k_{19} & 0 \\ -k_1 & 0 & 0 & 0 & 0 & (k_{20} + k_1) \end{bmatrix}_{6\times 6} \tag{30}$$

$$\boldsymbol{K}_{21} = \begin{bmatrix} 0 & & \cdots & & 0 \\ 0 & & & & \\ 0 & & \ddots & & \vdots \\ 2k_l r_4 & 2k_l r_4 & & & \\ 0 & 0 & 0 & \cdots & 0 \end{bmatrix}_{5\times 6} \tag{31}$$

$$\boldsymbol{K}_{22} = \begin{bmatrix} k_2 & -k_2 & 0 & 0 & 0 \\ -k_2 & (k_2 + k_3) & -k_3 & 0 & 0 \\ 0 & -k_3 & \left(k_3 + \frac{1}{n^2}k_4\right) & -\frac{k_4}{n} & 0 \\ 0 & 0 & -\frac{k_4}{n} & (k_4 + k_5 - 2k_l {r_4}^2) & -k_5 \\ 0 & 0 & 0 & -k_5 & (k_5 + k_9) \end{bmatrix}_{5\times 5} \tag{32}$$

$$\boldsymbol{K}_{31} = \begin{bmatrix} 0 & 0 & \cdots & & \cdots & 0 \\ 0 & 0 & \vdots & & & \vdots \\ 0 & 0 & \cdots & & \cdots & 0 \\ 2k_r r_9 & 0 & -2k_r r_9 & 0 & 0 & 0 \end{bmatrix}_{4\times 5} \tag{33}$$

$$\boldsymbol{K}_{32} = \begin{bmatrix} 0 & & \cdots & 0 & 0 \\ \vdots & \ddots & & & 0 \\ & & \ddots & \vdots & 0 \\ 0 & \cdots & & 0 & -k_9 \end{bmatrix}_{4\times 5} \tag{34}$$

$$\boldsymbol{K}_{33} = \begin{bmatrix} k_6 & -k_6 & 0 & 0 \\ -k_6 & (k_6 + k_7) & -k_7 & 0 \\ 0 & -k_7 & \left(k_7 + \frac{1}{n^2}k_8\right) & -\frac{k_8}{n} \\ 0 & 0 & -\frac{k_8}{n} & (k_8 + k_9 + 2k_r {r_9}^2) \end{bmatrix}_{4\times 4} \tag{35}$$

$$\boldsymbol{K}_{42} = \begin{bmatrix} 0 & \cdots & 0 & k_l r_h r_4 & 0 \\ & \ddots & 0 & -k_l r_h r_4 & 0 \\ \vdots & & 0 & 0 & 0 \\ & & & \ddots & \vdots \\ 0 & & \cdots & & 0 \end{bmatrix}_{6\times 5} \tag{36}$$

$$\boldsymbol{K}_{43} = \begin{bmatrix} 0 & \cdots & 0 & 0 \\ & \ddots & \vdots & 0 \\ \vdots & & 0 & k_r r_h r_9 \\ 0 & & 0 & -k_r r_h r_9 \\ & \ddots & \vdots & 0 \\ 0 & \cdots & 0 & 0 \end{bmatrix}_{6\times 4} \tag{37}$$

$$K_{44} = \begin{bmatrix} 2k_l{r_h}^2 & 0 & 0 & 0 & 0 & k_l{r_h}^2 \\ 0 & 2k_l{r_h}^2 & 0 & 0 & 0 & k_l{r_h}^2 \\ 0 & 0 & 2k_r{r_h}^2 & 0 & k_r{r_h}^2 & 0 \\ 0 & 0 & 0 & 2k_r{r_h}^2 & k_r{r_h}^2 & 0 \\ 0 & 0 & k_r{r_h}^2 & k_r{r_h}^2 & 2k_r{r_h}^2 & 0 \\ k_l{r_h}^2 & k_l{r_h}^2 & 0 & 0 & 0 & 2k_l{r_h}^2 \end{bmatrix}_{6\times 6} \tag{38}$$

Since $\boldsymbol{K}$ is a symmetric matrix. Thus $\boldsymbol{K}_{12} = {\boldsymbol{K}_{21}}^T, \boldsymbol{K}_{13} = {\boldsymbol{K}_{31}}^T, \boldsymbol{K}_{23} = {\boldsymbol{K}_{32}}^T, \boldsymbol{K}_{24} = {\boldsymbol{K}_{42}}^T$, $\boldsymbol{K}_{34} = {\boldsymbol{K}_{43}}^T$.

$\boldsymbol{Q}$, the generalized force matrix, is defined in Equation (39).

$$\boldsymbol{Q} = \begin{bmatrix} \boldsymbol{O}_{1\times 3} & -m_4 g & -m_7 g & 0 & Te_1 & -T_2 & \boldsymbol{O}_{1\times 3} & T_{e6} & -T_7 & \boldsymbol{O}_{1\times 8} \end{bmatrix}^T \tag{39}$$

where T_{e1} and T_{e6} are the driving torque of motors 1 and 6, respectively, and T_2 and T_7 are the braking torque of brakes 2 and 7, respectively.

According to the dynamic model of the hoisting mechanism, the load torque of the motor can be obtained as follows:

$$T_{m1} = c\left(\dot{\theta}_1 - \dot{\theta}_2\right) + k(\theta_1 - \theta_2) \tag{40}$$

$$T_{m6} = c\left(\dot{\theta}_6 - \dot{\theta}_7\right) + k(\theta_6 - \theta_7) \tag{41}$$

where c is the damping coefficient, θ and $\dot{\theta}$ are the rotation angle and the rotational speed of the rotor, respectively, and T_{m1} and T_{m6} are the load torque of the motor 1 and motor 6, respectively.

According to the coupling relationship between the motor control model and the mass-stiffness-damping model of the hoisting mechanism, the following equations can be obtained as:

$$\omega_{r1} = \dot{\theta}_1 \tag{42}$$

$$\omega_{r6} = \dot{\theta}_6 \tag{43}$$

Since both the motor driving torque and the brake braking torque are the result of the coupling between the motor and the mechanical structure, they are not constant values. It is difficult to describe these torques by a single time-varying function. The motor driving torque is the real time output signal of the motor, and the brake braking torque is the real time resistance torque after the brake start to act. Therefore, as to accurately reflect the dynamic response of the hoisting mechanism under various working conditions, it is necessary to simultaneously solve for the motor driving torque and the brake braking torque.

4.4. Calculation of the Structural Parameters of the Hoisting Mechanism

To solve the abovementioned dynamic Equations, it is necessary to first solve for the parameters in Table 8, such as the equivalent mass, equivalent moment, damping and stiffness of the main beam.

4.4.1. Calculation of Equivalent Mass of the Main Girder

Because of the long length of the main girder, the mass distributed in the direction of the length of the main beam is equivalent to the position of the lifting trolley. When the lifting trolley is at a certain distance from the left end of the main girder, the mass of the main beam is equivalent to:

$$m_8 = \frac{3L^4 m}{\pi^4 a^2 (L-a)^2} \tag{44}$$

where a is the distance between the hoisting trolley and the left end of the main girder, L is the length of the main girder, and m is the total mass of the main beam steel structure.

4.4.2. Calculation of the Equivalent Moment of Inertia

When the hoisting mechanism is running, the rotating shafts of brakes, motors, reducers, drums and other structural parts were not on the same axis. Therefore, it is necessary to convert the inertia moment of each component to the power takeoff shaft.

The kinetic energy of the moving parts in the transmission system is:

$$E_1 = \frac{1}{2}\sum_{i=1}^{m} J_i {\omega_i}^2 + \frac{1}{2}\sum_{j=1}^{n} m_j {v_i}^2 \tag{45}$$

Assuming that the equivalent inertia moment of load is J_L, the equivalent kinetic energy on the motor shaft is:

$$E_2 = \frac{1}{2} J_L {\omega_L}^2 \tag{46}$$

From the law of energy conservation law, $E_1 = E_2$, the equivalent inertia moment of load is:

$$J_L = \sum_{i=1}^{m} J_i \left(\frac{\omega_i}{\omega_L}\right)^2 + \sum_{j=1}^{n} m_j \left(\frac{v_j}{\omega_L}\right)^2 \tag{47}$$

4.4.3. Calculation of the Damping Coefficient

According to viscous damping theory, the damping force is proportional to the moving speed, and the absolute value of the ratio of the damping force to the speed is the damping coefficient. Additionally, the damping factor is usually calculated by multiplying the damping ratio and the critical damping factor. That is, the damping coefficient of the part can be obtained as:

$$c_i = \zeta c_{ci} = \zeta \times 2\sqrt{k_i m_i} (i = 1, 2 \ldots 20) \tag{48}$$

where c_i, ζ, c_{ci}, k_i and mi are the actual damping coefficient, damping ratio, critical damping coefficient, stiffness and mass of the part, respectively. The damping ratio ζ is taken as 0.1.

4.4.4. Calculation of Stiffness Coefficient

In the hoisting mechanism, the calculation of stiffness mainly involves the main girder, the rotating shaft and the wire rope.

The stiffness of main girder is calculated by using the equivalent calculation method. When the lifting trolley is at a certain distance from the left end of the main girder, the stiffness of the main girder can be obtained as:

$$k_8 = \frac{3EIL}{a^2 (L-a)^2} \tag{49}$$

where E is the elastic modulus of the main girder and I is the inertia moment of cross-sectional.

The rotational stiffness of the rotating shaft is calculated using Equation (50):

$$k_t = \frac{T}{\varphi} = \frac{GI_p}{l} \tag{50}$$

where T is the torque acting on the rotating shaft, φ is the torsion angle of the rotating shaft, G is the shear modulus of elasticity of the torsion shaft, and I_p is the length of the torsion shaft.

The stiffness of the wire rope is related to its length, and the stiffness of a single wire rope can be calculated using Equation (51).

$$k_i = \frac{E_z A}{L_0 - \sum y_i} \tag{51}$$

where E_z is the elastic coefficient of the wire rope, E_z = 110 × 109 N^2/m, A is the area of the wire rope cross-section, A = 6.154 × 10^{-4} m^2, L_0 is the initial length of the wire rope, L_0 = 6 m, and $\sum y_i$ is the displacement of the shortened length of the unilateral wire rope.

According to the above method, the values of the parameters in Table 7 were calculated, as shown in Table 10.

Table 10. The values of the parameters in Table 7.

Parameters	Value	Unit	Parameters	Value	Unit
m_1	44,800	kg	m_j	3500	kg
m_4	25,000	kg	m_7	25,000	kg
m_8	90,000	kg	J_2,J_7	0.049	kg·m^2
J_1,J_6	7.22	kg·m^2	J_4,J_9	2158	kg·m^2
J_3,J_8	0.1633	kg·m^2	J_{10}~J_{15}	0.04	kg·m^2
J_5	26.45	kg·m^2	k_1	8 × 106	N/m
k_{20}	9 × 106	N/m	k_3,k_7	4.26 × 104	N/m
k_2,k_6	1.26 × 104	N/m	k_4,k_8	2 × 105	N/m
k_r,k_l	4.1 × 105	N/m	k_5,k_9	2 × 105	N/m
k_{18},k_{19}	2.5 × 105	N/m	c_1	80,000	N·s/m
c_{20}	90,000	N·s/m	c_3,c_7	5000	N·s/m
c_2,c_6	5000	N·s/m	c_4,c_8	5000	N·s/m
c_r,c_l	4080	N·s/m	c_5,c_9	2000	N·s/m
r_4,r_9	0.7	m	c_{18},c_{19}	6.5 × 104	N·s/m
r_i (i = 10,11, ... ,15)	0.09	m	N	173.03	

Thus, by substituting the system parameters into the mathematical model of the motor system and transmission mechanisms, the dynamic load distribution of the bilaterally driven hoisting mechanism can be solved.

4.5. Comparison of Dynamic Simulation and Experiment

According to the procedure shown in Figure 11, the dynamic simulation of the hoisting mechanism under different working conditions was conducted.

Taking the test condition shown in Figure 10a as an example, when the hoisting weight (40.44 t) of the hoisting mechanism was in the process of hoisting and braking in the 1st gear, the torque at measuring point 2 on the load drum shaft was tested and simulated. The simulation and test torques are shown in Figure 13.

Figure 13 shows the torque at point 2 on the spool shaft before or after braking. The vibration amplitude of the simulated torque was larger than it in test, the mean value of the simulated torque was smaller than the tested torque, and the residual torque after braking in simulation was also smaller than it in test. These results may have been related to the assumptions such as ignoring friction in dynamic modeling, and further quantitative analysis was needed.

However, the tested torque and simulated torque periodic oscillated with a period of 13 s during the stable lifting process. Additionally, the tested torque and simulated torque at the measuring point vibrated during braking and retained stable residual torque after braking. These meant that the dynamic responses obtained by the simulation and test were basically the same. Thus, using the dynamic model to further analyze the fatigue fracture failure of the load drum shaft was credible.

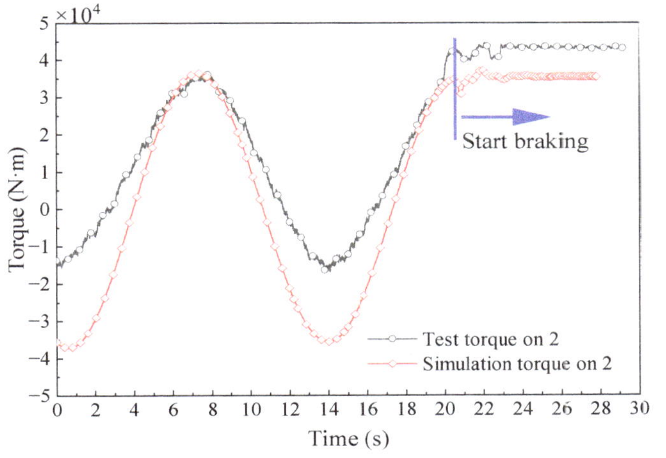

Figure 13. Comparison of simulation and experimental data for the torque at test point 2.

5. Simulation Results and Application Analysis

5.1. Effects of the Motor and Its Controls

According to the mathematical model, the simulation model of the three-phase asynchronous motor and its speed control system was built in the MATLAB/Simulink, as shown in Figure 14. The motor model mainly included the stator module, rotor module, flux linkage module and torque module. The motor speed control module mainly included the thyristor stator speed regulation module with voltage regulation, the rotor speed regulation module with connecting resistance, and the braking control module with energy consumption.

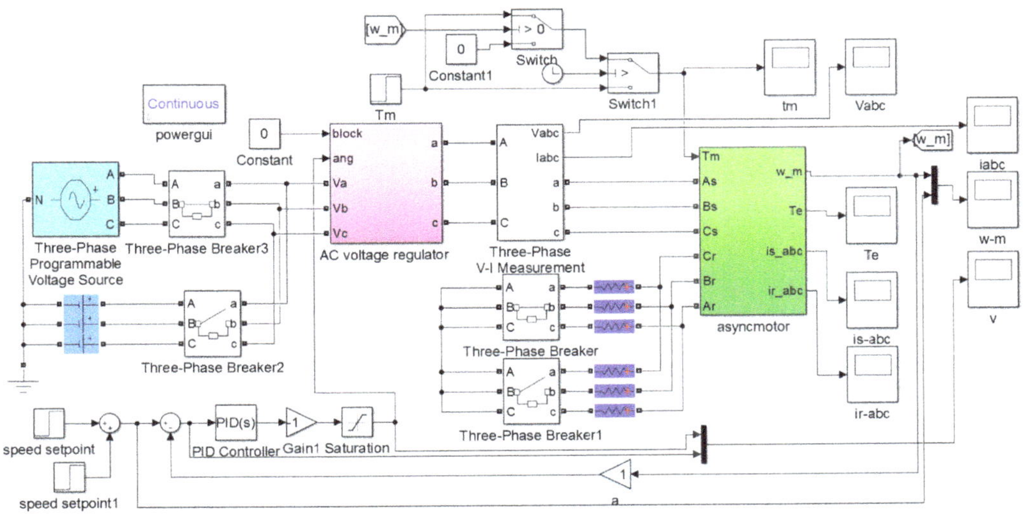

Figure 14. Three-phase asynchronous motor and its speed control model.

As to analyze the electromechanical coupling effect more clearly, the load was directly connected to the motor in Section 4.1. The parameters used in the simulation are shown in Table 11.

Table 11. Motor performance and its values.

Performance	Value	Performance	Value
Phase voltage/V	380	Rotor resistance/Ω	0.027
Power frequency/Hz	50	Rotor leakage inductance/H	0.000462
DC voltage/V	15	Mutual inductance/H	3.6
Rotor resistance at startup/Ω	0.873	Pole pairs	4
Rotor resistance at steady state/Ω	0.209	Rotor moment/kg·m^2	7.22
Stator resistance/Ω	0.042	Friction coefficient	0.0
Stator leakage inductance/H	0.0000296		

The given motor speed and load torque during the simulation are shown as solid lines in Figures 15 and 16, and the total simulation time was 12 s. When starting, the given motor speed was 15 rad/s, and the load was 300 N m. At 3 s and 5 s, the given speed was increased to 30 rad/s and 75 rad/s, respectively. At 8 s, the speed remained unchanged, and the load was increased to 500 N m. At 10 s, the brake was turned on until the rotor speed dropped to zero, that is, the motor load became zero. The dotted line in Figure 15 shows the rotational speed obtained by the simulation.

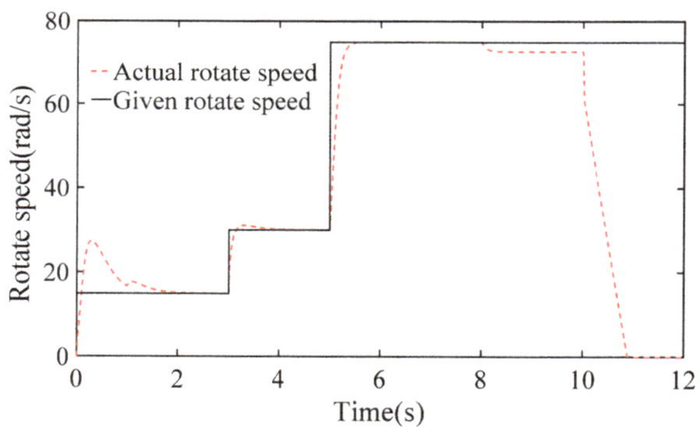

Figure 15. Motor given speed and actual speed.

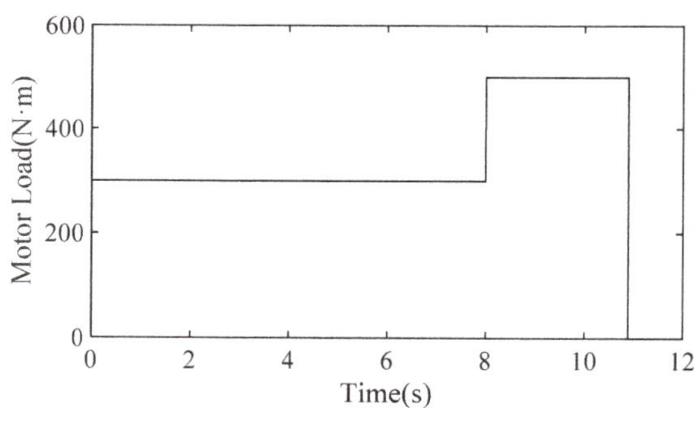

Figure 16. The given load torque of motor.

Figure 15 shows that the actual speed of the motor tracked the change in the given speed, and it decreased slightly and stabilized quickly when the load increased. In addition, when the motor started to brake, the rotate speeds decreased rapidly, and there was a sudden change in acceleration for the electric braking torque was not stable.

Under the given speed and load, as shown in Figures 15 and 16, the current and electromagnetic torque of motor are shown in Figures 17 and 18. A comparison showed that the changes in currents of stator and rotor were synchronous with the changes in electromagnetic torque. Additionally, the three-phase current of the rotor followed the change in the three-phase current of the stator, and due to the load, there was a slip rate between them. The frequency of three-phase current in the rotor was lower than it in the stator.

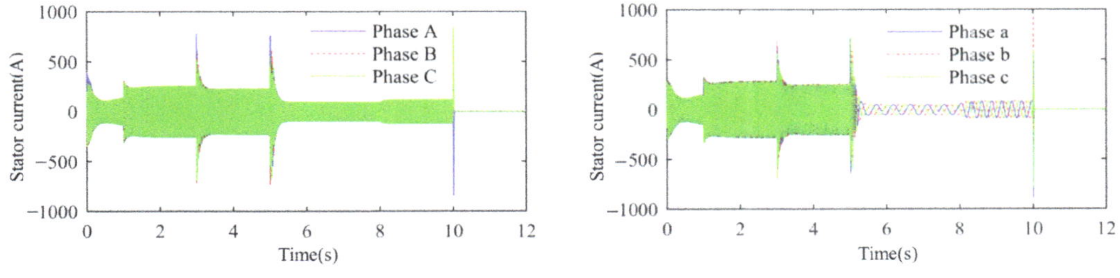

(a) (b)

Figure 17. Current curves of motor: (**a**) three-phase current of stator; (**b**) three-phase current of rotor.

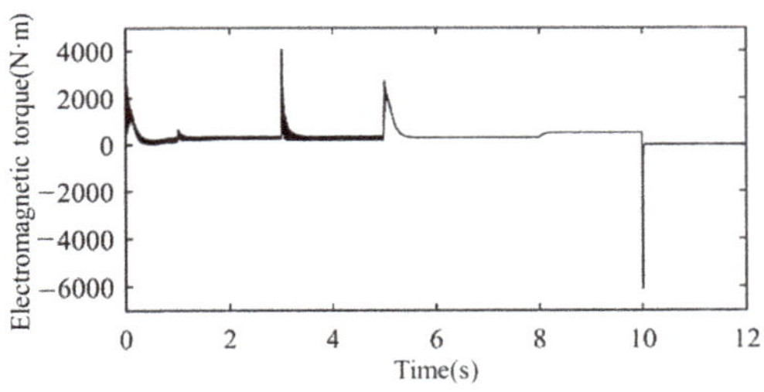

Figure 18. Electromagnetic torque of motor.

It is well known that the electromagnetic torque of motor depends on its stator current, which is generated by the stator voltage in the stator winding. In reality, the loads also had influences on the induced electromotive force and induced current on the rotor, which finally affected the stator current. This was the principle of coupling between the motor and the mechanical load, which made the electromagnetic torque of motor undergo complex dynamic changes.

Figure 18 shows that in the processes of starting, regulating the speed, braking and changing the load, the motor ensured a relatively stable speed, but the output torque obviously underwent overshoots and fluctuations. This inevitably resulted in large dynamic forces on the mechanical transmission and result in its fatigue fracture, including the load drum shaft.

5.2. Effects of Nonsynchronous Control

According to design requirements, the structures of the bilateral drive hoisting mechanism should be symmetrical, and the control should be synchronized. However, the asynchronous start of the motor and the asynchronous braking of the brakes often happened, which could result in the motor towing.

5.2.1. Single-motor Towing

If the right motor (motor 6) in Figure 12 was damaged, then, only the left motor (motor 1) would drive the entire hoisting mechanism at 1st gear speed to raise the object, and the operation time would be 20 s. The variations in the parameters of the hoisting mechanism from the simulation are shown in Figures 19–22.

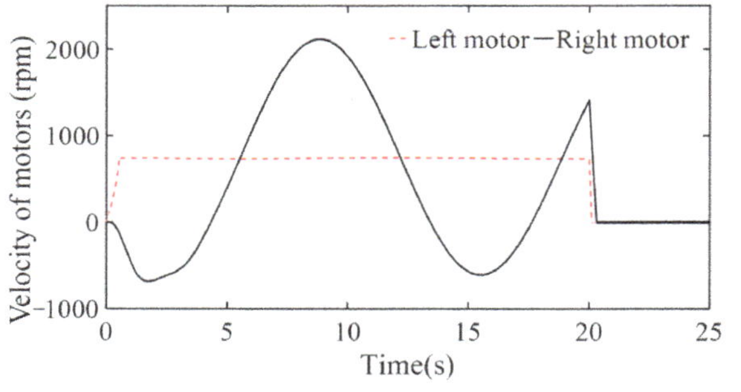

Figure 19. Speed of the motor shaft when driving a single motor.

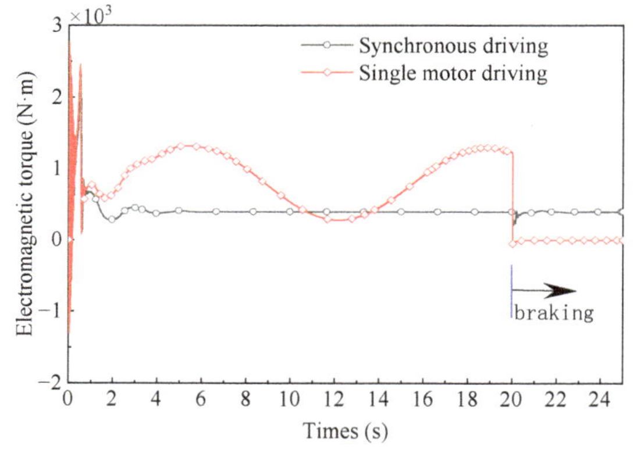

Figure 20. Electromagnetic torque comparison.

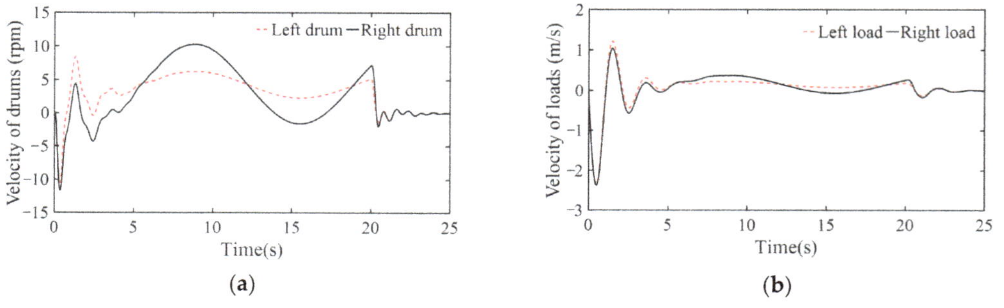

Figure 21. Drum speed and lifting speed when a single motor is driving: (**a**) left and right drum speeds; (**b**) lifting speed.

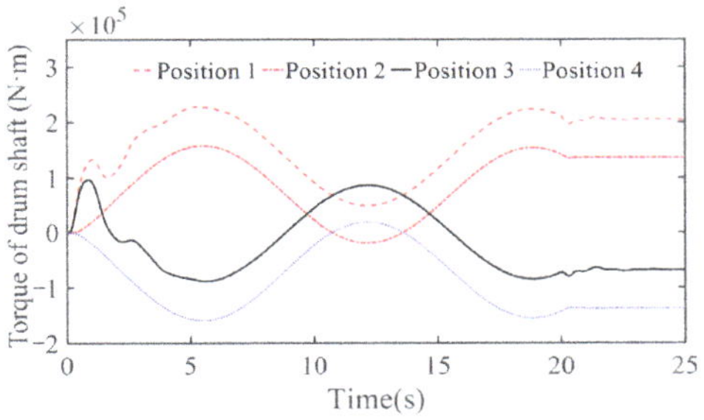

Figure 22. Torque of drum shaft when a single motor is driving.

As shown in Figure 19, under the motor speed control system, the speed of the left motor changed smoothly. However, the right motor 6 was towed, and its rotational speed oscillated with a period of 13 s. Additionally, the average speed of right motor was equal to that of the left motor.

As shown in Figure 20, the electromagnetic torque of the left motor exhibited overshoot behavior that was basically consistent with the oscillation of the electromagnetic torque during synchronous driving. Additionally, the electromagnetic torque of the left motor was twice the average when the hoisting mechanism was in synchronous drive. In addition, comparison of Figures 19 and 20 shows that the electromagnetic torque of left motor oscillated periodically, with a 90-degree out-of-phase difference from the rotational speed.

As shown in Figures 21 and 22, single-motor towing also affected the dynamic behavior of the two load drums. Figure 21a shows that the rotational speed amplitude of the left load drum, 5 r/min, was less than that of the right load drum, 15 r/min, but their oscillation tendencies were basically the same. The speeds of the two load drums decreased rapidly after braking and stopped after short-term oscillation. It can be concluded from Figure 21b that the rotational speeds of the two load drums oscillated with a period of 13 s after 5 s, and the average values were both 4 r/min, which ensured that the lifting speeds of the left and right sides were basically the same.

As shown in Figure 22, when the left motor towed the right motor, the torque was significantly larger on the input shaft of the left load drum than the input shaft of the right load drum after startup. After 5 s, the torques of the shafts of the left and right drums both fluctuated with a period of 13 s. Additionally, after braking, there were constant residual torques on the shafts of the left and right drums, and the torques on the left and right drum shafts had equal absolute values and opposite directions. Therefore, when a single motor was towed in the bilateral propelled hoisting mechanism, it generated a large alternating load on the drum shaft during starting, stable operation and braking. This was another load source for fatigue fracture of the spool shaft.

However, in actual operation, there is usually no damage to a single motor for a long time. More working conditions lead to motor drag caused by unsynchronized motor startup or brake braking.

5.2.2. Motor startup Time Difference

When the hoisting mechanism lifted a heavy object in 1st gear and the startup time difference Δt of the two motors was 0.05 s, 0.1 s, 0.3 s, and 0.5 s, the torque of the left and right drum shafts changed, as shown in Figure 23. According to Figure 23, the asynchronous start of the motor generated a large torque on the load drum shaft, and the torque was positively correlated with the time difference, Δt. When Δt decreased from 0.5 s to 0.05 s, the absolute value of the torque decreased from 2.7×10^4 N m to 2.5×10^3 N·m. The torque decreased slowly with running time increased.

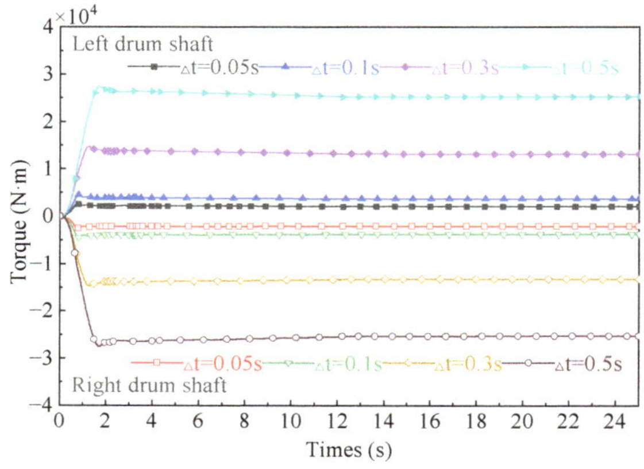

Figure 23. Torque on the left and right drum shafts under different power-on time.

5.2.3. Brake starting Time Difference

In the design condition, the two motors of the hoisting mechanism were powered on at the same time, heavy objects were lifted in 1st gear for 12 s and then the motors were braked at different times. When the brake starting time difference was 0.05 s, 0.1 s, 0.3 s, and 0.5 s, the torques on the output shafts of the left and right drums are shown in Figure 24. According to Figure 24, when the braking times of the left and right brakes differed, the torque on the output shaft of the load drum suddenly changed and a large residual torque was retained after braking. In addition, the absolute value was equal for the residual torques on the left and right load drum shafts, and both were positively correlated with the brake time difference. When $\Delta t'$ is 0.5 s, the absolute value of the moment reached a maximum of 1.25×10^4 N·m.

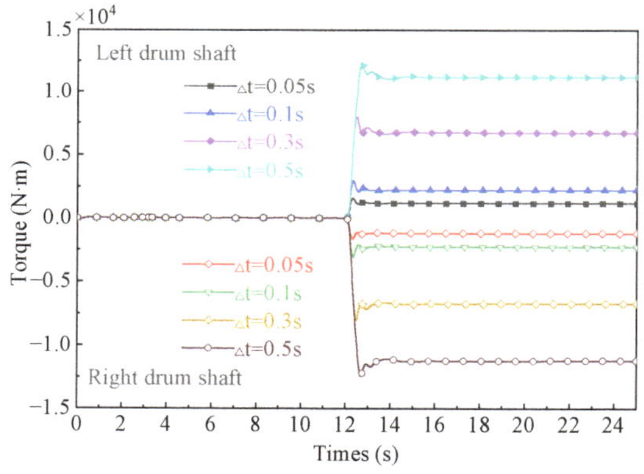

Figure 24. Torques on the left and right drum shafts under different brake closing time.

From the above analysis, it can be concluded that although coupling realized the synchronous lifting of hoisting weights, asynchronous control of motors and brakes changed the force distribution in the hoisting mechanism and generated periodic vibrations or constant dynamic loads on the left and right load drum shafts. The instantaneous torque value reached 2×10^5 N·m, causing shock and fatigue damage to the structure and reducing the service life of the spool shaft.

6. Conclusions

Considering fracture morphology and failure factors such as material structure, manufacturing process, and load, the fatigue fracture of the load drum shaft of the bilateral drive casting bridge crane was investigated. According to the test results, dynamic modeling and simulations, the dynamic loading of the crane during operation was a key factor in the fatigue fracture of load drum shafts. The main conclusions are as follows:

(1) The dynamic characteristics testing showed that starting the motor, braking and motor dragging caused by service conditions and control behavior caused the left and right load drum bearings to undergo oscillating torques with the cycles of approximately 13 s. Additionally, starting, shifting, and braking caused sudden changes in the torque on the load drum shaft, affecting the dynamic performance of the hoisting mechanism.

(2) The dynamic simulation results of the hoisting mechanism quantified the influence of the motor speed and load changes on the motor output electromagnetic torque and current changes and the influence of asynchronous control on the torque of the load drum shaft. The results showed that although the coupling can realize synchronous lifting of the left and right hoisting weights, the asynchrony of the left and right speed controls changed the load distribution in the hoisting mechanism and generated additional periodic or constant torques on the left and right load drum shafts. The torque value reached 2×105 N·m, giving rise to impact and fatigue damage to the structure and reducing the service life of the load drum shaft.

(3) Because several assumptions were made in the formulation of the dynamic simulation model, the simulation results and the test results differed in amplitude, but other dynamic responses were basically the same. Therefore, this dynamic simulation can be used to explain the origins of the fatigue fracture of spool shafts. However, further refinement of the model is required for applications in fatigue-resistant design.

Author Contributions: Conceptualization, D.X. and Y.L.; methodology, D.X.; software, Y.L.; validation, Y.Z. and F.X.; formal analysis, D.X. and Y.L.; resources, F.X.; data curation, Y.Z. and F.X.; writing—original draft preparation, Y.L.; writing—review and editing, Y.Z.; project administration, D.X. All authors have read and agreed to the published version of the manuscript.

Funding: This research was funded by the National Natural Science Foundation of China under grant number 51975323.

Conflicts of Interest: The authors declare no conflict of interest. The funders had no role in the design of the study; in the collection, analyses, or interpretation of data; in the writing of the manuscript; or in the decision to publish the results.

References

1. Wu, H.; Guo, H.; Gao, D. Generalization of life characteristic investigation for probabilistic fatigue failure. *Struct. Environ. Eng.* **2002**, *29*, 38–44.
2. Zerbst, U.; Madia, M.; Klinger, C.; Bettge, D.; Murakami, Y. Defects as a root cause of fatigue failure of metallic components. I: Basic aspects. *Eng. Fail. Anal.* **2019**, *97*, 777–792. [CrossRef]
3. Zerbst, U.; Madia, M.; Klinger, C.; Bettge, D.; Murakami, Y. Defects as a root cause of fatigue failure of metallic components. II: Non-metallic inclusions. *Eng. Fail. Anal.* **2019**, *98*, 228–239. [CrossRef]
4. Zerbst, U.; Madia, M.; Klinger, C.; Bettge, D.; Murakami, Y. Defects as a root cause of fatigue failure of metallic components. III: Cavities, dents, corrosion pits, scratches. *Eng. Fail. Anal.* **2019**, *97*, 759–776. [CrossRef]
5. Liu, Z.F.; Tsang, K.S.; Liu, Y.; Pang, J.H.L. Finite element and experimental study on multiaxial fatigue analysis of rail clip failures. *Fatigue Fract. Eng. Mater. Struct.* **2020**, *43*, 2390–2401. [CrossRef]
6. Farrahi, G.H.; Ahmadi, A.; Kasyzadeh, K.R. Simulation of vehicle body spot weld failures due to fatigue by considering road roughness and vehicle velocity. *Simul. Model. Pract. Theory* **2020**, *105*, 102168. [CrossRef]
7. Sedrakian, A.; Zineb, T.B.; Billoet, J.L. Contribution of industrial composite parts to fatigue behaviour simulation. *Int. J. Fatigue* **2002**, *24*, 307–318. [CrossRef]
8. Chan, K.S. Roles of microstructure in fatigue crack initiation. *Int. J. Fatigue* **2010**, *32*, 1428–1447. [CrossRef]
9. Nasr, A.; Nadot, Y.; Bouraoui, C. Fatigue initiation in C35 steel: Influence of loading and defect. *Int. J. Fatigue* **2010**, *32*, 780–787. [CrossRef]

10. Shirani, M.; Härkegård, G. Fatigue life distribution and size effect in ductile cast iron for wind turbine components. *Eng. Fail. Anal.* **2011**, *18*, 12–24. [CrossRef]
11. Das, R.; Jones, R. Fatigue life enhancement of structures using shape optimization. *Theor. Appl. Fract. Mech.* **2009**, *52*, 165–179. [CrossRef]
12. Zheng, Y.; Wang, Z.G. Historical review of fatigue study. *Mater. Sci. Eng.* **1993**, *11*, 1–6.
13. Li, W.; Yan, Q.; Xue, J.H. Analysis of a crankshaft fatigue failure. *Eng. Fail. Anal.* **2015**, *55*, 139–147. [CrossRef]
14. Roy, A.; Palit, P.; Das, S.; Mukhyopadyay, G. Investigation of torsional fatigue failure of a centrifugal pump shaft. *Eng. Fail. Anal.* **2020**, *112*, 759–776. [CrossRef]
15. Seifoori, S.; Parrany, A.M.; Khodayari, M. A high-cycle fatigue failure analysis for the turbocharger shaft of BELAZ 75131 mining dump truck. *Eng. Fail. Anal.* **2020**, *116*, 104752. [CrossRef]
16. Hoddinott, D.S. Railway axle failure investigations and fatigue crack growth monitoring of an axle. *Proc. Inst. Mech. Eng. Part F J. Rail Rapid Transit* **2004**, *218*, 283–292. [CrossRef]
17. Asi, O. Fatigue failure of a rear axle shaft of an automobile. *Eng. Fail. Anal.* **2006**, *13*, 1293–1302. [CrossRef]
18. Ognjanovic, M.; Simonovic, A. Research of rail traction shafts and axles fractures towards impact of service conditions and fatigue damage accumulation. *Eng. Fail. Anal.* **2010**, *17*, 1560–1571. [CrossRef]
19. *GB 6067.5-2014*; Safety Rules for Lifting Appliances—Part 5. Bridge and Gantry Cranes. China Quality and Standards Publishing: Beijing, China, 2014.
20. Li, Y. *Fracture Failure Analysis of Load Drum Shaft for Casting Overhead Crane Considering Dynamic Factors*; Tsinghua University: Beijing, China, 2018.
21. Feng, P. Analysis and improvement on crane drum shaft crack. *Metall. Equip.* **2017**, *233*, 74–75.
22. Wang, J.D.; Zhang, J. Analysis of stress concentration for load drum shaft fracture. *Spec. Equip. Saf. Technol.* **2012**, *2*, 12–14.
23. Bai, Y.; Du, S.H. Analysis on cracks in the roller bearing shaft of ladle cranes. *Gui Gu* **2012**, *5*, 135–140.
24. Ma, J.X.; Wang, S.W.; Wei, J.J. Analysis and precautions of cracks on main winding barrel shaft of crane. *Metall. Equip. Manag. Maint.* **2011**, *29*, 64–65.
25. Wu, Y.M. Analysis of metallurgical crane load drum shaft broken accident and discussion on maintenance measures. *Sci. Technol. Innov.* **2019**, *16*, 150–160.
26. Li, J.C.; Mao, G.X.; Luo, Y.D. Accident analysis and maintenance measures of metallurgical crane load drum broken shaft. *Mech. Eng. Autom.* **2005**, *133*, P81–P82. [CrossRef]
27. Gao, D.L. *Design and Implementation of Speed Sensorless Vector Control System for Induction Motor*; University of Chinese Academy of Sciences: Beijing, China, 2014.
28. Guo, W.R.; Li, Y. Application of MATLAB in power electronic circuit simulation. *Electron. Des. Appl.* **2013**, *42*, 26–29.

electronics

Article

Dynamic Response and Energy Absorption Characteristics of a Three-Dimensional Re-Entrant Honeycomb

Jun Zhang, Boqiang Shi * and Tian Han

School of Mechanical Engineering, University of Science and Technology Beijing, Beijing 100083, China
* Correspondence: shiboqiang@ustb.edu.cn

Abstract: In this paper, we design a new three-dimensional honeycomb with a negative Poisson's ratio. A honeycomb cell was first designed by out-of-plane stretching a re-entrant honeycomb and the honeycomb is built by spatially combining the cells. The in-plane response and energy absorption characteristics of the honeycomb are studied through the finite element method (FEM). Some important characteristics are studied and listed as follows: (1) The effects of cell angle and impact velocity on the dynamic response are tested. The results show that the honeycomb exhibits an obvious negative Poisson's ratio and unique platform stress enhancement effect under the conditions of low and medium velocity. An obvious necking phenomenon appears when the cell angle parameter is 75°. (2) Based on the one-dimensional shock wave theory, the empirical formula of the platform stress is proposed to predict the dynamic bearing capacity of the honeycomb. (3) The energy absorption in different conditions are investigated. Results show that as the impact velocity increases, the energy absorption efficiency gradually decreases. In addition, with the increase of cell angle, the energy absorption efficiency is gradually improved. The above study shows that the honeycomb has good potential in using in vehicle industry as an energy absorption material. It also provides a new strategy for multi-objective optimization of mechanical structure design.

Keywords: negative Poisson's ratio; impact response; deformation mode; energy absorption; platform stress

Citation: Zhang, J.; Shi, B.; Han, T. Dynamic Response and Energy Absorption Characteristics of a Three-Dimensional Re-Entrant Honeycomb. *Electronics* **2022**, *11*, 2725. https://doi.org/10.3390/electronics11172725

Academic Editor: Je-Hyeong Bahk

Received: 17 July 2022
Accepted: 24 August 2022
Published: 30 August 2022

1. Introduction

Re-entrant honeycombs have attracted considerable attention due to their excellent mechanical properties, including high stiffness and specific strength [1,2] and superior heat dissipation capabilities [3]. Because of their distinctive energy absorption abilities, they are associated with lightweight material [4–6]. Thus they are extensively used in the field of transportation [7], aerospace and construction [8]. Many studies have recently been published that seek to investigate the in-plane compressing properties [9,10] and out-of-plane compressing properties [2,11] of honeycombs.

Meanwhile, some scholars have used experimental, numerical and theoretical methods [12] to research the compression properties of the re-entrant honeycombs [13–16]. They found that adding ribs in the cell of re-entrant honeycomb can improve Young's modulus and the energy absorption capacity [17,18]. Some researchers have investigated the deformation modes under different compress velocities and found that the re-entrant honeycomb has greater impact resistance than hexagonal honeycomb [5]. Some researchers also introduce a hierarchy into re-entrant honeycombs to investigate the in-plane crashworthiness performance [19]. In recent years, gradient honeycomb has also attracted the attention of many scholars. Studies have found that this honeycomb has better ability by changing the parameters to enhance energy absorption capacity [20–22]. There are many studies on the cell configuration of re-entrant honeycomb and the cell configuration at different compression speeds. Most re-entrant honeycomb is designed by directly stretching 2D configuration to 3D configuration and such design commonly has limited abilities.

In this study, a three-dimensional re-entrant honeycomb is proposed and its in-plane compress performance is investigated. The deformation of the honeycomb under different conditions is calculated by the finite element method, and then the stress formula of the platform is fitted according to the stress–strain curve. Finally, the energy absorption abilities of the honeycomb under different conditions are discussed.

2. Finite Element Model and Parameters

2.1. Digital Model

The new three-dimensional re-entrant honeycomb is formed by the rotation shown in Figure 1. The cell of traditional re-entrant honeycomb in an out-of-plane tensile is shown in Figure 1a, the representative structural cell (RSC) of the three-dimensional re-entrant honeycomb in orthogonal space is shown in Figure 1b and the array of the re-entrant honeycomb in orthogonal space is shown in Figure 1c. The dimensions of the RSC are shown in Figure 2, where the length of upper and lower cell walls of the re-entrant honeycomb structure is 2L, the length of the ligament connecting the adjacent honeycomb structure is L, t represents the thickness of cell wall of the honeycomb, d represents the width outside the cell wall and α represents the angle between the oblique edge of the re-entrant honeycomb and the horizontal plane (i.e., the cell angle). For the traditional re-entrant honeycomb, the range of α is from 0° to 90°. The internal edges of the traditional re-entrant honeycomb structure are overlapped when α is 0°, and when α is 90°, the re-entrant honeycomb structure becomes a square honeycomb structure.

(a) (b) (c)

Figure 1. Evolution of three-dimensional re-entrant honeycomb structure. (**a**) Traditional re-entrant honeycomb; (**b**) Representative structural cell (RSC); (**c**) Array of the RSC.

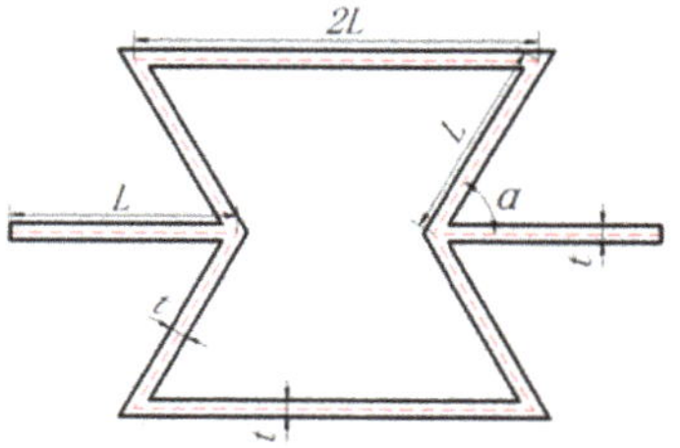

Figure 2. Structural parameters of the traditional re-entrant honeycomb ($2L$ is the length of upper and lower cell walls, L is the length of the ligament, α is the angle between the cell walls and t is the thickness of cell wall.).

According to the theory of porous materials [1], the relative density of honeycomb materials can be calculated by the ratio of the volume of RSC in Figure 1b to the total volume of three-dimensional space. Therefore, the relative density of three-dimensional honeycomb materials, $\Delta\rho$, can be written as Formula (1)

$$\Delta\rho = \frac{V_{RVE}}{V_{Total}} = \frac{td(16L - d)}{8(L\sin\alpha + \frac{t}{2})(2L - L\cos\alpha)^2} \quad (1)$$

where V_{RVE} is the volume of three-dimensional honeycomb structure, and V_{Total} is the total volume of representative structural cells in three-dimensional space in Formula (1).

2.2. Model Parameters and Constraints

The schematic diagram of the three-dimensional honeycomb calculation model is illustrated in Figure 3, where the direction setting in the simulation model is shown. The specimen formed by the three-dimensional honeycomb is placed between the upper and lower plate. In the test, the lower plate is fixed and the upper plate is driven by the external velocity, and it compresses the honeycomb specimen along the negative direction of the y-axis. The cell wall length (L) is 5 mm, the thickness (t) is 0.3 mm and the cell wall width (d) is 1 mm. By changing the cell angle and impact velocity, the dynamic response characteristics of the model in y-axis direction are calculated by Abaqus/Explicit dynamic finite element method. The matrix material of honeycomb is selected to be aluminum, assuming that the material is an ideal elastic–plastic material model, which conforms to the Mises yield criterion. The material parameters are given in Table 1. In order to facilitate the calculation of the simulation model, the upper compressed and the lower fixed plate are both regarded as a rigid plate.

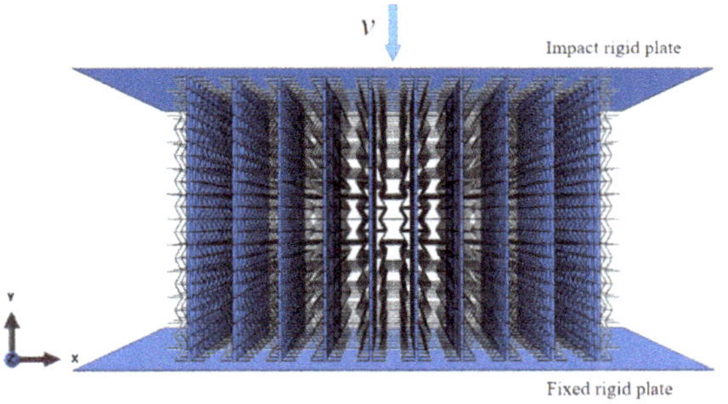

Figure 3. The model of simulation.

Table 1. Mechanical properties of the aluminum.

Material	ρ/(Kg·m^{-3})	E/GPa	σ_s/MPa	v
Aluminum	2700	69	76	0.3

The Abaqus software is used. In order to ensure the convergence of the calculation process, each cell wall of the three-dimensional honeycomb is discretized by S4R shell element, and five integral points are taken along the direction of cell wall thickness. The general contact of each element of the simulation model is set as an automatic contact and rigid plate, and this can greatly reduce the workload of defining different contact pairs. Therefore, multiple contacts are defined in the calculation, such as the general contact between the rigid plates and the honeycomb specimen and the self-contact between the internal elements of the specimen. Since the contact surfaces cannot be completely smooth, the friction coefficient is set to 0.02 for calculation accuracy [23]. In order to verify the model accuracy, we set simulation parameters as the same as the ones in article [24] and make a comparison. The results show that the deformation is completely consistent in Figure 4, which proves the simulation model is correct. In addition, this article also compares the performances in convergence with different mesh sizes. The force over calculation time is shown in Figure 5. Considering the calculation time and the accuracy of the results, the mesh size is set to 0.5.

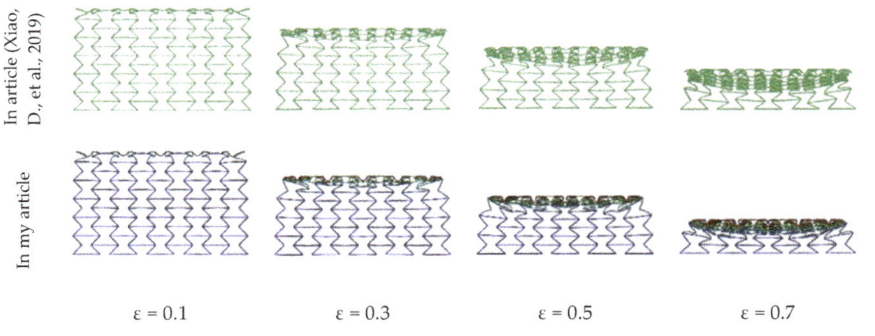

Figure 4. Comparison of deformation at speed of 100 m/s. Reprinted with permission from ref. [24]. Copyright 2022 Elsevier.

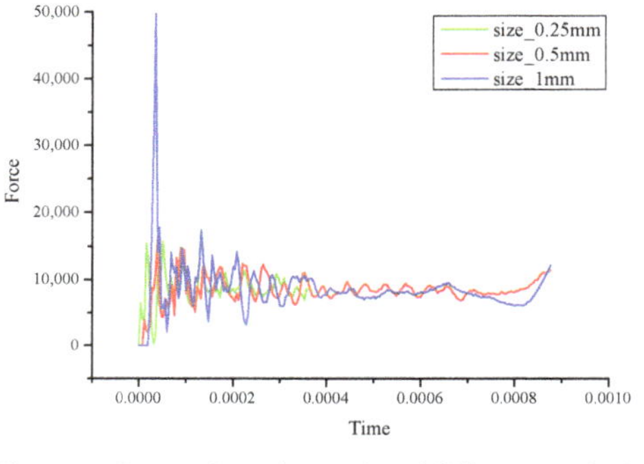

Figure 5. Comparison the results of different mesh sizes.

According to the calculation results in [23], the dynamic response of honeycomb specimens can be stabilized when the number of cells in each axial direction exceed 10. Therefore, the specimen has 10 structural cells in the *x*-axis and *z*-axis directions, and the number of cells in the *y*-axis direction change according to the cell angle. The number of cells set in each axis direction are marked in Figure 6. In order to facilitate the comparison of energy absorption efficiency, the height of specimen in *y*-axis is maintained at about 95 mm.

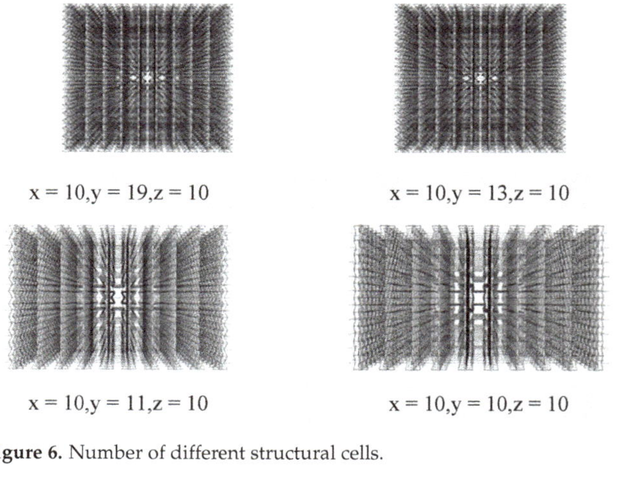

Figure 6. Number of different structural cells.

2.3. Calculation Critical Velocity

The impact velocity is a significant index affecting the dynamic response characteristics of materials. The dynamic response process is discussed under three conditions:

1. When the impact velocity is lower than the first critical impact velocity (i.e., the notch wave velocity), the whole specimen is slowly compressed, and the force is relatively uniform during the impact process. The material undergoes quasi-static deformation.
2. When the impact velocity exceeds the first critical impact velocity (i.e., notch wave velocity), the impact process transits from the overall deformation to the local deformation and the local deformation band is formed. With the increase of impact velocity, the local deformation of the upper end of the specimen is more obvious.
3. When the impact velocity is higher than the second critical impact velocity, the local deformation zone will propagate from the upper end to the lower end of the specimen in the mode of a shock wave.

The impact velocity of honeycomb material with local deformation zone during impact is called the first critical impact velocity, and its calculation formula is as follows:

$$v_{c1} = \int_0^{\varepsilon_1} \sqrt{\frac{\sigma(\varepsilon)}{\Delta\rho\rho_A}} d\varepsilon \quad (2)$$

In Formula (2), ε_1 is defined as the corresponding nominal strain (i.e., initial strain) when the stress reaches the stress peak for the first time in the process of impact fluctuation. $\sigma(\varepsilon)$ represents the elastic modulus of honeycomb material in the online elastic stage, and $\Delta\rho$ is the relative density of honeycomb material. ρ_A is the density of the honeycomb material.

The shock velocity when honeycomb material deformation is compressed with shock wave deformation characteristics is known as the second critical shock velocity. The calculation formula is as follows:

$$v_{c2} = \sqrt{\frac{2\sigma_p\varepsilon_3}{\Delta\rho\rho_A}} \quad (3)$$

σ_p is the plateau stress of honeycomb materials under quasi-static compression in Formula (3), and ε_3 is the locking strain, that is, the strain value at the beginning of the densification stage of honeycomb materials.

According to the above Formulas (2) and (3), when the cell parameters are as follows: t = 0.3 mm, d = 1 mm, α = 45° and L = 5 mm, the first critical impact velocity $V_{c1} \approx 11$ m/s and the second critical impact velocity $V_{r2} \approx 62$ m/s are calculated. This paper selects the impact velocity V_1 = 3 m/s ($V_1 < V_{cr1}$), V_2 = 20 m/s ($V_{cr1} < V_2 < V_{cr2}$) and V_3 = 200 m/s ($V_{cr2} < V_3$) to study the impact deformation in order to observe the influence of different impact velocities on the dynamic response of three-dimensional re-entrant honeycombs.

3. The Result of Simulation and Discussion

3.1. Deformation Mode

Under different impact velocities, the deformation of model is an important characteristic of the dynamic response of honeycomb. The reason for the deformation is that the wall of cell inside the specimen is rotation and buckling under external loads.

When the angle α is 45°, the deformation of the three-dimensional honeycomb under three different impact velocities of low speed (V_1 = 3 m/s), medium speed (V_2 = 20 m/s) and high speed (V_3 = 200 m/s) are shown in Figures 7–9, respectively. The nominal strain (ε) in the Figures is the ratio of the displacement of the specimen in *y*-axis direction to the initial height.

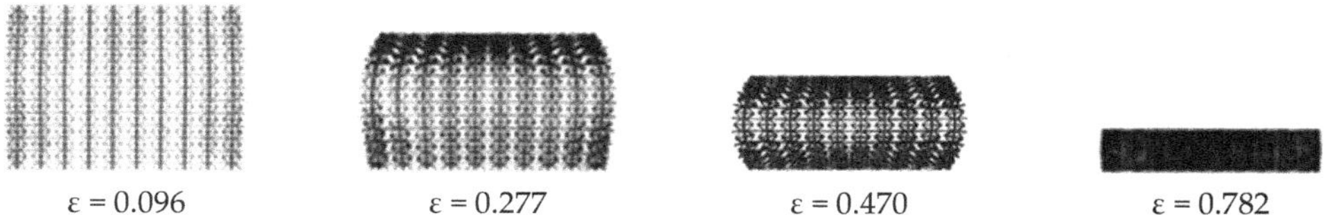

Figure 7. Deformation of specimen at low speed (3 m/s and α = 45°).

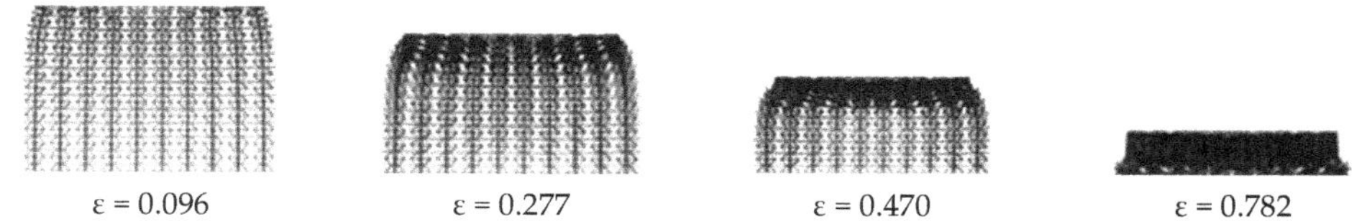

Figure 8. Deformation of specimen at medium speed (20 m/s and α = 45°).

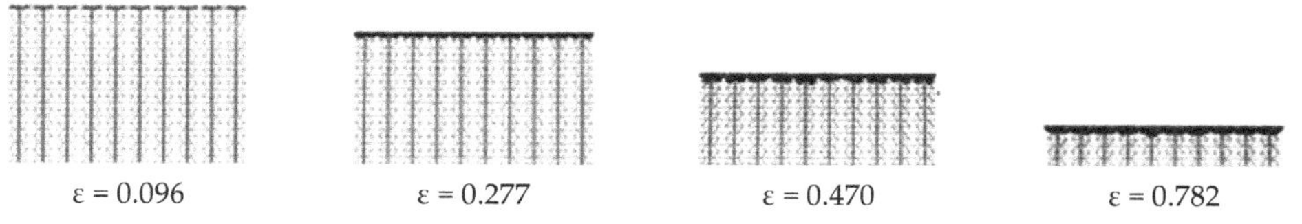

Figure 9. Deformation of specimen at high speed (200 m/s and α = 45°).

In the case of low velocity (V_1 = 3 m/s), the deformation process of three-dimensional honeycomb can be roughly divided into four stages. Phase I (ε = 0.096) is mainly the rotation of the inclined cell wall inside the three-dimensional honeycomb. The results show that the more the impact velocity is close to the velocity of quasi-static compression, the more uniform the stress is in the compression process. The specimen is uniformly deformed all the time in this stage and the upper and lower ends are close to the middle under the action of the x-axis force generated by the rotation of the cell wall, which shows that the specimen has a specific negative Poisson's ratio. The middle part of the specimen in the y-axis direction has little force and almost no deformation, so the middle part of the specimen is convex during compression. In Phase II (ε = 0.277), the deformation is mainly caused by the continuous rotation of the inclined cell wall in the upper structure of the specimen in order to withstand the compression deformation in the y-axis direction. Therefore, under the action of the cohesion in the x-axis direction, the upper end of the specimen has obvious concave phenomenon. When the upper end is compressed to a certain extent, it enters Phase III (ε = 0.470). The internal inclined cell wall of the upper end of the specimen will maintain a certain angle in the process. The pressure is transferred from the upper end of the specimen to the lower end, which leads to the rotation of the internal inclined cell wall of the lower end of the specimen. The cohesive force begins to contract to the middle, and the concave shape appears to bear the impact force transmitted from the upper end. When the upper and lower ends are basically symmetrical and the upper and lower ends of the specimen are concave and the middle part is convex, forming a 'barreling' state, this stage is completed. Then the deformation enters Phase IV (ε = 0.782), the specimen continues compressing in the y-axis direction. Because the upper and lower ends of the specimen have been compressed to a certain extent, the inclined cell wall of the middle part of the specimen in the y-axis direction will rotate. Under the action of transverse force in the x-axis direction, the inclined cell wall begins to converge to the middle vertical plane. When the inner cell wall of the structure basically parallel, the adjacent cell walls reach full contact density, and the compression is stopped.

In the case of medium velocity (V_2 = 20 m/s), the deformation can be divided into two stages. The first stage includes the deformations where the strain is between 0 to 0.47. When ε = 0.096, the impact energy cannot be transmitted to the lower end of the specimen, resulting in the compression of the upper end of the specimen. The inclined cell wall begins to rotate inside the specimen. Because the impact velocity is higher than the one at low speed, the horizontal cell wall will produce buckling and the external impact energy is absorbed. Due to the rotation and buckling of the inner cell wall of the specimen structure, the upper part of the specimen will produce an obvious concave deformation, where negative Poisson's ratio characteristics are shown. When ε = 0.277 and ε = 0.470, the deformation still presents in the first stage. The specimen is compressed layer by layer from top to bottom. In the second stage (ε = 0.782), the compression of the specimen is passed layer by layer to the lower of the specimen, and the internal cell walls begin to contact each other and produce dense compression. At this stage, the bottom layer of the specimen cannot bear the force in the x-axis direction due to the lessened friction force, so there is a slight rollover phenomenon.

At high speeds (V_3 = 200 m/s), the compression deformation can be seen as one stage. The inertia effect plays a leading role due to the fast impact speed. The inclined cell wall in the specimen structure cannot produce rotation and only buckling deformation occurs. The specimen is compressed layer by layer from upper to lower, until the compression reaches to the bottom and the inner cell wall of the structure is fully contacted and dense. Negative Poisson's ratio can be hardly shown in this process.

It can be concluded that when the speed is low, the whole specimen is more evenly deformed from top to bottom. Due to the effect of friction, when the speed is low, the specimen presents 'barreling'. With the speed increases, this phenomenon gradually disappears, inertial force plays a major role, the specimen deforms from the upper end and the bottom deformation becomes smaller.

The above research discusses the influence of different impact velocities on the deformation of specimen at the same cell angle. The deformations of specimen with different cell angles under the same impact velocity is discussed next. In the test, the cell wall length L = 5 mm is unchanged, the impact velocity is 3 m/s and the compression deformation is set as ε = 0.186. The deformation of specimens with different cell angles α = 30°, 45°, 60° and 70° are shown in Figure 10a–d, respectively.

(**a**) α = 30° (**b**) α = 45° (**c**) α = 60° (**d**) α = 75°

Figure 10. Deformation at different angles at speed of 3 m/s (ε = 0.186).

When the cell angle α = 30°, the impact energy is first transferred from the upper of the specimen to the lower. Shrinkage deformations in the vertical direction are produced in the middle part. The shrinkage deformation of the lower part of the specimen is greater than that of the upper part of the specimen. The deformation mode of this case is that the deformation is small near the upper end and the deformation is great next to the lower end.

When the cell angle is α = 45°, the impact energy is also transferred from the upper to the lower of the specimen. In the process of the compression deformation of the specimen, due to the rotation of the inclined cell wall inside the specimen, the transverse deformation in the x-axis direction is produced. Both ends of the specimen shrink and the middle position of the y-axis direction is relatively small, so it finally presents the 'barreling' mode.

When the cell angle is α = 60°, the impact energy is transferred from the upper to the lower parts of the specimen. In the process of compression deformation, the situation is basically the same as that of α = 45°, so it finally presents the 'barreling' state. Compared with α = 45°, the deformation at the lower part of the specimen is smaller, and the deformation is mainly concentrated in the upper part of the specimen.

When the cell angle $\alpha = 75°$, the internal cell wall of the specimen is sparser due to the larger angle of α. When the impact begins from the upper part of the specimen, it shows that the upper part of the specimen is prone to deformation, while the lower part does not produce deformation. A specific position at the upper part of the specimen shows the necking phenomenon, and the specimen has obvious negative Poisson's ratio characteristics.

By studying the impact deformation of different cell angles under the same impact velocity, it is found that when the cell angle $\alpha = 30°$, the deformation of the negative Poisson's ratio mainly occurs at the lower part of the specimen, and the deformation morphology is different from that with other angles where the lower shrinkage is larger than the upper shrinkage. When the cell angle is $\alpha = 45°$ and $\alpha = 60°$, the deformation modes are basically the same, showing the shape of 'barreling'. Besides the phenomenon of negative Poisson's ratio still exists at the upper and lower ends of the specimen. However, with the increase of the cell angle, the deformation at the lower part of the specimen gradually weakens and mainly concentrates on the upper part of the specimen. When the cell angle is $\alpha = 75°$, the upper end of the specimen displays an obvious 'necking' phenomenon, and the lower end almost has no deformation. It can be concluded that under the same impact velocity, with the increase of the cell angle, the deformation position gradually transits from the lower of the specimen to the upper end and shows different deformation modes.

When the impact velocity is low ($V_1 = 3$ m/s), the nominal stress–strain curve of the specimen is shown in Figure 11. When the three-dimensional honeycomb parameters are $\alpha = 45°$, L = 5 mm and t = 0.3 mm, the ε represents the nominal strain in the horizontal coordinate, that is, the ratio of the compression reaction of the upper rigid plate to the initial contact area of the specimen, and the σ represents the nominal stress in the vertical coordinate.

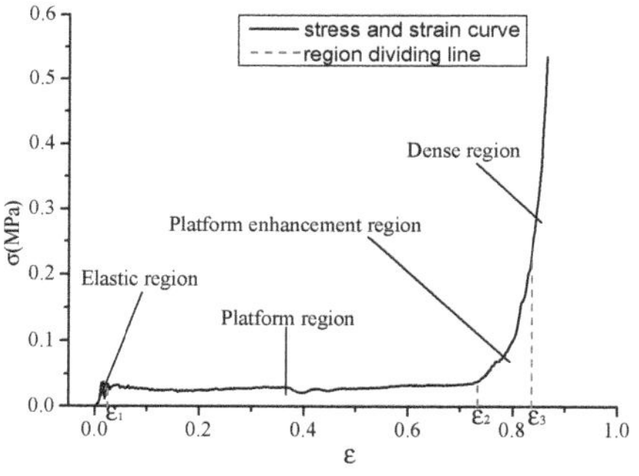

Figure 11. Nominal stress–strain curve of specimen under in-plane impact.

When honeycomb is subjected to in-plane compression, it is studied according to [25]. The compression process of traditional honeycomb is generally divided into three regions, as shown in Figure 11, which are the linear elastic region, platform region and dense region. However, compared with the traditional honeycomb, the compression process of the three-dimensional re-entrant honeycomb is divided into four regions, namely, the linear elastic region, platform region, platform enhancement region and dense region.

The linear elastic region is a process, where the compression stress of the upper rigid plate suddenly increases and reaches the initial stress peak in a very short time. After that ($\varepsilon = \varepsilon_1$), the stress begins to fluctuate and finally tends to be stable. In the platform region, the compressive stress of the specimen after reaching the initial strain ε_1 fluctuates around a certain value and remains relatively stable. The specimen undergoes great compressive deformation in this stage, so it is the main stage of energy absorption. After the end of the platform stage, it enters the platform enhancement region. With the continuous increase of the compressive strain of the specimen, the stress no longer remains relatively stable, but gradually increases with a specific slope and exceeds the platform stress value to a certain extent. After the strain at the end of the enhancement stage reaches ε_3, the cells in the specimen begin to contact with each other in dense region. At this region, the stress value of the specimen rises sharply in a small strain stage until the inner wall of the cells in the specimen is completely bonded together and the dense stage ends.

3.2. Platform Stress

When the stress remains in a relatively stable region from ε_1 to ε_2, the stress in this region is called the platform stress (σ_p). It is an important indicator for describing the dynamic response characteristics of the honeycomb and can be calculated by the following Formulas (4) and (5):

$$\sigma_P = \frac{\int_{\varepsilon_1}^{\varepsilon_3} \sigma(\varepsilon) d(\varepsilon)}{\varepsilon_1 - \varepsilon_3} \tag{4}$$

$$\sigma(\varepsilon) = \frac{F(\varepsilon)}{L_x \times L_z} \tag{5}$$

In Formula (4), ε_1 is the initial strain, that is, the corresponding strain value when the initial stress is just stable and reaches the platform stress, so the value of ε_1 is very small. In order to achieve a high calculation accuracy, the value of ε_1 in this paper is set as 0.013. ε_3 is dense strain, that is, the strain corresponding to the contact between adjacent cell walls within the specimen. In Formula (5), the value of $F(\varepsilon)$ is derived from the average value of the force of the upper rigid plate in the platform area obtained by the simulation. L_x is the length of the specimen in the x-axis direction, and L_z is the length of the specimen in the z-axis direction.

According to the one-dimensional shock wave theory [1,25], the formula of platform stress is obtained as follows:

$$\sigma_P = m\sigma_s \Delta\rho^2 + \frac{\Delta\rho\rho_s v^2}{1 - n\Delta\rho} \tag{6}$$

where σ_s represents the yield stress of the matrix material, $\Delta\rho$ represents the relative density of the designed honeycomb material, ρ_s represents the density of the matrix material and m and n are the coefficients to be calculated or fitted.

The stress over different velocities are calculated using the FEM method and listed in Table 2. According to Formula (6) and the data points in Table 2, three curves are fitted and plotted in Figure 12, and the formula of the three curves are obtained by linear regression, so as to solve the parameters m and n in the formula. By verifying the results, it has been found that the value of n is too large, mainly because the value of ρs in the formula leads to the inapplicability of the formula. It is found that the value of n conforms to the linear distribution by observation, so the calculation formula of platform stress is modified by fitting the value of n again. The modified formula is as follows:

$$\sigma_p = 2.188\sigma_s \Delta\rho^2 + \left(0.0023\Delta\rho + 6.948e^{-9}\right) v^2 \tag{7}$$

The comparison between the results of the three-dimensional honeycomb platform stress under different densities by FEM and the curves of the formula are shown in the Figure 12 as well. It can be seen that the fitting of the result is better, and the smaller the relative density is, the higher the fitting degree is. Therefore, the rationality of the correction formula is verified.

Table 2. Platform stress in different condition.

v/(m/s)	σp/(MPa)			
	Δρ = 0.037	Δρ = 0.02	Δρ = 0.012	Δρ = 0.008
3	0.037	0.026	0.023	0.019
7	0.038	0.031	0.028	0.022
20	0.081	0.053	0.039	0.035
35	0.167	0.109	0.058	0.055
70	0.623	0.303	0.192	0.171
100	1.450	0.615	0.378	0.314
200	3.778	2.158	1.403	1.079

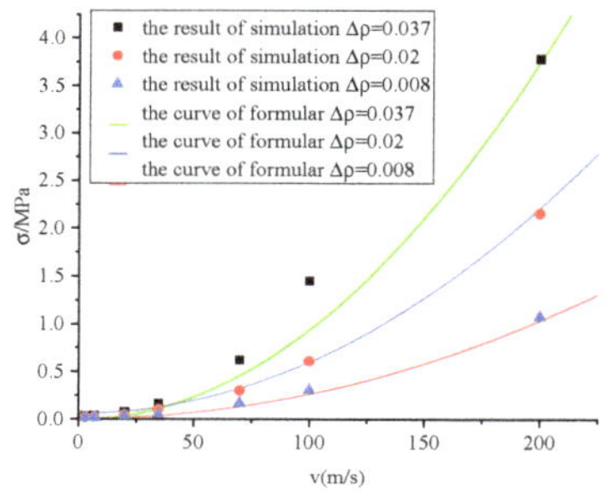

Figure 12. Comparison of accuracy under different densities.

3.3. Energy Absorption

The effects of the impact velocity and cell angle on the stress and strain of specimen during impact are studied below. Firstly, under the condition of the constant cell angle (relative density), the nominal stress and strain curves of three-dimensional re-entrant honeycomb are obtained by simulation. It can be concluded from Figure 13a that under this condition, the stress increases with the increase of velocity. In addition, the stress and strain of different cell angles (different densities) under the same impact velocity can be obtained in Figure 13b. It can be seen from Figure 13b, the stress decreases with the increase of the cell structure angle.

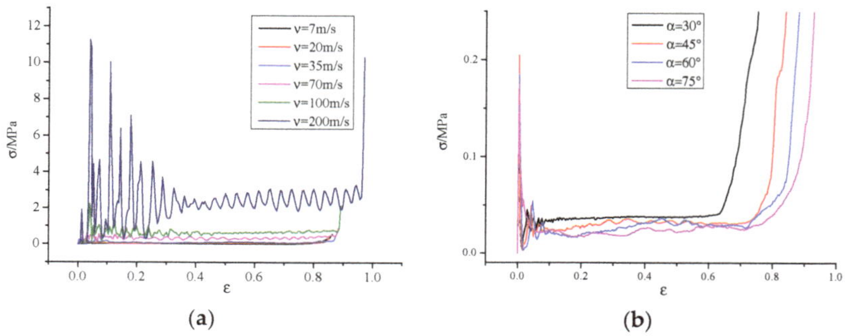

Figure 13. The relationship between nominal stress and strain of specimen. (**a**) The stress and strain of different velocities. (**b**) The stress and strain of different cell angles.

Energy follows the first principle of thermodynamics under external loads, which can be expressed by Formula (8). Ignoring the energy of friction loss and the energy of the damping dissipation of surrounding media in Formula (8), the external work is mainly converted into kinetic energy and the internal energy absorbed by the impact object, so the sum of the two is regarded as the total energy absorbed by the material.

$$E_w + E_{qb} = E_u + E_k + E_f \quad (8)$$

where E_u is the internal energy of the material, E_k is the kinetic energy of the material, E_f is the energy of the contact friction loss, E_w is the work done by the external load and E_{qb} is the energy dissipated by the surrounding medium damping.

The curve of total energy over strain is shown in Figure 14. Figure 14a shows that when the cell angle (relative density) unchanged, the ability to absorb energy during compression increases with the increase of velocity. In addition, when the velocity is constant (V_1 = 3 m/s), the ability to absorb energy increases with the increase of cell angle shown Figure 14b. Therefore, the energy absorption ability of three-dimensional honeycomb can be improved by changing the impact velocity and cell angle.

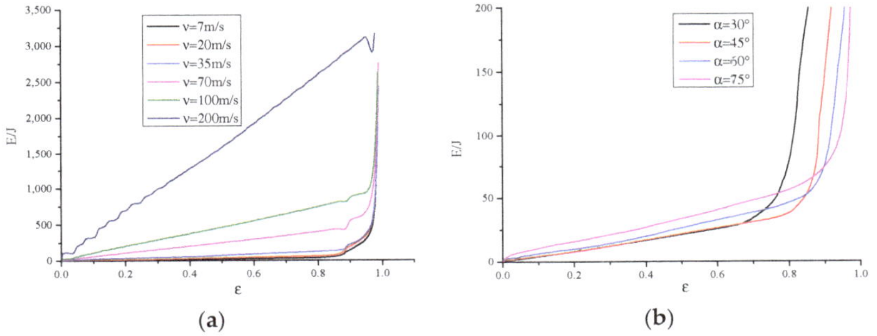

Figure 14. Relationship between energy absorption and strain of specimens. (**a**) E (total energy) with different velocities. (**b**) E (total energy) with different angles.

In order to investigate the energy absorption distribution of three-dimensional honeycomb structure under in-plane impact, the internal energy distribution coefficient Φ (the proportion of internal energy in total absorbed energy) is defined. The formula is as follows:

$$\Phi = \frac{Eu}{Ek + Eu} \quad (9)$$

The influence of impact velocity and cell angle on the internal energy distribution coefficient Φ during the impact of specimen is studied below. The variation of the internal energy distribution coefficient Φ with the nominal strain is shown in Figure 15. The condition that relative density (cell angle) of the honeycomb is constant and the impact velocity is different in Figure 15a. The result show that the impact velocity has a great influence on the internal energy distribution coefficient Φ. With the increase of the impact velocity, the internal energy distribution coefficient Φ decreases accordingly, and its value gradually decreases from 0.95 at a low speed impact (V = 7 m/s) to 0.45 at a high speed impact (V = 200 m/s). It can be concluded that when the impact velocity is lower than the second critical impact velocity, the honeycomb absorbs most of the internal energy. With the increase of impact velocity, the proportion of internal energy distribution decreases due to the increase of inertial effect. In addition, when the impact velocity (V = 3 m/s) is constant and the relative density (cell angle) changes, the internal energy distribution coefficient Φ also depends on the cell angle, as shown in Figure 15b. Under the same impact velocity, the internal energy distribution coefficient Φ increases slightly with the increase of cell angle. It can be concluded that the effect of impact velocity on the absorption of impact energy is greater than the cell angle.

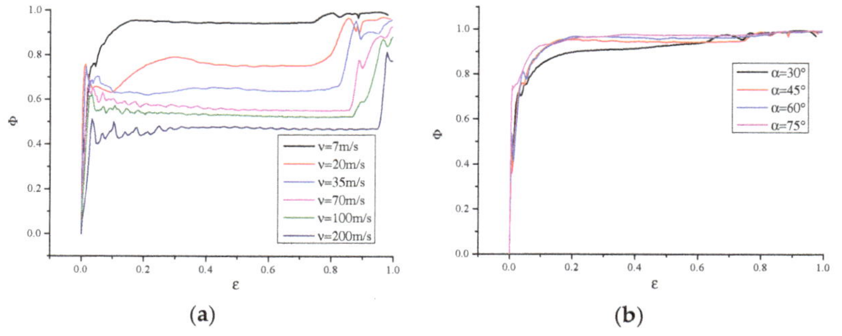

Figure 15. Relationship between energy distribution coefficient and strain. (**a**) Φ (internal energy distribution coefficient) with different velocities. (**b**) Φ (internal energy distribution coefficient) with different cell angles).

4. Conclusions

Based on the traditional re-entrant honeycomb, a 3D honeycomb through a spatial combination is designed in this paper. Then, the dynamic response characteristics of three-dimensional re-entrant honeycomb are numerically analyzed by the explicit dynamic finite element method, and the following conclusions are drawn.

Three-dimensional honeycomb with different cell angles exhibits different deformation modes at the same speed. When the cell angle is 30°, it presents the different compression deformation morphology where the shrinkage of the lower part is larger than that of the upper part. When the cell angles were 45° and 60°, the compressive deformation pattern of 'barreling' was presented. When the cell angle is 75°, the specimen shows the necking phenomenon, which conform to the negative Poisson's ratio material under axial compression. The rotation and bending deformation of the cell wall are the main reasons for the negative Poisson's ratio of the honeycomb. With the increase of impact velocity, the deformation localization is obvious, the inertial effect is gradually enhanced and the negative Poisson's ratio characteristic is weakened. In addition, the stress–strain curve of three-dimensional re-entrant honeycomb adds the platform enhancement region compared with the traditional honeycomb.

Based on the one-dimensional shock wave theory, the empirical formula of the platform stress of three-dimensional re-entrant honeycomb is given, which is proved to be in good agreement with the FEM calculation results. In addition, it can be seen from the fitting results that the smaller the relative density, the higher the fitting degree.

The impact velocity has a great influence on the internal energy distribution coefficient Φ. With the increase of the impact velocity, the internal energy distribution coefficient Φ accordingly decreases, and its value gradually decreases from 0.95 at low speed (V = 7 m/s) to 0.45 at high speed (V = 200 m/s). Therefore, it can be concluded that when the impact velocity is lower than the second critical impact velocity, the material in this paper mainly absorbs internal energy. With the increase of impact velocity, the proportion of internal energy distribution will decrease due to the increase of the inertial effect.

Author Contributions: This research was supervised by B.S. and T.H.; the investigation, software and writing were carried out by J.Z. All authors have read and agreed to the published version of the manuscript.

Funding: This work was supported by National Key R&D Program of China (2018YFC0810500).

Data Availability Statement: The raw/processed data required to reproduce these findings cannot be shared at this time as the data also forms part of an ongoing study.

Acknowledgments: The authors would like to thank the support from the National Key R&D Program of China (2018YFC0810500).

Conflicts of Interest: The authors declare no known competing financial interest or personal relationship that could have appeared to influence the work reported in this paper.

References

1. Gibson, L.J.; Ashby, M.F. *Cellular Solids: Structure and Properties*; Cambridge University Press: Cambridge, UK, 1997.
2. Zen, S.; Ava, K.; Gedikli, H.; Alver, M.; Aslan, M. Low-energy impact response of composite sandwich panels with thermoplastic honeycomb and re-entrant cores. *Thin-Walled Struct.* **2020**, *156*, 106989.
3. Hong, J.; Ma, F.; Xu, X.; Yang, J.; Zhang, H. A novel mechanical-electric-hydraulic power coupling electric vehicle considering different electrohydraulic distribution ratios. *Energy Convers. Manag.* **2021**, *249*, 114870. [CrossRef]
4. Qi, C.; Jiang, F.; Yang, S.; Remennikov, A. Multi-scale characterization of novel re-entrant circular auxetic honeycombs under quasi-static crushing. *Thin-Walled Struct.* **2021**, *169*, 108314. [CrossRef]
5. Wang, H.; Lu, Z.; Yang, Z.; Xiang, L. A novel re-entrant auxetic honeycomb with enhanced in-plane impact resistance. *Compos. Struct.* **2019**, *208*, 758–770. [CrossRef]
6. Evans, K.E.; Alderson, K.L. Auxetic materials: The positive side of being negative. *Eng. Sci. Educ. J.* **2000**, *9*, 148–154. [CrossRef]
7. Hong, J.; Wang, Z.; Ma, F.; Yang, J.; Xu, X.; Qu, C.; Zhang, J.; Shan, T.; Hou, Y.; Zhou, Y. Thermal Runaway Prognosis of Battery Systems Using the Modified Multi-Scale Entropy in Real-World Electric Vehicles. *IEEE Trans. Transp. Electrif.* **2021**, *7*, 2269–2278. [CrossRef]
8. Tan, H.L.; He, Z.C.; Li, K.X.; Li, E.; Cheng, A.G.; Xu, B. In-plane crashworthiness of re-entrant hierarchical honeycombs with negative Poisson's ratio. *Compos. Struct.* **2019**, *229*, 111415. [CrossRef]
9. Xiao, D.; Dong, Z.; Li, Y.; Wu, W.; Fang, D. Compression behavior of the graded metallic auxetic re-entrant honeycomb: Experiment and finite element analysis. *Mater. Sci. Eng. A* **2019**, *758*, 163–171. [CrossRef]
10. Djemaoune, Y.; Krstic, B.; Rasic, S.; Radulovic, D.; Dodic, M. Numerical Investigation into In-Plane Crushing of Tube-Reinforced Damaged 5052 Aerospace Grade Aluminum Alloy Honeycomb Panels. *Materials* **2021**, *14*, 4992. [CrossRef]
11. Wang, Y.; Yu, Y.; Wang, C.; Zhou, G.; Karamoozian, A.; Zhao, W. On the out-of-plane ballistic performances of hexagonal, re-entrant, square, triangular and circular honeycomb panels. *Int. J. Mech. Sci.* **2020**, *173*, 105402. [CrossRef]
12. Shen, L.; Wang, Z.; Wang, X.; Wei, K. Negative Poisson's ratio and effective Young's modulus of a vertex-based hierarchical re-entrant honeycomb structure. *Int. J. Mech. Sci.* **2021**, *206*, 106611. [CrossRef]
13. Fu, M.H.; Chen, Y.; Hu, L.L. A novel auxetic honeycomb with enhanced in-plane stiffness and buckling strength. *Compos. Struct.* **2017**, *160*, 574–585. [CrossRef]
14. Yu, R.; Luo, W.; Yuan, H.; Liu, J.; He, W.; Yu, Z. Experimental and numerical research on foam filled re-entrant cellular structure with negative Poisson's ratio—ScienceDirect. *Thin-Walled Struct.* **2020**, *153*, 106679. [CrossRef]
15. Mukhopadhyay, T.; Adhikari, S. Effective in-plane elastic properties of auxetic honeycombs with spatial irregularity. *Mech. Mater.* **2016**, *95*, 204–222. [CrossRef]
16. Günaydin, K.; Eren, Z.; Kazanci, Z.; Scarpa, F.; Grande, A.M.; Türkmen, H.S. In-plane compression behavior of anti-tetrachiral and re-entrant lattices. *Smart Mater. Struct.* **2019**, *28*, 115028. [CrossRef]
17. Li, D.; Yin, J.; Dong, L.; Lakes, R.S. Strong re-entrant cellular structures with negative Poisson's ratio. *J. Mater. Sci.* **2018**, *53*, 3493–3499. [CrossRef]
18. Lu, Z.X.; Li, X.; Yang, Z.Y.; Xie, F. Novel structure with negative Poisson's ratio and enhanced Young's modulus—ScienceDirect. *Compos. Struct.* **2016**, *138*, 243–252. [CrossRef]
19. Ingrole, A.; Hao, A.; Liang, R. Design and modeling of auxetic and hybrid honeycomb structures for in-plane property enhancement. *Mater. Des.* **2017**, *117*, 72–83. [CrossRef]
20. Biswas, A.N.; Mahesh, N.; Sreekanth, P. Aluminium 6061 & HDPE incorporated study on gradient honeycomb-hybrid auxetic fused structure for energy absorption. *Mater. Today Proc.* **2022**, *56*, 1035–1042.
21. Chen, S.; Tan, X.; Hu, J.; Zhu, S.; Wang, B.; Wang, L.; Jin, Y.; Wu, L. A novel gradient negative stiffness honeycomb for recoverable energy absorption. *Compos. Part B Eng.* **2021**, *215*, 108745. [CrossRef]
22. Wei, Y.; Zhang, Y.; Song, Q.; Zhou, X.; Zhou, Y.; Shen, Y. Effects of different conFigureurations and gradients on compression responses of gradient honeycombs via selective laser melting—ScienceDirect. *Thin-Walled Struct.* **2022**, *170*, 108462. [CrossRef]
23. Zhang, X.C.; Zhu, X.Y.; Li, N. A study of the dynamic response characteristics of hexagonal chiral honeycombs. *J. Vib. Shock* **2016**, *35*, 1–7.
24. Xiao, D.; Kang, X.; Li, Y.; Wu, W.; Fang, D. Insight into the negative Poisson's ratio effect of metallic auxetic reentrant honeycomb under dynamic compression. *Mater. Sci. Eng. A* **2019**, *763*, 138151. [CrossRef]
25. Yan, X.; Wang, R.; Chen, Y.; Cheng, Y.; Liang, X. Energy absorption characteristics of honeycomb materials with negative poisson's ratio*. *IOP Conf. Ser. Mater. Sci. Eng.* **2021**, *1040*, 012002. [CrossRef]

electronics

MDPI

Article

An Intelligent Breast Ultrasound System for Diagnosis and 3D Visualization

Yuanyuan Lu [1,2,†], Yunqing Chen [3,†], Cheng Chen [3], Junlai Li [1,*], Kunlun He [2,*] and Ruoxiu Xiao [3,4,*]

1 Department of Ultrasound, The Second Medical Center, Chinese PLA General Hospital, Beijing 100853, China; lff791207@163.com
2 Medical Big Data Research Center, Chinese PLA General Hospital, Beijing 100853, China
3 School of Computer and Communication Engineering, University of Science and Technology Beijing, Beijing 100083, China; serein7z@163.com (Y.C.); b20170310@xs.ustb.edu.cn (C.C.)
4 Shunde Graduate School, University of Science and Technology Beijing, Foshan 100024, China
* Correspondence: li_jl@yeah.net (J.L.); kunlunhe@plagh.org (K.H.); xiaoruoxiu@ustb.edu.cn (R.X.)
† These authors contributed equally to this work.

Abstract: Background: Ultrasonography is the main examination method for breast diseases. Ultrasound imaging is currently relied upon by doctors to form statements of characteristics and locations of lesions, which severely limits the completeness and effectiveness of ultrasound image information. Moreover, analyzing ultrasonography requires experienced ultrasound doctors, which are not common in hospitals. Thus, this work proposes a 3D-based breast ultrasound system, which can automatically diagnose ultrasound images of the breasts and generate a representative 3D breast lesion model through typical ultrasonography. Methods: In this system, we use a weighted ensemble method to combine three different neural networks and explore different combinations of the neural networks. On this basis, a breast locator was designed to measure and transform the spatial position of lesions. The breast ultrasound software generates a 3D visualization report through the selection and geometric transformation of the nodular model. Results: The ensemble neural network improved in all metrics compared with the classical neural network (DenseNet, AlexNet, GoogLeNet, etc.). It proved that the ensemble neural network proposed in this work can be used for intelligent diagnosis of breast ultrasound images. For 3D visualization, magnetic resonance imaging (MRI) scans were performed to achieve their 3D reconstructions. By comparing two types of visualized results (MRI and our 3D model), we determined that models generated by the 3D-based breast ultrasound system have similar nodule characteristics and spatial relationships with the MRI. Conclusions: In summary, this system implements automatic diagnosis of ultrasound images and presents lesions through 3D models, which can obtain complete and accurate ultrasound image information. Thus, it has clinical potential.

Keywords: ultrasonography; breast lesions; neural network; system design; diagnosis; visualization

Citation: Lu, Y.; Chen, Y.; Chen, C.; Li, J.; He, K.; Xiao, R. An Intelligent Breast Ultrasound System for Diagnosis and 3D Visualization. *Electronics* **2022**, *11*, 2116. https://doi.org/10.3390/electronics11142116

Academic Editor: Domenico Mazzeo

Received: 10 June 2022
Accepted: 27 June 2022
Published: 6 July 2022

1. Introduction

As female sex organs, breasts have both internal function and external beauty, which are pivotal to women [1–5]. Usually, breasts are continuously undergoing developmental changes, and the structure of their glands is complex. Thus, treatment of breast nodules is always confusing for doctors [6–8]. At present, doctors mainly rely on the observation of breast images to find the developmental law of lesions. This can not only correctly recognize diseases and treat them reasonably, but also improve the understanding of breast nodule growth.

Ultrasonography has the characteristics of a real-time observation, low cost and high diagnostic level, and it has become the most important method of breast examination [9–13]. Hospitals generate many ultrasound images of the breast every day, which requires experienced ultrasound doctors to analyze the ultrasound images. However, most breast ultrasound images are tumor-free, so ultrasound doctors spend a lot of time every day analyzing tumor-free ultrasound images, which is a waste of ultrasound doctor resources. At the same time, the ultrasound image results of tumor patients mainly depend on the doctor's statement of the image characteristics, which depends largely on the doctor's professional level and expression ability. In addition, the subjective will of the patient is often the main factor in determining the treatment plan. Therefore, how to diagnose ultrasound images efficiently and quickly, and how to more accurately express the lesion structure, are urgent problems to be solved. It is of great scientific significance for patients to understand the results of breast ultrasound and explore the potential for breast disease [14] more intuitively.

Researchers have proposed several medical image nodule diagnosis algorithms based on convolutional neural networks, such as DenseNet [15], GoogLeNet [16], and ResNet [17] used in this paper. Due to the low resolution of the breast ultrasound image and the diversity of shape and texture features of lesions, a single network model is not generally effective for benign and malignant diagnosis of lesions, and has the disadvantage of poor generalization.

In addition, 3D models of breasts obtained by reconstruction allow patients to have more intuitive understanding and communication with doctors. The current methods for obtaining 3D breast structure include computed tomography (CT) and MRI imaging. Due to space limitations of the scanning machine, patients can only adopt a lying-down position when acquiring CT and MRI, which is different from the reclining posture on the operating table, resulting in greater deformation of the breast. Therefore, reconstructed 3D models cannot correctly realize surgical navigation. Although there are already some methods that enable doctors to collect 3D ultrasound data of patients, they require cumbersome positioning and take a long time [18].

In this work, we proposed a breast ultrasound system, which can simultaneously diagnose and visualize breast ultrasound nodules. The remainder of this paper is organized as follows: In Section 2, we introduce the method of neural network ensemble and 3D breast model generation. In Section 3, the experimental results and discussion of the ensemble network and 3D breast nodule model are presented. Section 4 provides a conclusion about the investigated methods, challenges, and future directions for the employment of the system in clinical applications.

2. Methods

2.1. 3D-Based Breast Ultrasound System

The breast ultrasound system consists of two parts, as shown in Figure 1. The first part is the intelligent diagnosis of breast ultrasound image by ensemble neural network. The ensemble method of neural network is shown in Figure 1a. The second part is the 3D visualization of breast locator, which is designed based on the structural characteristics of female breasts to realize the positioning of female breast nodules. It includes ultrasound scanning equipment, workstation (Figure 1b), and locator (Figures 1c and 2). GE LOGIQ E9 Ultrasound Machine is adopted as ultrasound scanning equipment provided by Ultrasound Supply. The workstation environment is Microsoft Workstation, Python 3.6. Operating software is 3D-based Breast Ultrasound Software, which is mainly developed based on The Visualization Toolkit (VTK) official application programming interface (version 7.1.1). Locator is designed based on the structural characteristics of female breasts to realize the positioning of female breast nodules.

The flowchart of breast ultrasound system proposed in this work is shown in Figure 3, which is divided into intelligent diagnosis and 3D visualization. In the intelligent diagnosis part, the breast ultrasound images are input into a pre-trained network, and the prediction results of the ensemble neural network are obtained by weighted ensemble, to realize the automatic benign and malignant diagnosis of breast ultrasound images and save medical resources. In the 3D visualization part, the ultrasound doctor firstly analyzes the ultrasound image and obtains the spatial position and geometric characteristics of breast nodules. After the analysis, the corresponding parameters are input into the 3D visualization software, and the software automatically generates the 3D breast nodule model.

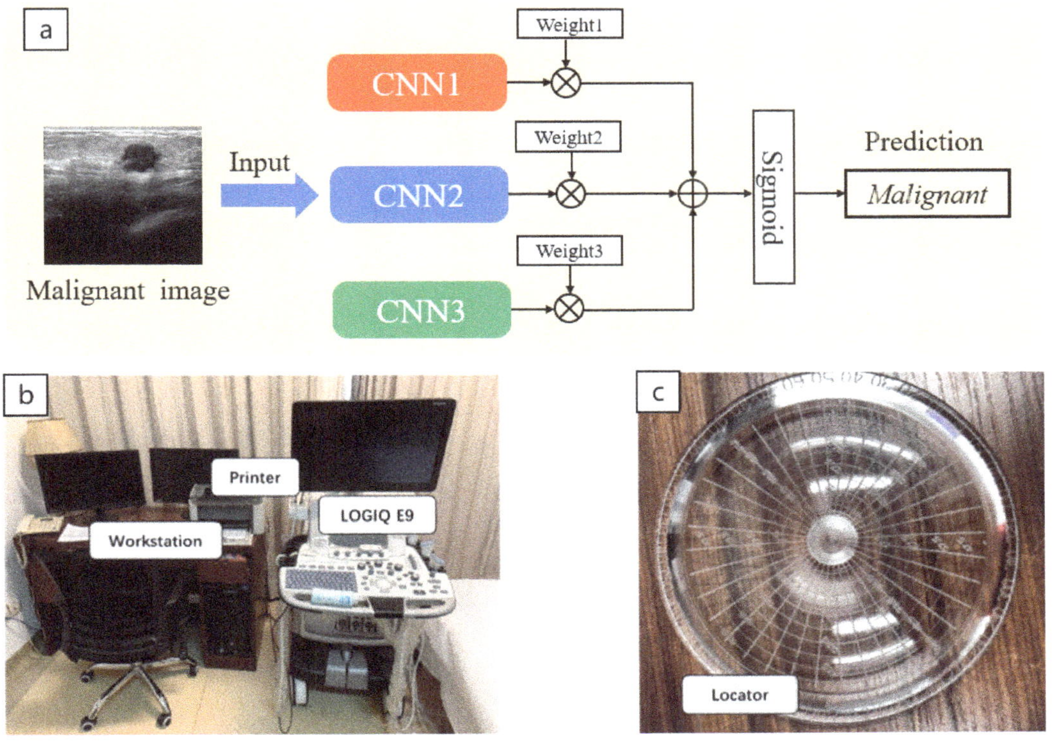

Figure 1. 3D-based breast ultrasound system. (**a**) Weighted ensemble network architecture. (**b**) Ultrasound scanning equipment workstation. (**c**) Locator proposed in this work.

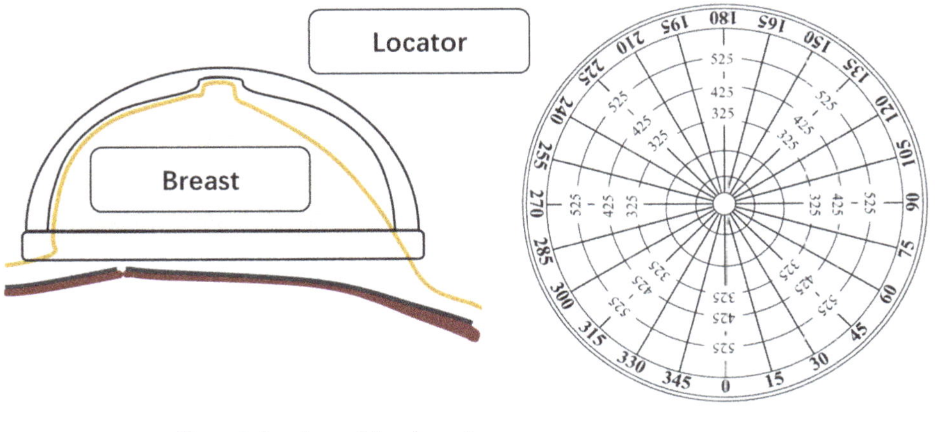

Figure 2. Locator and its schematic.

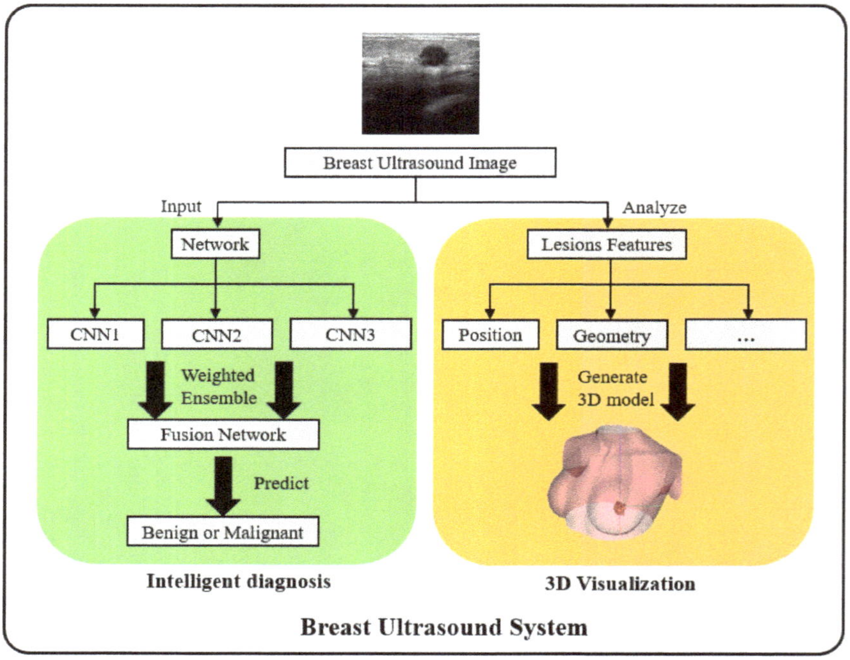

Figure 3. Flowchart showing the breast ultrasound system for intelligent diagnosis and 3D visualization.

2.2. Ensemble Neural Network

Neural networks usually require a large amount of data to train the network. However, the amount of ultrasound image data in the breast is generally low. To achieve good performance in a small dataset, we select three networks from widely recognized network: DenseNet [15], GoogLeNet [16], ResNet [17], VGG16 [19], ResNeXt [20], AlexNet [21], MobileNet [22], and NasNet [23]. Then, we combine them via the weighted ensemble method. In our work, we use a public breast ultrasound image dataset [24] (647 in total) to train these three neural networks and test the performance of a single neural network and ensemble neural network.

The dataset is divided into 80% training set and 20% testing set. After training the three neural networks, the trained model is obtained. To take advantage of different models, we use a weighted ensemble strategy (Figure 4) to combine their results. It is assumed that the benign probability values of the neural network outputs are P_1, P_2 and P_3, respectively. By multiplying these three results by their respective weights (W_1, W_2 and W_3) and summing them up, the sum result is then mapped to the range 0–1 by Sigmoid function to make sure the ensemble result is a probability distribution. The benign and malignant probability value of the ensemble neural network is shown in the following formula:

$$P_{benign} = \text{Sigmoid}(W_1 \times P_1 + W_2 \times P_2 + W_3 \times P_3) \tag{1}$$

$$P_{malignant} = 1 - P_{benign} \tag{2}$$

In our work, $W_1 = W_2 = W_3 = 0.33$. The parameters above are evaluated by experienced breast ultrasound doctors according to the performance of a single neural network.

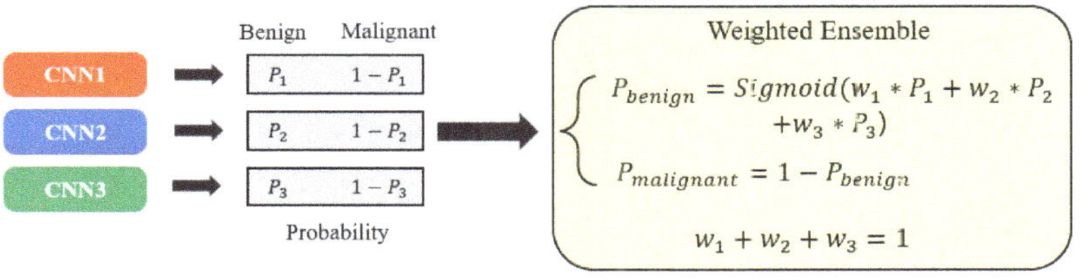

Figure 4. Weighted ensemble strategy we proposed in this work. Choosing three pre-trained models, the ensemble result is obtained via the strategy.

2.3. Locator

Locator is mainly worn on the patient's breast. Its schematic is displayed in Figure 2. It is designed as a spherical structure that matches the shape of a human breast. The center of the inner side of the locator is equipped with a round hole-shaped groove that matches the position of the nipple, and the outer side is marked with a plurality of first- and second-measurement lines for measuring the breast of the wearer. The first-measurement lines with 15 degrees unit distance are used to obtain the angle of the breast, and the second-measurement lines with 0.1 mm unit distance measure the length of the breast. Each first-measurement line, respectively, extends from the center to the edge of the spherical structure. They are equal in length, and each is marked with an angle. Each second-measurement line is an annular line with the center of the spherical structure as the circle center. Second-measurement lines are parallel to each other, and each one is marked with a distance. Thus, locator can realize the structured measurement of patient's breast, including measuring the height of the breast axis and the diameter of the breast base.

Conversion method of spatial information is as follows: Firstly, a straight line to connect origin and nodule center is made. Arc length from intersection of the straight line and skin to center of the nipple is defined as l, measured by locator's second-measurement lines. Angle of nodule center on horizontal plane is θ, measured by the first-measurement lines of locator. It ranges from 0 to 360 degrees, increasing in counterclockwise direction. In addition, vertical distance between nodule center and skin is defined as d, which can be measured by an ultrasound scanner. Then, sphere center of locator is taken as origin, and surface of locator is sphere. Thus, a coordinate system is established and its z-axis is from the center of the sphere to the nipple. Horizontal direction from right to left side of the chest is uniformly defined as x-axis, which satisfies the left-hand coordinate system. $\theta = 0$ is satisfied on x-axis. Therefore, the ball radius is radius of curvature of nipple horizontal section.

Assuming that breast height is H and width is $2W$, $(R-H)^2 + W^2 = R^2$ is satisfied and $R = \frac{H^2+W^2}{2H}$ is obtained (Figure 5a). H and W can be measured by doctors.

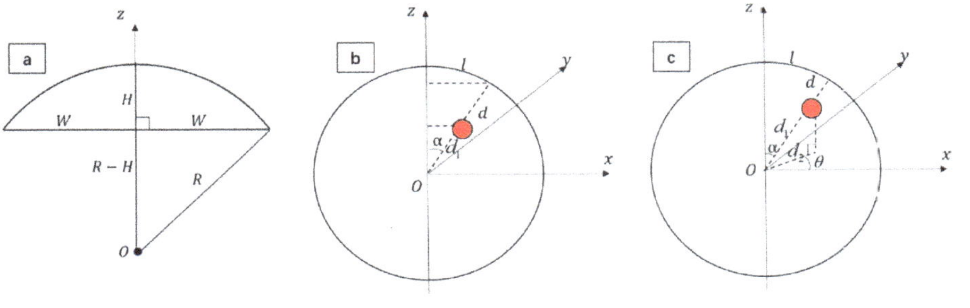

Figure 5. Spatial transformation of the locator. (**a**) Calculate the radius of the breast sphere. (**b**) Obtain the nodule z-axis coordinates (**c**) Obtain the nodule x-axis and y-axis coordinates.

To obtain z-axis coordinates of the nodule position, α and d_1 are defined as shown in Figure 5b, where red area represents the nodule, which satisfies $\alpha = \frac{l}{R}$. Because of $d_1 = R - d$, the nodule z coordinate can be expressed as:

$$\begin{aligned} z &= d_1 \times cos\alpha \\ &= (R-d) \times cos\left(\frac{l}{R}\right) \end{aligned} \tag{3}$$

Calculation of the x-axis and y-axis coordinates is as follows: projection length on the XOY plane of the line from the nipple center to the origin is defined as d_2 (Figure 5c), which satisfies:

$$\begin{aligned} d_2 &= d_1 \times cos(90^\circ - \alpha) \\ &= (R-d) \times sin\alpha \\ &= (R-d) \times sin\left(\frac{l}{R}\right) \end{aligned} \tag{4}$$

Therefore, x and y coordinates of the nodule are:

$$\begin{cases} x = d_2 cos\theta = (R-d) sin\left(\frac{l}{R}\right) cos\theta \\ y = d_2 sin\theta = (R-d) sin\left(\frac{l}{R}\right) sin\theta \end{cases} \tag{5}$$

Since $z > 0$ and the range of θ are 0–360 degrees, positive and negative changes of x and y coordinates are consistent with Equation (3). Therefore, (x, y, z) of the breast nodule in the 3D coordinate system can be converted through actual measurement parameters.

2.4. Model Preparation and Report Generation

Preparation of the models is mainly based on signs of breast ultrasonography. Referring to ultrasound images, four basic nodule models of smooth, burr, horned, and lobed cases are established. Dealing with nodules of different sizes, horizontal and vertical length, and height of nodules measured by ultrasound scanner correspond to the size in x-, y-, and z-axis. Thus, through size transformation of models, they are contracted or expanded. For a smoothing effect, noise and subdivision are performed on the basic geometric model. Some textures realize calcification (microcalcification, arc-shaped calcification, etc.) and blood flow (poor blood supply, rich blood supply) through basic geometric combinations. In addition, the setting of outer cover establishes a fuzzy high-echo halo feature. Some models need to be provided with materials. Cinema 4D model materials are applied to control color and transparency. According to input information, size transformation and morphological combination of models are carried out. Finally, three typical body types are established.

Generating module is composed of three parts, including 3D space positioning information of lesions, internal information of 3D lesions, and external information of 3D lesions. The 3D space positioning information marks the center position of the nodules, internal information shows textures and characteristics of nodules, and external information expresses the relative position of nodules in the breast. Thus, a 3D ultrasound diagnosis report is generated, which has the patients' name and number, 3D models of the nodules, and diagnosis information. The 3D model is adjusted to appropriate angle, viewed through manual interaction of visualization module, and obtained through real-time screenshots. Finally, it is output according to the current report format commonly applied in hospitals and printed by the writing function.

3. Experiment Results and Discussion

3.1. Ensemble Network

To evaluate the performance of the network, we use the accuracy, recall, F1 score and the AUC (area under curve) of ROC (receiver operating characteristic curve) metrics [25,26]. Results of different neural networks and ensemble networks are shown in Table 1. We first train all individual neural networks and calculate the metrics, then select the model with higher F1 score and combine the models through weighted ensemble strategy. Finally, we determine four groups of neural network combinations. They are (NasNet, AlexNet, DenseNet), (NasNet, GoogLeNet, AlexNet), (VGG16, GoogLeNet, NasNet), (VGG16, NasNet, DenseNet). By using the weighted ensemble strategy, the ensemble neural network is obviously improved compared with that of a single network. The performance and metrics of these networks have listed in the Table 1, Figures 6 and 7.

Table 1. Results of different neural and ensemble networks on a testing set.

Neural Network	Accuracy	Recall	F1 Score	AUC
DenseNet	0.7752	0.5952	0.6329	0.8328
GoogLeNet	0.8294	0.7381	0.7381	0.8760
AlexNet	0.7984	0.7143	0.6977	0.8410
ResNet	0.7597	0.5238	0.5867	0.8240
MobileNet	0.7287	0.4524	0.5205	0.7466
NasNet	0.8062	0.6190	0.6753	0.8481
ResNeXt	0.7829	0.5952	0.6410	0.8106
VGG16	0.8372	0.6905	0.7342	0.8547
(NasNet, AlexNet, DenseNet)	0.8295	0.6667	0.7179	0.8719
(NasNet, GoogLeNet, AlexNet)	0.8450	0.7381	0.7561	0.8790
(VGG16, GoogLeNet, NasNet)	0.8527	0.6905	0.7532	**0.8831**
(VGG16, NasNet, DenseNet)	**0.8992**	**0.7619**	**0.8312**	0.8711

The bold numbers indicate that the network performs best on this metric.

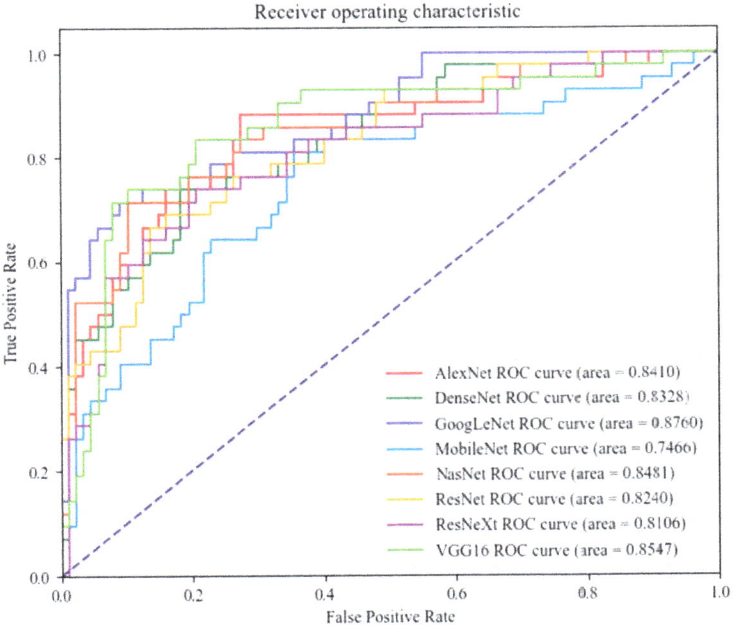

Figure 6. ROC of classical neural networks on a testing dataset.

Among combination groups, the (VGG16, NasNet, DenseNet) group achieved the best result, with acc of 0.8992 and F1 score of 0.8312. Compared with the best single neural network, VGG16, accuracy increased by 6.2%, and F1 score increased by 9.7%. All of the above indicate that the ensemble network obtained by the weighted ensemble method has better performance and is more suitable for benign and malignant diagnosis of breast ultrasound images.

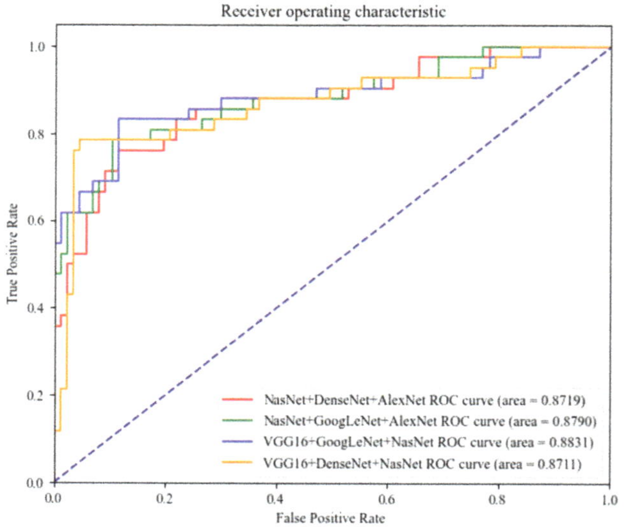

Figure 7. ROC of different ensemble neural networks on a testing dataset.

3.2. Breast Lesion 3D Visualization

Two senior sonographers (both with over 25 years of experience) conducted experiments on three subjects. First, lesion space information of three patients was obtained through the locator. Then, 3D-based Breast Ultrasound Software realized the generation of models and reports. Information representation of the three patients is shown in Table 2. In addition, three subjects underwent MRI scans at the same time, and the acquired MRI volume data were analyzed by the operating imaging physician in Mimics Research 19.0 to complete segmentation of the skin layer, lesion, and pectoralis major in sequence.

Table 2. Characterization information of three ultrasound volume data.

Data	Characterization Information Description
Subject 1	Medium breast shape; upper right half, 5 × 15 × 23 mm, smooth, lengthwise, clear and complete, and position (51 mm, 80°, 10 mm). Upper left half, 46 × 35 × 63 mm, burr, horizontal length, rich blood supply, position (0 mm, 0°, 30 mm).
Subject 2	Medium breast shape, upper right side, 42 × 31 × 44 mm, burr, horizontal and long, clear and complete border, lack of blood supply, position (0 mm, 0°, 27 mm). The upper left half, 50 × 58 × 60 mm, burr, rich blood supply, clear and complete border, position (0 mm, 0°, 35 mm).
Subject 3	Medium breast shape, upper right side, 41 × 30 × 39 mm, burr, lengthwise, clear and complete, rich blood supply, location (26 mm, 340°, 20 mm).

To unify the structure, all models used green nodules. Pectoralis major muscle tissues were shown as dark red with a certain degree of transparency, and white with high transparency were used as skin tissue. In the ultrasound reports of the first group, the most typical patient feature was that the size of the left thoracic nodule was 46 × 35 × 63 mm (Figure 8a), which was about 58.8 times larger than that of the right thoracic nodule (Figure 8b). Therefore, there were obvious differences in individual size between the two lesions, which are clearly shown in the green nodules in Figure 8c,d. Using the 3D-based breast ultrasound system, a basic burr nodule model was established on the left side of the chest, and the size was transformed according to the horizontal length and size. Basic geometry was added to show the state of rich blood supply (Figure 8e). On the right side of the chest, a smooth basic nodule model was established, and the size was transformed according to length and size (Figure 8f). Finally, all positionings were completed by the spatial information. Comparing the reconstruction results from MRI and models generated by the 3D-based breast ultrasound system, shape, size, and spatial location basically matched, meeting the clinical requirements of the doctor–patient information guidance.

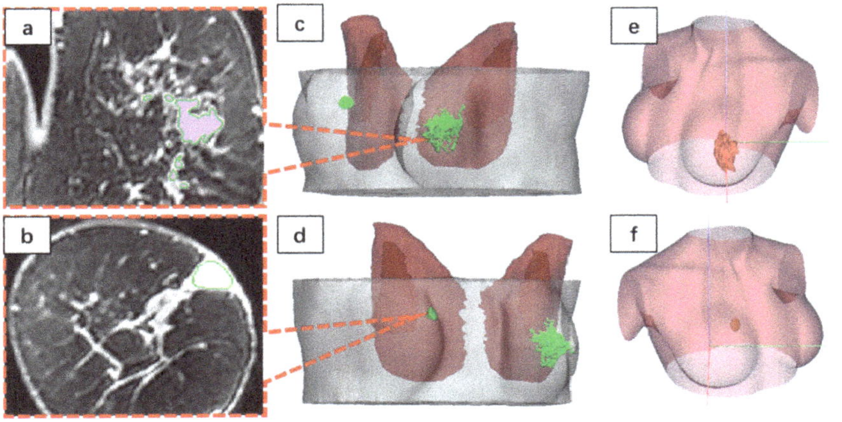

Figure 8. Result of Subject 1. (**a**,**b**) Representation of the MRI imaging parts of the left and right breast of Subject 1. (**c**,**d**) Reconstruction structure based on the MRI. (**e**,**f**) Generative models of 3D-based breast ultrasound system.

Figure 9a,b reveal MRI imaging parts of the second patient, which are different from the first patient. Through a 3D model reconstructed by the MRI, it can be observed that nodules on the left and right sides of the chest are large and have obvious burr characteristics (Figure 9c,d). In the 3D-based breast ultrasound system, the left and right sides of the chest were both imported with the burr nodule model, and transformation was computed from the actual size. A basic geometric model to the left thoracic nodule was added to simulate the state of a rich blood supply (Figure 9e). In the end, the generated model basically matched the size and the spatial location of the reconstructed model's structural features (Figure 9e,f).

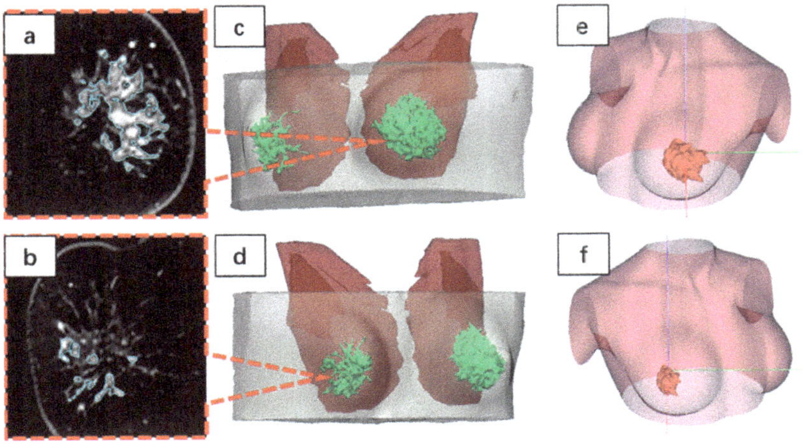

Figure 9. Result of Subject 2. (**a**,**b**) MRI imaging parts of the left and right breast of Subject 2. (**c**,**d**) Reconstruction structure based on the MRI. (**e**,**f**) Generative models of the 3D-based breast ultrasound system.

Figure 10a introduces parts of the MRI imaging of the third group of cases. Based on the reconstruction structure, only the right side of the chest contained medium-sized nodules, showing burrs and a rich blood supply (Figure 10b). The 3D-based breast ultrasound system imported the burr nodule model and added basic geometric structure as blood structure. Finally, it was stretched according to scale. The generated model was basically the same as the reconstructed result (Figure 10c). It is worth mentioning that muscle fiber length characteristics of the third group of cases are obvious. However, the model generated in the 3D-based breast ultrasound system cannot reflect this feature. This is because the 3D-based breast ultrasound system only contains a small, medium, and large chest that can be contracted in proportion, so it cannot reflect the patient's weight, chest width, and chest length. This will bring an error to the relative position of nodules and thoraxes.

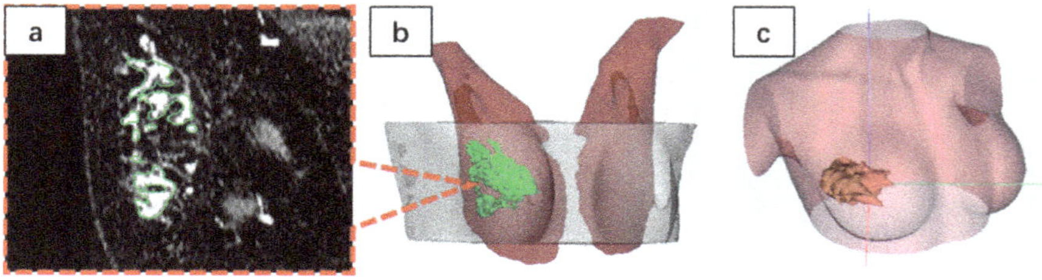

Figure 10. Result of Subject 3. (**a**) Parts of the MRI imaging of the right breast of Subject 3. (**b**) Reconstruction structure based on the MRI. (**c**) Generative model of the 3D-based breast ultrasound system.

3.3. Discussion

In breast nodule diagnosis, compared with the neural network models used in this paper (DenseNet, GooGleNet and AlexNet), the model generated by the weighted ensemble method improved in accuracy, recall and F1 score metrics to varying degrees. Among the experiments, the combination of VGG16, NasNet and DenseNet, with accuracy of 0.8992, recall of 0.7619 and F1 score of 0.8312, obtained the best score and verified that the ensemble neural network generated by the weighted ensemble method can effectively improve the performance of the neural network and better diagnose breast ultrasound images.

In breast nodule visualization, the size and location of nodules generated by the 3D breast ultrasound system basically matched the reconstructed structure. This is because the ultrasound probe and locator can accurately measure the size and position of the nodule. The error mainly comes from three aspects: (1) the measurement error of the ultrasonic breast nodule size may be due to human error; (2) when locating breast nodules on ultrasound, the selection of reference points (usually the area near the center of the nodules) may lead to localization errors; (3) structural errors, including calcification, blood supply and hyperechoic halo, may occur when the simulated edge shapes of basic nodules are different. It is worth mentioning that the 3D visualization of the breast ultrasound system aims to provide instructive 3D ultrasound reports to help doctors construct surgical navigation and facilitate communication with patients. Therefore, the 3D visualization of the breast ultrasound system does not need to establish a high-precision model system. However, how to more accurately simulate the patient's body shape, simulate more accurate characterization methods, and enrich the basic model of nodules are still problems worth exploring. This not only deals with common nodule representations, but also generates a guiding model structure for complex situations. This also provides a direction for the next generation of 3D-based breast ultrasound systems.

4. Conclusions

This paper proposes an intelligent diagnosis and 3D visualization system based on ultrasonic breast images. The main work includes: first, through training widely recognized classical networks, the ultrasonic image experts evaluate the network performance and give the weight for the following ensemble. The ensemble neural network is obtained by the weight. Second, based on the texture characteristics of ultrasonic images, the basic model and geometric combination rules of breast nodules were used to construct a three-dimensional model of typical breast nodules. Third, a locater-based method for locating breast nodules was proposed to achieve spatial calibration of nodular information in ultrasound images. Fourthly, the performance of the network was evaluated on the test set with 129 samples on three separate networks and on the ensemble neural network. Lastly, to verify the 3D visualization performance of the system, three groups of clinical patients were tested. After comparison, the model generated based on the 3D breast ultrasound system was basically consistent with the reconstructed 3D nodular structure. Therefore, the 3D-based breast ultrasound system proposed in this paper can realize automatic diagnosis of breast ultrasound images, meet certain clinical needs, and solve the serious problems existing in doctor–patient communication and preoperative breast navigation. It has clinical benefits.

Author Contributions: Conceptualization, Y.L. and R.X.; investigation, C.C. and J.L.; resources, Y.C. and K.H.; writing—original draft preparation Y.L. and C.C.; writing—review and editing, Y.C. and R.X.; supervision, R.X.; project administration, R.X.; funding acquisition, R.X. All authors have read and agreed to the published version of the manuscript.

Funding: The research was funded by National Natural Science Foundation of China (62176268), Non-profit Central Research Institute Fund of Chinese Academy of Medical Sciences (2020-JKCS-008), Major Science and Technology Project of Zhejiang Province Health Commission (WKJ-ZJ-2112), and Scientific and Technological Innovation Foundation of Shunde Graduate School of USTB (BK19BF004).

Institutional Review Board Statement: The studies involving human participants were reviewed and approved by the ethic committee of the PLA General Hospital. Each patient obtained informed consent before examination.

Informed Consent Statement: Informed consent was obtained from all subjects involved in the study.

Data Availability Statement: Not applicable.

Conflicts of Interest: The authors declare no conflict of interest.

References

1. Cheng, H.-D.; Shan, J.; Ju, W.; Guo, Y.; Zhang, L. Automated breast cancer detection and classification using ultrasound images: A survey. *Pattern Recognit.* **2010**, *43*, 299–317. [CrossRef]
2. Raafat, S.S.; Ezzat, S.Z.; Khachaba, Y.A.; Aboul-Nasr, L.A. Autologous mastopexy and autoaugmention of the breast. *Plast. Reconstr. Surg. Glob. Open* **2020**, *8*, e3126. [CrossRef] [PubMed]
3. Jeng, C.-J.; Hou, M.-F.; Liu, H.-Y.; Wang, L.-R.; Chen, J.-J. Construction of an integrated sexual function questionnaire for women with breast cancer. *Taiwan. J. Obstet. Gynecol.* **2020**, *59*, 534–540. [CrossRef] [PubMed]
4. Nguyen, Q.T.; Tsien, R.Y. Fluorescence-guided surgery with live molecular navigation—A new cutting edge. *Nat. Rev. Cancer* **2013**, *13*, 653–662. [CrossRef] [PubMed]
5. Tu, C.-C.; Hsu, P.-K. Global development and current evidence of uniportal thoracoscopic surgery. *J. Thorac. Dis.* **2016**, *8*, S308. [PubMed]
6. Bezerra, L.; Ribeiro, R.; Lyra, P.; Lima, R. An empirical correlation to estimate thermal properties of the breast and of the breast nodule using thermographic images and optimization techniques. *Int. J. Heat Mass Transf.* **2020**, *149*, 119215. [CrossRef]
7. Zhang, Z.; Zhang, X.; Lin, X.; Dong, L.; Zhang, S.; Zhang, X.; Sun, D.; Yuan, K. Ultrasonic diagnosis of breast nodules using modified faster R-CNN. *Ultrason. Imaging* **2019**, *41*, 353–367. [CrossRef] [PubMed]
8. Patel, S.; Jong, T.; Haden, A. Pruritic Nodules on the Breast. *Cutis* **2019**, *103*, E3–E5. [PubMed]
9. Froelich, M.F.; Kaiser, C.G. Cost-effectiveness of MR-mammography as a solitary imaging technique in women with dense breasts: An economic evaluation of the prospective TK-Study. *Eur. Radiol.* **2021**, *31*, 967–974. [CrossRef]
10. Legrand, J.; Kirchgesner, T.; Sokolova, T.; Berg, B.V.; Durez, P. Early clinical response and long-term radiographic progression in recent-onset rheumatoid arthritis: Clinical remission within six months remains the treatment target. *Jt. Bone Spine* **2019**, *86*, 594–599. [CrossRef] [PubMed]
11. Van Zelst, J.C.; Platel, B.; Karssemeijer, N.; Mann, R.M. Multiplanar reconstructions of 3D automated breast ultrasound improve lesion differentiation by radiologists. *Acad. Radiol.* **2015**, *22*, 1489–1496. [CrossRef] [PubMed]
12. Cui, S.; Chen, M.; Liu, C. DsUnet: A new network structure for detection and segmentation of ultrasound breast lesions. *J. Med. Imaging Health Inform.* **2020**, *10*, 661–666. [CrossRef]

13. Graziano, L.; Barbosa, P.N.V.P.; Travesso, D.J.; de Lima Tourinho, T.; Tyng, C.J.; Bitencourt, A.G.V. CT-guided biopsy of breast lesions: When should it be considered? *Breast J.* **2019**, *25*, 1050–1052. [CrossRef]
14. Rashmi, R.; Prasad, K.; Udupa, C.B.K.; Shwetha, V. A comparative evaluation of texture features for semantic segmentation of breast histopathological images. *IEEE Access* **2020**, *8*, 64331–64346. [CrossRef]
15. Huang, G.; Liu, Z.; Van Der Maaten, L.; Weinberger, K.Q. Densely connected convolutional networks. In Proceedings of the IEEE Conference on Computer Vision and Pattern Recognition, Honolulu, HI, USA, 21–26 July 2017; pp. 4700–4708.
16. Szegedy, C.; Liu, W.; Jia, Y.; Sermanet, P.; Reed, S.; Anguelov, D.; Erhan, D.; Vanhoucke, V.; Rabinovich, A. Going deeper with convolutions. In Proceedings of the IEEE Conference on Computer Vision and Pattern Recognition, Boston, MA, USA, 7–12 June 2015; pp. 1–9.
17. He, K.; Zhang, X.; Ren, S.; Sun, J. Deep residual learning for image recognition. In Proceedings of the IEEE Conference on Computer Vision and Pattern Recognition, Las Vegas, NV, USA, 27–30 June 2016; pp. 770–778.
18. Lu, Y.; Li, J.; Zhao, X.; Li, J.; Feng, J.; Fan, E. Breast cancer research and treatment reconstruction of unilateral breast structure using three-dimensional ultrasound imaging to assess breast neoplasm. *Breast Cancer Res. Treat.* **2019**, *176*, 87–94. [CrossRef]
19. Simonyan, K.; Zisserman, A. Very deep convolutional networks for large-scale image recognition. *arXiv* **2014**, arXiv:1409.1556.
20. Xie, S.; Girshick, R.; Dollár, P.; Tu, Z.; He, K. Aggregated residual transformations for deep neural networks. In Proceedings of the IEEE Conference on Computer Vision and Pattern Recognition, Honolulu, HI, USA, 21–26 July 2017; pp. 1492–1500.
21. Krizhevsky, A.; Sutskever, I.; Hinton, G.E. Imagenet classification with deep convolutional neural networks. In *Advances in Neural Information Processing Systems*; MIT Press: Cambridge, MA, USA, 2012; Volume 25.
22. Howard, A.G.; Zhu, M.; Chen, B.; Kalenichenko, D.; Wang, W.; Weyand, T.; Andreetto, M.; Adam, H. Mobilenets: Efficient convolutional neural networks for mobile vision applications. *arXiv* **2017**, arXiv:1704.04861.
23. Zoph, B.; Vasudevan, V.; Shlens, J.; Le, Q.V. Learning transferable architectures for scalable image recognition. In Proceedings of the IEEE Conference on Computer Vision and Pattern Recognition, Salt Lake City, UT, USA, 18–23 June 2018; pp. 8697–8710.
24. Al-Dhabyani, W.; Gomaa, M.; Khaled, H.; Fahmy, A. Dataset of breast ultrasound images. *Data Brief* **2020**, *28*, 104863. [CrossRef] [PubMed]
25. Chen, C.; Zhou, K.; Zha, M.; Qu, X.; Guo, X.; Chen, H.; Wang, Z.; Xiao, R. An effective deep neural network for lung lesions segmentation from COVID-19 CT images. *IEEE Trans. Ind. Inform.* **2021**, *17*, 6528–6538. [CrossRef]
26. Chen, C.; Xiao, R.; Zhang, T.; Lu, Y.; Guo, X.; Wang, J.; Chen, H.; Wang, Z. Pathological lung segmentation in chest CT images based on improved random walker. *Comput. Methods Programs Biomed.* **2021**, *200*, 105864. [CrossRef] [PubMed]

electronics MDPI

Review

Advances in Modeling and Suppression Methods of EMI in Power Electronic Converters of Third-Generation Semiconductor Devices

Xiaogang Wu [1,2], Xinjia Gao [2], Jiulong Wang [2,*], Zheng Li [2], Shirui Du [2], Shuchun Gao [2], Feiqiang Li [3,*], Jiuyu Du [4], Nickolay I. Shchurov [5] and Xinyang Zhang [6,7]

1 National Rail Transit Electrification and Automation Engineering Technique Research Center, Southwest Jiaotong University, Chengdu 611756, China; xgwu@hrbust.edu.cn
2 School of Electrical and Electronic Engineering, Harbin University of Science and Technology, Harbin 150080, China; gao_xinjia@163.com (X.G.); lizhengmr@163.com (Z.L.); dsr2000dsr@163.com (S.D.); gscsdust1024@163.com (S.G.)
3 Beijing Sinohytec Co., Ltd., Beijing 100084, China
4 State Key Laboratory of Automotive Safety and Energy, Tsinghua University, Beijing 100084, China; dujiuyu@tsinghua.edu.cn
5 Department of Electrical Engineering Complexes, Novosibirsk State Technical University, Novosibirsk 630073, Russia; nischurov@mail.ru
6 School of Mechanical Engineering, University of Science and Technology Beijing, Beijing 100083, China; zhangxyustb@163.com
7 Shunde Innovation School, University of Science and Technology Beijing, Foshan 528399, China
* Correspondence: wangjiulong1517@163.com (J.W.); lifeiqiang@autcht.com (F.L.); Tel.: +86-45-186-391-699 (J.W.)

Abstract: With the development of high-frequency, miniaturized, and lightweight power electronic devices, third-generation semiconductor devices are more and more used in the main circuits of power electronic converters. The electromagnetic interference (EMI) generated by their fast switching can affect the performance of power electronic converters. Therefore, it is necessary to investigate the modeling and suppression methods of conducted noise in power electronic converters of third-generation semiconductor devices. This paper describes the EMI sources and coupling paths of EMI in third-generation semiconductor devices used in power electronic converters. The modeling methods of EMI are summarized from the perspectives of power devices and coupling paths. The suppression methods of conducted noise are summarized by suppressing EMI sources and improving coupling path characteristics. This paper provides a reference for the electromagnetic compatibility design of power electronic converters for third-generation semiconductor devices.

Keywords: electromagnetic interference; EMI source; coupling path; EMI model; EMI suppression

Citation: Wu, X.; Gao, X.; Wang, J.; Li, Z.; Du, S.; Gao, S.; Li, F.; Du, J.; Shchurov, N.I.; Zhang, X. Advances in Modeling and Suppression Methods of EMI in Power Electronic Converters of Third-Generation Semiconductor Devices. *Electronics* **2023**, *12*, 2348. https://doi.org/10.3390/electronics12112348

Academic Editor: Giulio Antonini

Received: 25 March 2023
Revised: 16 May 2023
Accepted: 20 May 2023
Published: 23 May 2023

1. Introduction

Traditional power electronic converters use semiconductor devices represented by silicon to realize energy conversion, but such devices have issues such as large loss, large volume, slow switching speed, and poor voltage and current resistance [1]. Therefore, power electronic devices composed of first-generation semiconductor devices have the issues of low conversion efficiency, large volume, and low operating frequency [2].

The third-generation semiconductor devices composed of silicon carbide (SiC) and gallium nitride (GaN) have higher blocking voltages, higher operating temperatures, and higher switching speeds than Si-based power electronic devices [3]. Taking a standard SiC device as an example, it is noteworthy that its blocking voltage is one order of magnitude higher than that of a Si device, its drift layer is thinner, its on-resistance is lower, there is no bipolar charge storage mechanism in it, and its switching dynamic behavior is greatly improved [4]. Compared with Si devices, SiC devices can reduce switching losses by 70% at the same switching voltage change rate. When third-generation semiconductor devices are used in new energy systems, their low loss characteristics can reduce the volume and weight of heat dissipation systems, and their high switching frequency characteristics can reduce the volumes and weights of passive devices. Therefore, third-generation semiconductor devices further promote the trend of high frequency usage and miniaturization in power electronics, and further improve the performance of power converters, which are considered to be the key driving forces for "carbon peak and carbon neutralization".

The third generation of semiconductors can greatly improve operating frequency, reduce the volumes of devices and radiators, and increase power density in converters [5]. However, it must be noted that the high-frequency switching device is the main high-frequency noise source in a converter. The switching device generates electromagnetic interference (EMI) through the capacitive loop formed between the insulating sheet, the radiator, and the ground, bringing threats to the stable operation of the new energy system [6]. Therefore, the questions of how to improve the performance of third-generation semiconductor devices, reduce EMI caused by high switching frequency, and improve the electromagnetic compatibility of these systems have become the focuses of recent research.

This paper analyzes the EMI sources and coupling paths used in the third-generation semiconductor devices within power electronic devices. The EMI measurement and modeling methods used on third-generation semiconductor devices are reviewed. On this basis, the methods of suppressing EMI are summarized.

The structure of this paper is as follows. Section 2 analyzes the EMI mechanism and testing and analysis methods used on third-generation semiconductor devices in power electronic devices. Section 3 summarizes the methods of EMI modeling. Section 4 summarizes the EMI suppression methods. In Section 5, the electromagnetic compatibility research on power electronic devices is explored.

2. EMI Mechanism and Testing and Analysis Methods Used on Third-Generation Semiconductor Devices

2.1. The Generation Mechanism and Coupling Path of EMI

The EMI caused by third-generation semiconductor devices is mainly concentrated in two aspects, noise sources and coupling paths, as shown in Figure 1.

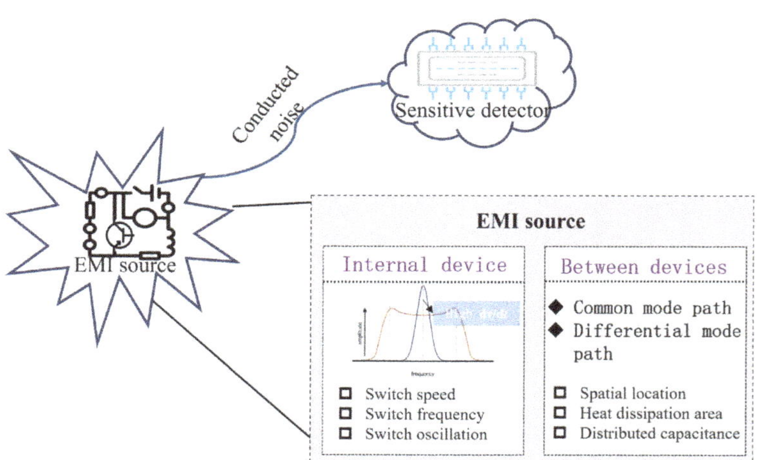

Figure 1. Main EMI sources.

2.1.1. The Generation Mechanism of EMI Source

The high $\mathrm{d}v/\mathrm{d}t$ caused by the high-frequency switching of a third-generation semiconductor device is the main source of EMI in a power electronic device. In addition, the parasitic capacitors between the ports and the loop stray inductors during the switching can cause the high-frequency oscillation of the gate source at high/low levels, as shown in Figure 2 [7].In these capacitors, the high-frequency oscillation accounts for a small proportion when the gate-source is at high or low level, and the dv/dt of the switching device is the main source of EMI.

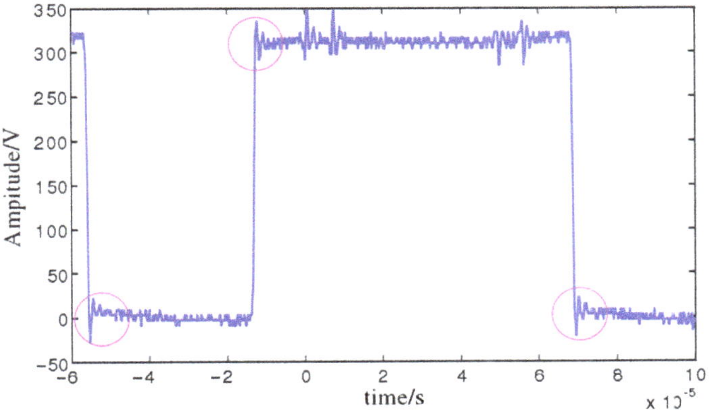

Figure 2. Measured voltage waveform between drain and source. (Pink circles indicate that parasitic capacitance during switching causes high-frequency oscillations of the gate source at high/low levels).

In their study on EMI sources, Zhang et al. [8] pointed out that the fast-switching speed, high switching frequency, and ringing of SiC devices will increase the spectrum of EMI sources. Because SiC devices have a higher switching speed, higher switching frequency, and higher ringing than Si devices, the EMI noise generated by SiC devices will be higher than that generated by Si devices. Han et al. [9] analyzed the influence of third-generation semiconductor devices on the conducted noise characteristics of a system when applied in three-phase inverters. It was found that the high switching speed characteristic mainly affected the conducted noise in the high-frequency band of the system, and that the high switching frequency characteristic would improve the EMI in the whole conducted noise band. Wen and Dalal et al. [10,11] pointed out that the displacement current may be increased by several orders of magnitude due to the fast-switching transient and increased voltage amplitude of SiC MOSFET. When the displacement current is equivalent to the rated current of SiC MOSFET, the EMI and switching loss performance of the converter will be seriously deteriorated.

2.1.2. Propagation Characteristics and Coupling Path Analysis

The EMI of third-generation semiconductor devices is not only related to the switching behavior of power devices, but is also related to the parasitic coupling paths of power electronic devices. The peak of the EMI test spectrum usually comes from coupling path impedance resonance. The impedance characteristics of the coupling path are related to the circuit topology, grounding mode, and impedance characteristics of the system devices. Third-generation semiconductor devices do not change the EMI coupling path impedance characteristics of converter systems. Therefore, the analysis of the converter coupling path is still based on the analysis method used on first-generation semiconductors in power electronic devices. Based on the different EMI coupling channels, EMI is usually divided into common mode (CM) noise and differential mode (DM) noise, which correspond to the CM EMI coupling path and DM EMI coupling path, respectively.

CM noise is mainly formed by the interaction between the instantaneous dv/dt during the switching process and the parasitic capacitance inside a system. The parasitic capacitors in the system are mainly composed of the power source, the cable from the new energy system to the ground, and the radiator from the power electronic device to the ground. At the moment the switch turns on or turns off, dv/dt will act on the parasitic capacitors in the system to generate a large charge and discharge current, forming a CM current, and its flow path will flow through the ground and ground capacitors. A schematic diagram of this process is shown in Figure 3.

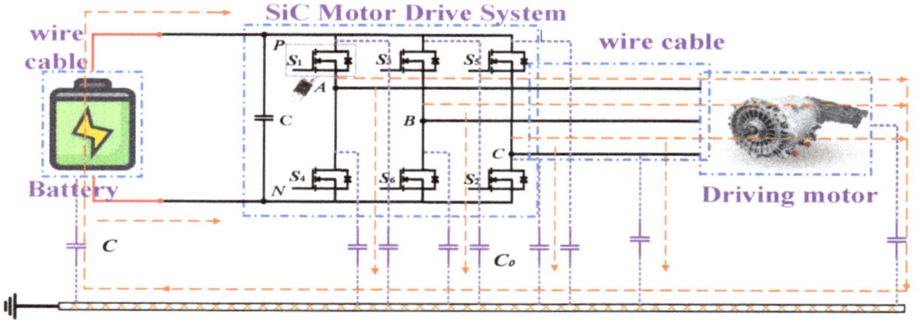

Figure 3. CM noise coupling path of an electric drive system.

The DM noise of a power electronic device is formed by the interaction between the di/dt generated during the turn-on and turn-off processes and the stray inductors in the system. The stray inductors in the system mainly include the parasitic inductors of the pin, the inductors of the cable, etc. A diagram of the coupling path is shown in Figure 4.

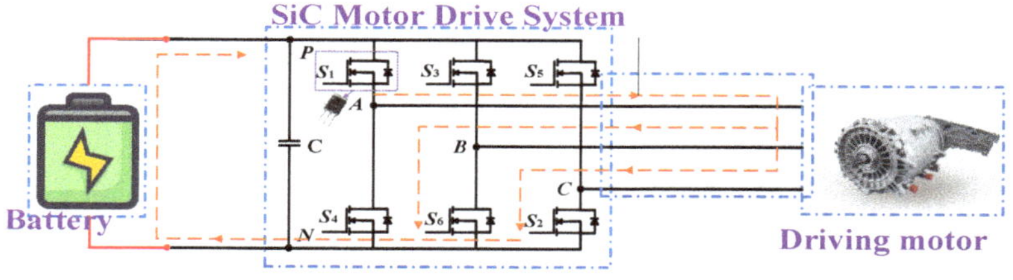

Figure 4. The DM noise coupling path of an electric drive system.

As part of the research on coupling paths in power electronic devices that use third-generation semiconductors, Xie and Han [12,13] analyzed the coupling path impedance characteristics of SiC devices in Buck and Boost converters by using the transfer function method. The transfer function analysis method does not need to simplify a coupling path into CM and DM equivalent circuits. Instead, the node current equation and loop voltage equation are used on the equivalent circuit of the coupling path to analyze the impact of the transfer function characteristics of the EMI noise source of the converter system on CM and DM noise. Marlier et al. [14] used two equivalent models and a defined switching function to analyze the EMI coupling path in a system. The final results could be obtained by calculating the convolution of the switching function and the discrete results of each equivalent model. However, these studies were carried out at the system level, and the influence of the parasitic parameters of wide-bandgap devices on EMI cannot be analyzed from the perspective of packaging. Jia et al. [15] studied the influence of switching frequency, switching speed, and switching ringing on the conducted CM noise of a SiC electric vehicle powertrain. In the low-frequency band, the CM noise of the studied system increased significantly with an increase in the switching frequency. The switch ringing mainly affected the CM noise spectrum in the high-frequency band.

2.2. Analysis Method of Conducted Noise Characteristics of Third-Generation Semiconductor Devices

The noises of third-generation semiconductor devices mainly include conduction noise and radiation noise. This paper mainly discusses conduction noise. Conductive noise research focuses on two aspects of test methods and analysis, as shown in Figure 5.

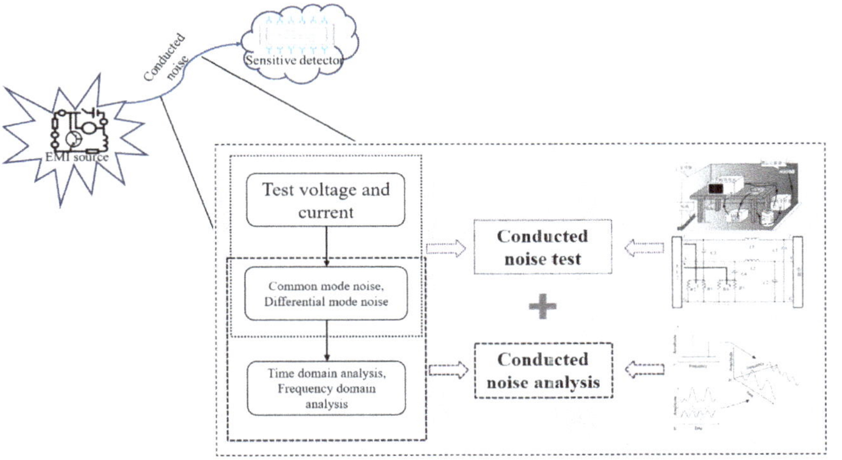

Figure 5. Conducted noise analysis and test methods.

2.2.1. Conducted Noise Test Methods

A typical conduction test platform mainly includes the following: a Line Impedance Stabilization Network (Line Impedance Stabilization Network, LISN), spectrum analyzer, impedance analyzer, and EMI voltage/current probe. The test platform structure of a typical electric drive system is shown in Figure 6.

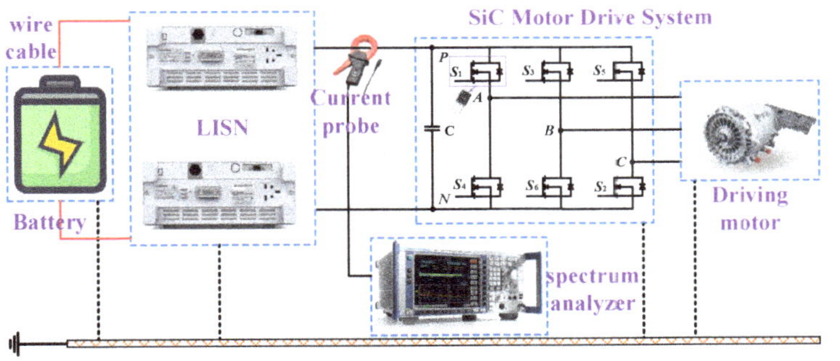

Figure 6. Test platform for a typical electric drive system.

Within research on the EMI test platform and methods, Adamowicz et al. [16] proposed a test technology to characterize and extract parameters of ferrite EMI suppressors by using vector network analyzers and micro-strip test fixtures. The extracted EMI suppressor parameters in their study were compared with the data table values. Bandara et al. [17] investigated CM- and DM-conducted emissions without using a LISN. Using the switching power source as the device under test, the CM and DM impedances of the actual measurement device, with and without a LISN, were extracted; the study results better illustrated the influence of LISNs. Li. et al. [18] extracted the online impedance of an electrical system by treating their inductively coupled probe and the monitored electrical system as three cascaded two-port networks. Their method simplified the setting and calibration process, and could monitor the real-time impedance of the key electrical system in the study without interrupting the normal operation of the system, providing convenience for EMI research. Aiello et al. [19] studied the EMI sensitivity of a Hall effect current sensor, and compared the EMI stability of this non-contact device with that of the resistive current sensing method. The measurement results showed that the Hall effect current sensor was more affected by EMI than the resistive current sensing method. Lemmon et al. [20] created a specially designed hardware platform for evaluating the EMI behaviors of high-power systems based on third-generation semiconductor devices using custom metrics, providing a reference for the development of peripheral structures and the design of EMI mitigation components in future research.

2.2.2. Conducted Noise Analysis Methods

Conducted noise analysis methods for power electronic devices are mainly divided into two types: the time domain analysis method and the frequency domain analysis method. Conducted noise modeling can be done in PSpice and Saber. Both analysis methods require the modeling of EMI interference sources and EMI coupling paths, with the main difference between the two being in the modeling of EMI interference sources. The time domain analysis method focuses on using a detailed physical model and conducting the time domain waveform analysis of all devices [21]. Li et al. [22] compared the experimental results of time domain analysis and frequency domain analysis. The results showed that both methods were effective as long as the modeling was appropriate. However, with the increasing demand for computing resources and simulation time, the frequency domain analysis method is recommended as the preferred method for EMI analysis. The specific discussion of this topic is covered in Section 3.2.

2.3. *The Effect of Temperature on EMI*

Among wide-bandgap semiconductor materials, SiC has a much wider bandgap than Si and can handle higher voltages and faster switching speeds and show better efficiency [23]. On the other hand, GaN is particularly attractive for high-pressure, high-temperature, and high-pressure applications due to its higher electric field breakdown capability, good thermal conductivity, and high electron mobility [24]. The performances of wide-bandgap materials with Si are compared in Table 1 [25].

Table 1. Comparison of wide-bandgap materials properties with Si.

Properties	Si	6H-SiC	4H-SiC	GaN
Thermal conductivity (W/cm K)	1.5	4.9	4.9	1.3
Band gap (eV)	1.12	3.03	3.26	3.45
Breakdown field (MV/cm)	0.3	2.5	2.2	3.3
Dielectric constant	11.9	9.66	9.7	8.5–10.4
Electron mobility (cm^2/Vs)	1500	400	800	2000

With the diversification of the working environment, researchers have been considering the influence of environmental factors on the circuit. The concept of electromagnetic robustness of chips was proposed by Ben in 2009 [26]. This researcher combined the problem of electromagnetic capacitance with the problem of semiconductor reliability. A microprocessor chip was subjected to experiments on high and low temperature aging and hot carrier injection. The relationships between negative bias, temperature instability, and hot carrier injection, caused by aging and the electromagnetic compatibility performance of the chip, were studied. Fernandez et al. [27] concluded that the electromagnetic sensitivity of CMOS was affected by negative-bias temperature instability. The mechanism by which electromagnetic interference and thermal stress affect semiconductor devices was explained in their paper. Dienot et al. [28] tested the PWM generation circuit and verified the effect of the combined stress of electromagnetism and heat on the function of the circuit. Wang et al. [29] proposed an efficient finite-difference time-domain (FDTD) algorithm for studying the frequency- and temperature-dependent characteristics of some graphene-based structures, with the auxiliary differential equation FDTD method and its conformal modification technique integrated together for handling atomically thin and electrically dispersive periodic geometries.

3. EMI Modeling Method

3.1. EMI Modeling Process

The basic process of conducting EMI modeling is as follows. (1) The circuit topology principle of the converter is clarified and the function of the converter is analyzed. (2) The high-frequency model of the component is established and the corresponding parameters are extracted. (3) The converter model is simulated in the simulation software. (4) The simulated EMI voltage and current spectrum are obtained via mathematical analysis. A flow chart of the modeling is shown in Figure 7. The flow chart mainly focuses on the modeling of the EMI source and the modeling of the coupling path of the parasitic parameter. On this basis, an equivalent circuit model is established. The technical difficulty of modeling lies in the extraction of the parameters of each component in the model.

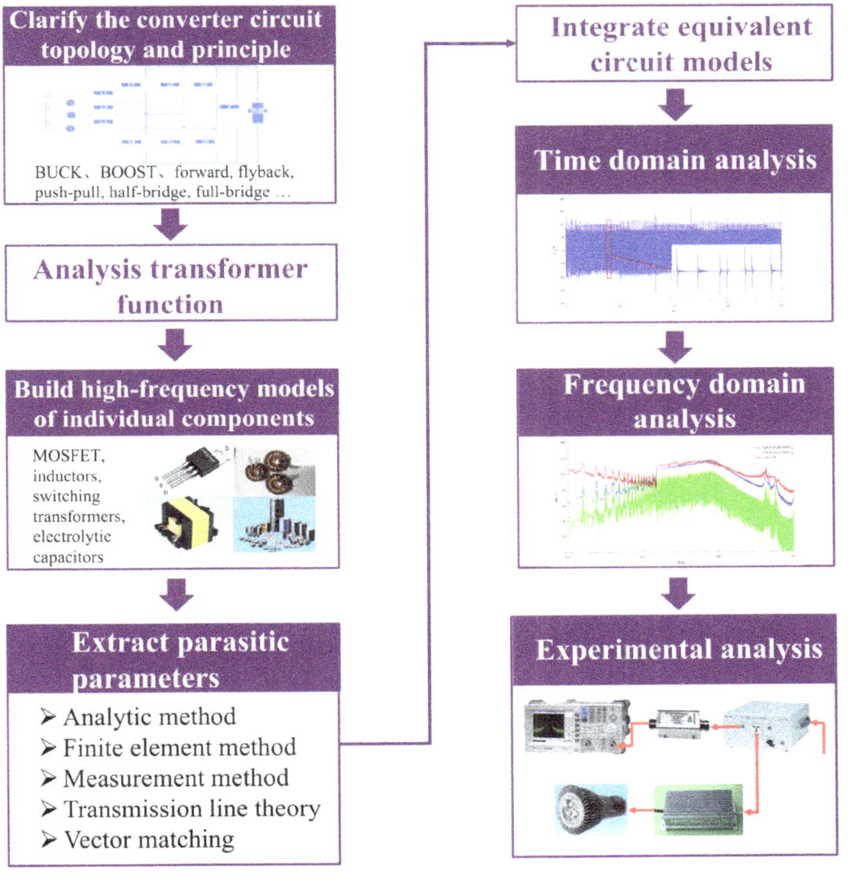

Figure 7. Flow of conducted noise modeling process.

3.2. Research Status of EMI Modeling

The high-frequency switching of third-generation semiconductor devices is the main factor required to generate switching EMI, and so the high-frequency modeling of the switching device must accurately simulate its transient characteristics [30]. The method used to model a switching device creates a circuit behavior model and a physical model. The physical model needs to partition the device according to the structure and function of the device. Depending on the carrier motion characteristics inside each sub-region, the device is described by physical equations. After that, under certain boundary conditions, the description equations are combined according to the current path to form a complete physical model of the device. An accurate physical model is helpful in order to describe the mechanism of the device [31,32]. The circuit model is modeled from the physical level of the device, and the parameters required for modeling here are mostly extracted via mathematical fitting. The relevant model parameters have no direct physical meanings. The circuit model can model a certain state of the device switching process, and can also model the whole device switching process. The circuit modeling method is relatively simple, and its modeling accuracy depends on the accuracy of parameter extraction. If the working conditions are inconsistent with the parameter-fitting conditions, the accuracy of the model will be significantly reduced. Willemen and Hsu [33,34] proposed an analytical model of MOSFET that used the parameters obtained from the device data tables and the parasitic parameters of the external circuit to simulate the high-frequency behavior of MOSFET. Y Yuan [35] proposed a MOSFET segmented behavior model that prioritizes runtime performance and convergence behavior, and is thus suitable for computer-based simulation analysis.

3.2.1. SiC Power Device Modeling

Device layer modeling is the basis of, and key to, SiC MOSFET research. Roccaforte and Smith [36,37] established the behavior model of SiC MOSFET, which is shown in Figure 8. The model consists of a drain-source resistance and three fixed parasitic inductors, and considers the influence of the parasitic inductors of the package on the model characteristics, but does not accurately simulate the SiC MOSFET switching process.

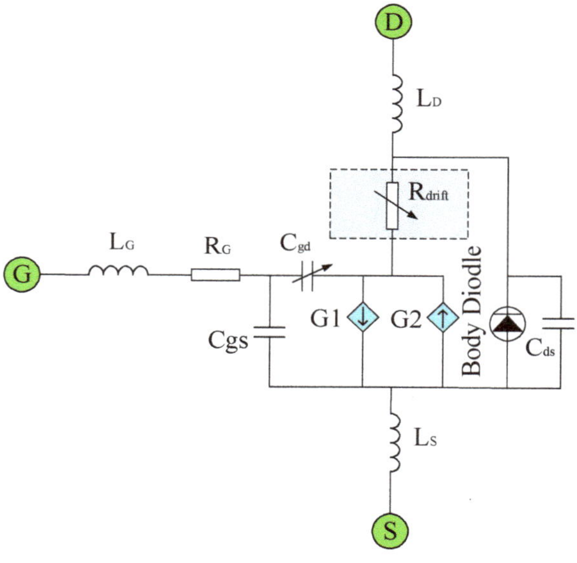

Figure 8. SiC MOSFET behavior model [38,39].

Merkert et al. [38] improved the behavior model of SiC devices by calculating the device loss, but their model did not consider the working mechanisms of the devices. Some scholars have tried to build a physical model of SiC MOSFET to clarify the working mechanism of these devices. The two described modeling methods are shown in Table 2.

Based on the theory of semiconductor physics, Wang et al. [40] used mathematical methods to express the internal mechanism of SiC MOSFET and constructed the device physical model. However, this model is complex, and its simulation takes a long time. Adamowicz and Johannesson [16,41] proposed a circuit-level model of SiC MOSFET based on the PSpice software by adjusting the parameters and modifying the model on the basis of the software's own MOSFET core. The modeling focus changed from the internal mechanism of the device to the external characteristics of the device, but the model did not consider the influence of temperature and power loss. Xin et al. [42] proposed a variable-temperature SiC MOSFET model. The model uses a temperature-controlled voltage source and a current source to compensate the static characteristics of SiC MOSFET, and analyzes the operating characteristics of the device at different temperatures, but does not consider real-time power loss.

Table 2. Comparison of power switch modeling methods [39].

Power Switch Modeling Method	Physical Model	Equivalent Circuit Model
Basic ideas	A method of semiconductor physics used to describe the internal structure of a device.	The circuit equivalent method is used to describe the circuit switching process.
Advantages	• Parasitic parameters can be explained. • The model is more accurate.	• The model is simple. • The model is suitable for the field of power electronics.
Disadvantages	• The model is more complex. • It involves a large amount of calculation.	• There are errors in parameter extraction. • The accuracy of the model is low.

Wang et al. [43] proposed a behavior-level modeling method. Their model is shown in Figure 9, and the device behavior model was established by using temperature compensation fitting. In the cited paper, the Miller capacitor between the gate and drain, which was modeled by the complex Siemens model with parameter extraction, required a large number of parameters, bringing difficulties to modeling. Sun et al. [44] added a conduction model under gate negative pressure and negative temperature based on reference [43].

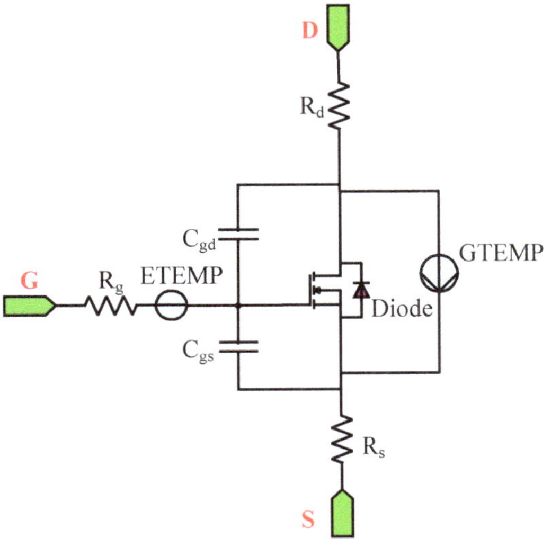

Figure 9. Temperature compensation behavior model [43].

3.2.2. GaN Power Device Modeling

The device equivalent circuit, which uses GaN, is shown in Figure 10 [45]. As emerging devices, there have been relatively few studies on high-frequency models involving GaN power devices.

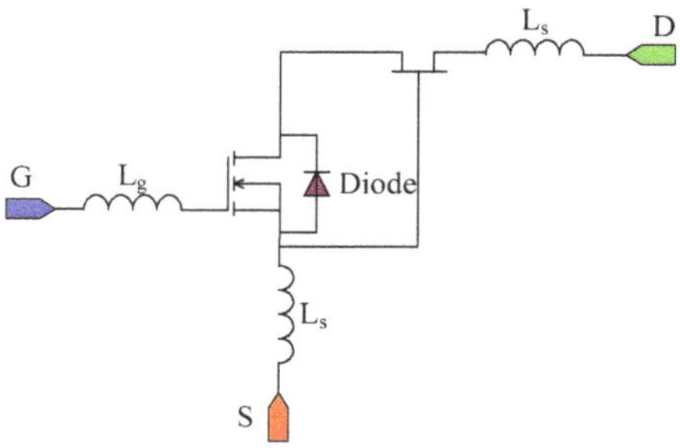

Figure 10. GaN equivalent circuit [45].

The high-frequency equivalent model that uses GaN is shown in Figure 11. Huang et al. [46] analyzed the steady-state parameters of the device in their study and compared them with Si MOSFETs at similar voltage and current levels. Pajnic et al. [47] analyzed the output characteristics of GaN devices. Their analysis results showed that GaN devices had the characteristics of high voltage, high switching frequency and low switching loss, and were thus suitable for high-voltage and high-current converters. Xie et al. [48,49] made a lot of simplifications in the switching process of the common gate-source structure device, and believed that the parasitic inductors between the low-voltage Si MOSFET and the high-voltage GaN HEMT were key to affect the switching characteristics of the device. Carrasco et al. [50] applied GaN HEMT to single-phase inverters. The switching process of the device in application was analyzed in their paper, but the expression of voltage/current and switching loss were not further analyzed. Garsed et al. [51] proposed a simplified analytical model of the switching process in GaN devices. However, the influence of stray parameters and reverse recovery—which are quite different from the actual current and voltage stress and switching loss— on the switching process were not considered.

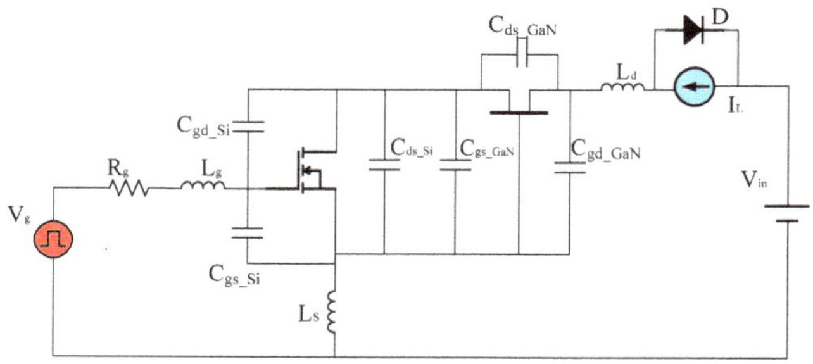

Figure 11. High-frequency equivalent circuit for GaN HEMT [46].

Liu et al. [52] established a loss analysis model for high-voltage GaN devices. Their model can accurately describe the switching process of a GaN device and can accurately analyze its output characteristics, transfer characteristics, and driving characteristics. The simulation and experiment in the paper were in good agreement. Huang et al. [53,54] analyzed the parasitic parameters inside the cascaded high-voltage GaN device in detail. By changing the installation position of the internal low-voltage Si MOSFET and the high-voltage depletion GaN device, the parasitic parameters inside the device were effectively reduced and the reliability of the device was improved in this study. Parikh and Chen [55,56] applied high-voltage GaN devices to Buck and LLC converters to evaluate their performance advantages. However, the influence of the differential packaging of GaN devices on the driving circuit was not analyzed in the paper, and the advantages of high voltage GaN devices and Si MOSFET in LLC resonant converters were not explained.

3.3. Conducted Noise Coupling Path Model

Conducted EMI propagates in the circuit in two modes: DM and CM. The coupling path includes not only the circuit composed of power devices, but also the circuit composed of parasitic elements and distributed elements [57].

The experimental results of a conducted EMI test are the sum of DM noise and CM noise. Due to the fact that EMI sources and coupling paths are different in DM noise and CM noise, it is necessary to analyze these two kinds of noise separately, and establish the DM and CM noise models of power converters. Since the modeling of the coupling path needs to be analyzed according to different circuit topologies, only the general modeling method is described here.

When modeling the conductive electromagnetic coupling loop, it is required to accurately find out the CM and DM noise coupling paths in the topology. When the converter contains multiple EMI sources and coupling paths, a multi-channel parallel analysis method should be adopted, as shown in Figure 12. Then, the circuit simulation software will be used to simulate the time domain, the time domain waveform of the noise voltage will be decomposed by Fourier, and finally, the spectrum of the conducted EMI will be obtained [58].

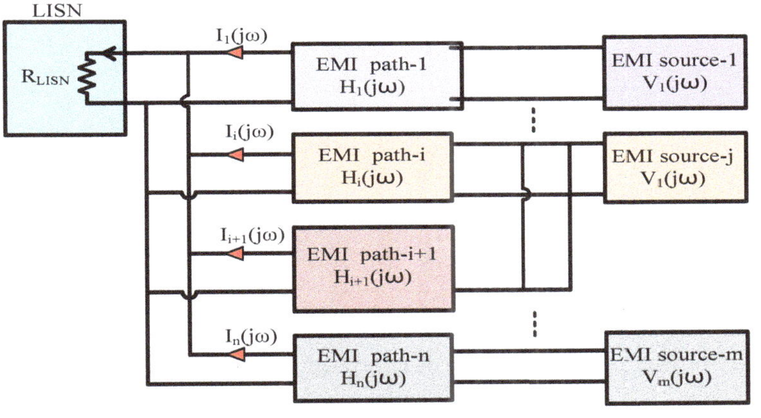

Figure 12. Schematic diagram of EMI conduction multi-channel parallel modeling.

The main models created via conduction coupling path modeling include the time domain model and frequency domain model. The aim of the time domain model is to put a high-frequency model of all devices into the circuit in order to simulate the transient waveform, and the obtained model is relatively complex. Since the time domain model does not deeply analyze the mechanism underlying the generation and propagation of conducted EMI, it is impossible to judge which parasitic devices are the main ones, and instead, the parasitic parameters of all devices are blindly extracted [59]. Therefore, the model accuracy of the time domain model is poor. Frequency domain modeling is a modeling method used to obtain the EMI spectrum via frequency domain calculation involving the EMI source. When the frequency domain model is simulated, the high-frequency model of the EMI source should be established, and then the coupling path of the EMI should be analyzed [47]. Considering that the parasitic parameters of passive devices have a great influence on the results of frequency domain prediction, the coupling path model [60] should be established based on the high-frequency model of passive devices. Then, according to the EMI source model and the coupling path model, the EMI value of the whole loop will be obtained. Finally, the obtained DM EMI and CM noise will be superimposed, yielding the EMI spectrum of the system.

3.3.1. Time Domain Model

The time domain analysis method is based on using the physical structure or behavior model of third-generation semiconductor devices to characterize EMI sources. In this method, circuit simulation software is used to simulate a whole converter system, and the EMI spectrum distribution of the system is obtained by using discrete Fourier transform. The accuracy of the time-domain simulation model mainly depends on the accuracy of the power device model. At present, a lot of research has been carried out on the simulation model of third-generation semiconductor devices. The mainstream method involves obtaining the behavioral-level circuit simulation model by fitting a manual chart of the device data or experimentally measured parameters by using software tools. However, circuit design engineers may not be able to obtain all the experimental test parameters required to establish this model for third-generation semiconductor devices, and the stability and convergence of the current device simulation model still need to be verified when the model is applied to the simulation of EMI in power electronic devices. Due to the fast-switching speeds of third-generation semiconductor devices, the time-domain-simulation step is usually set in a few nanoseconds to simulate the transient process of switching, and so, the time-domain simulation method still has the problem of long simulation time. In summary, the time-domain simulation analysis method has the disadvantages of non-convergence and long simulation time, and so it is difficult to use it for the parametric analysis of EMI characteristics and suppression measures in power electronic devices. Within research on the time domain analysis of power electronic devices, Duan and Dillan [61,62] proposed a simple modified SiC MOSFET behavior model using the SPICE language. In their paper, the key parameters of the model were analyzed and determined in detail, and the main parameters of the switch dynamic characteristic in the model were compared with the measured results. The results showed that the modified model had higher accuracy than the actual measurement results. Duan et al. [63] modeled the EMI source and coupling path of the conducted noise of a full SiC three-phase inverter, and predicted the conducted noise at the power port by using time-domain simulation and fast Fourier transform. In the frequency range of 10 kHz–30 MHz, the simulation results were basically consistent with the measured results. On the basis of establishing a system-level conducted electromagnetic interference model for an entire SiC-based electric vehicle powertrain, Jia et al. [64]. analyzed the influence of AC cable length, AC cable type, DC cable type, and other system layout characteristics on common-mode electromagnetic interference noise through the time-domain simulation of the system-level conducted electromagnetic interference model. Zhang et al. [65]. established a conducted noise model using the measured output voltage and the model of passive devices. This model used a time-domain approach. The measured output voltage of one of the Zhangbei projects was used as the excitation source for the simulation. The model of the inductor and transformer is depicted in Figure 13, where the conducted noise amplitude on the AC side is seen to be reduced by approximately 44 dBμV/Hz.

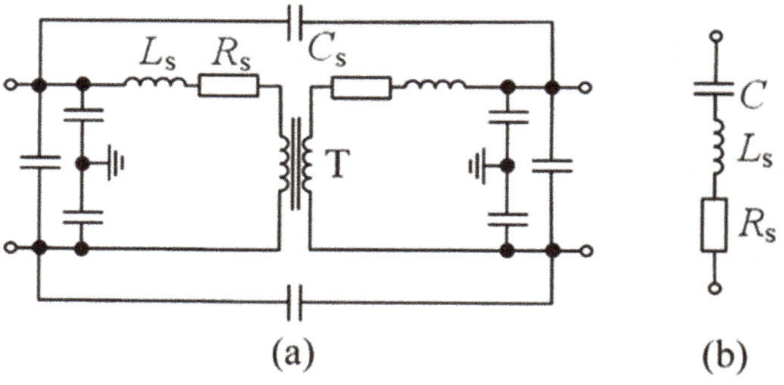

Figure 13. Wideband models of the passive devices: (**a**) transformer; (**b**) capacitor.

3.3.2. Frequency Domain Analysis

The frequency domain analysis method applies the superposition principle to linearize the nonlinear power electronic circuit. The time-domain switching waveforms of the power device are obtained by conducting simple experimental tests, and the spectrum description of the EMI source is obtained via discrete Fourier transform. The frequency domain modeling method, which has the advantage of fast simulation speed, can be used to analyze the EMI characteristics of a system and optimize the design of its EMI filter. The disadvantage of this modeling method is that it allows only one EMI source to be analyzed at a time.

As part of the research on the frequency domain analysis of power electronic devices, Revol et al. [66] replaced the switches of an inverter with current or voltage sources defined in the frequency domain. Their method considered all passive devices and achieved fast and stable simulation. Their experimental results showed that the calculation results of the model were in good agreement with the experimental results in the bandwidth range of 150 kHz–30 MHz, and that the frequency domain analysis method significantly shortened the simulation time. Li et al. [67] proposed a hybrid packaging structure SiC half-bridge power module with ultra-low parasitic inductors and low parasitic capacitors to ground. The low parasitic inductors could improve the switching speed, and the low parasitic capacitors could suppress the grounding current. Liu et al. [68] proposed a new frequency-domain method to predict conducted EMI in DC-AC converters with varying switching conditions during operation. Their method was based on the equivalent module terminal behavior frequency domain EMI source model of a switching cycle at a given operating point, and the module terminal behavior models of different operating regions were superimposed in the frequency domain to predict the conducted EMI of the entire operating cycle. Duan et al. [69] proposed a frequency-domain method for predicting the DM EMI of three-phase SiC inverters. Their calculation and experimental results showed that the proposed frequency domain calculation results were in good agreement with the experimental results in the switching frequency range and the resonant frequency range. Zhou et al. [70] proposed a new frequency-domain EMI modeling technique. In their paper, based on the frequency domain model, a three-terminal behavior model for analyzing mixed model EMI was derived. Differing from the traditional behavior model, the DM and CM EMI sources were independent of each other, and this helped readers to intuitively understand the mixed model phenomenon.

3.4. *Comparison of Time-Domain and Frequency-Domain Modeling Methods*

The two modeling methods are shown in Table 3.

Table 3. Comparison of conducted EMI modeling methods [71,72].

Modeling Method	Time Domain Modeling	Frequency Domain Modeling
Modeling approach	The high-frequency model of the EMI source is put into the circuit to simulate the transient waveform.	The superposition principle is used to linearize the circuit, and the spectrum description of the EMI source is obtained by using discrete Fourier transform.
Advantages	• It performs the simultaneous analysis of multiple EMI sources.	• The model is simple. • Only one EMI source can be analyzed by the model.
Disadvantages	• The model is complex. • Parameter extraction is difficult. • The simulation time is long.	• The simulation speed is fast. • Multiple EMI sources need to be equivalently processed. • The model accuracy is low.

Elrayyah et al. [73] studied the EMI paths of conducted EMI in DC/AC inverters, using fast Fourier transform (FFT) to simulate the EMI spectra. However, when analyzing CM noise, the error between the high-frequency band and measurement result was large, and there was a convergence problem. Hedayati et al. [74] directly used the frequency domain analytical method to establish a prediction model: that is, the circuit equation of the EMI source and the coupling path was obtained in the paper, and this equation was converted into a transfer function by using the mathematical method. However, for more complex circuit topologies, it would be difficult to directly obtain the transfer function of EMI, and so this method is only suitable for simple topology. Tang et al. [75] proposed a prediction model of conducted EMI based on Thevenin's theorem for a three-phase motor drive circuit and constructed the equivalent circuit of the CM conduction path and DM conduction path. The frequency domain analysis results from their study show that their method had high accuracy in predicting conducted EMI. Han et al. [9] used the spectral analysis method to analyze the CM noise of a motor drive system. A frequency domain prediction model in the range of 10 KHz–30 MHz was established. It simultaneously solved the convergence problem in the time-domain simulation and shortened the simulation time.

4. Study on EMI Suppression Methods for Power Electronic Devices

The existence of high dv/dt EMI sources and couplings in the conduction paths are the root causes of EMI in power electronic devices [76]. Therefore, the research on EMI suppression methods is mainly divided into two categories: one aims to suppress EMI sources, and the other aims to improve coupling path characteristics. The main measures to suppress an EMI source are the use of soft-switching technology, the optimization of the modulation mode, and the optimization of converter topology. The main measures to improve a coupling path involve active filtering technology, passive filtering technology, packaging optimization layout, and bridge balancing technology. These EMI suppression methods are classified as shown in Figure 14 [77].

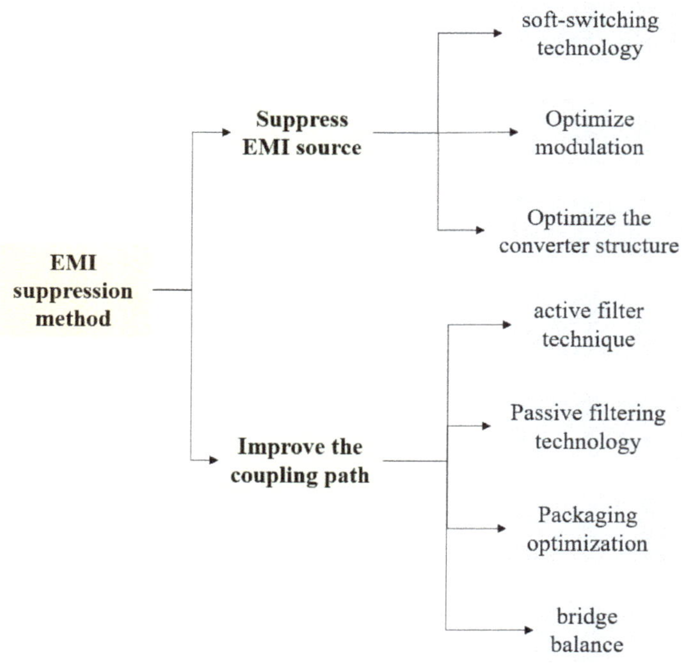

Figure 14. Classification of EMI suppression methods.

4.1. EMI Suppression Strategy for Suppressing EMI Sources

4.1.1. Soft-Switching Technology

Soft switching improves EMI noise by reducing the dv/dt and di/dt generated by system power devices. Soft-switching technology requires a corresponding soft-switching scheme according to the circuit topology of the application. Some soft-switching characteristics in power conversion systems are susceptible to perturbations by system parameters, which can affect EMI suppression [78] Third-generation semiconductor devices work at high frequencies. Their turn-on will have higher voltage and current overshoot and oscillation than Si-based semiconductor devices. Soft-switching technology can suppress the amplitude of voltage and current overshoot and oscillation, and effectively reduce the DM noise in the high-frequency bands of power electronic devices, but it has no obvious effect on the DM noise in low-frequency bands. Kim et al. [79] proposed a closed-loop gate driver to control di/dt and dv/dt that reduces the amplitude of high-frequency conducted EMI. Xie et al. [12] proposed a soft-switching circuit to suppress conducted EMI: that is, parallel RLC circuits were added at both ends of switches to reduce the switching loss in order to achieve the purpose of suppressing EMI. It was concluded that the amplitude of low-frequency current noise under CCM was small, but the amplitude of high-frequency current noise was higher than that under DCM.

4.1.2. Optimization of Modulation

Pulse width modulation is the main factor that determines power electronic noise. Modulation technology can improve the output voltage waveform quality of the inverter and suppress the conduction noise in a system. The optimization of the modulation method aims to directly adjust the generation process of the EMI source and thereby reduce the noise in the low-frequency band. Xiang et al. [80] proposed that EMI current can be suppressed by using CCM and TCM. By comparing the influences of the two methods on EMI, it was concluded that the current noise amplitude of CCM was smaller in the low-frequency range, but the current amplitude was higher than that of TCM in the high-frequency range. Xie et al. [81] used the RPWM strategy instead of the traditional SVPWM strategy. In their study, the volume of the filter inductor was effectively reduced by combining the passive filter. Therefore, the EMI current of the whole frequency band of the controller met the requirements. Omar et al. [82] proposed the use of different pulse random modulation techniques to increase the power spectral density in order to suppress conducted EMI. Natarajan et al. [83] derived a formula for calculating the power spectral density to predict EMI. Mihalic et al. [84] compared the effects of four methods—random pulse position modulation, random pulse width modulation, fixed duty cycle random carrier frequency modulation, and variable duty cycle random carrier frequency modulation—on the power spectral density and EMI of a DC-DC synchronous rectifier. It was concluded that random pulse width modulation and fixed duty cycle random carrier frequency modulation had the best effects in terms of suppressing the DC-DC converter. Dove et al. [85] transferred harmonic power with high amplitude to other frequencies by utilizing random pulse width modulation technology, and reduced the power of the previous harmonics. According to the probability distribution function, the switching behavior of the DC-DC converter was designed to eliminate EMI optimally. Hasan et al. [86] proposed a hybrid pulse modulation technique that modulated pulse width and pulse position in each pulse modulation period. EMI was effectively suppressed in a quasi-Z source converter composed of an impedance source network and a GaN device. Vedet et al. [87] proposed a pulse width modulation strategy to eliminate the CM voltage of the three-phase AC-DC-AC converter. In this strategy, the inverter and the rectifier worked at the same switching frequency, and the inverter and the rectifier switched synchronously to offset the CM voltages of the two converters.

Compared with the filter method, the pulse width modulation method has the advantages of not increasing system cost and design difficulty, and has high versatility for systems with different power levels. Due to the change in control mode that it incorporates, this method will adversely affect inverter output voltage.

4.1.3. Converter Structure Optimization

In a converter, the PCB layout, and device structure optimization, suppress EMI by varying the noise transmission path and system impedance, respectively. By optimizing the main circuit topology of the power electronic converter, the CM EMI source of dv/dt can be offset, thereby reducing the conduction EMI. The method of counteracting the EMI source fundamentally suppresses EMI, and its EMI suppression effect is more significant than those of other methods. Based on the existing research on DAB converters, a parallel dual DAB structure based on active neutral point clamping was proposed by Xie et al. [84]. The dv/dt in their paper was offset by controlling the on-off setting of the switch to reduce the EMI caused by the CM current. However, this increased the cost and reduced the power density of the system. Kumar et al. [88] adopted a split-winding structure. The inductance windings were redistributed at the input and output terminals to optimize the potential of the nodes in the circuit. The inverse dv/dt node was established to eliminate CM noise. Xie et al. [89] proposed a CM voltage offset method that eliminated the CM voltage of the converter by inserting a compensation voltage source into the input power cable of the converter. Li et al. [90] analyzed and determined that the parasitic capacitance of power devices to the ground was the main factor affecting common-mode interference. They increased the thickness of the thermal grease between the thermal conductive sheet of the power device and the heat sink, thereby increasing the parasitic impedance of the switching device to the ground and effectively reducing the low-frequency conduction noise of the converter system. Their method can reduce the volume of the converter, and its implementation cost is low; thus, it has engineering application value.

4.2. EMI Suppression Strategy to Improve Coupling Path Characteristics

4.2.1. Active Filtering Technology

Active filtering technology detects CM current in real time. The emitter follower is used to generate reverse CM voltage or CM current. The reverse CM current is injected into the converter to suppress the conducted EMI. There are three main parts of an AEF circuit. These are the noise-sensing circuit, noise-processing active circuit, and noise-injection circuit. Conventionally, the noise-processing active circuit involves an amplifier stage that is capable of driving the injection stage. Zhang et al. [91] proposed an active common noise canceller. Their method first detected the CM voltage or current from the coupling path, then used the amplifier circuit to generate reverse voltage or current. Finally, the reverse signal was injected into the circuit. The suppression of CM noise was realized. Fan and Bendicks [92,93] solved the problem of signal path delay in an active filter system by using a synthetic cancellation signal. Their experimental results showed that the synthetic cancellation signal had an obvious suppression effect on EMI in a specific frequency range. Adapa and Mueller [94,95] proposed an active filter technology, based on the generalized predictive pulse compensation method, wherein the EMI generated by a DC-DC converter was suppressed by the half-bridge gate drive circuit. The effectiveness of the method was verified on a Buck converter with a simulated output controller.

A combination of different noise-sensing and noise-cancellation methodologies yields the conventional AEF as shown in Figure 15 [76]. In addition, according to whether noise detection is completed on the source side or on the load side, the control scheme can use feedforward or feedback. AEF itself is a single-order filter. Due to the bandwidth limitations of the Detection, Processing, and Elimination stages, AEF is designed to provide noise attenuation from the EMI frequency range (150 kHz, according to the International Radio Interference Standards Special Committee) to several MHz. In order to provide noise attenuation in a higher frequency range, another passive component must be used. This passive element is combined with AEF to form HEF. The higher the attenuation and bandwidth provided by AEF are, the smaller the additional passive components required to form HEF become. The choice of specific AEF topology depends on the source impedance and load impedance [96].

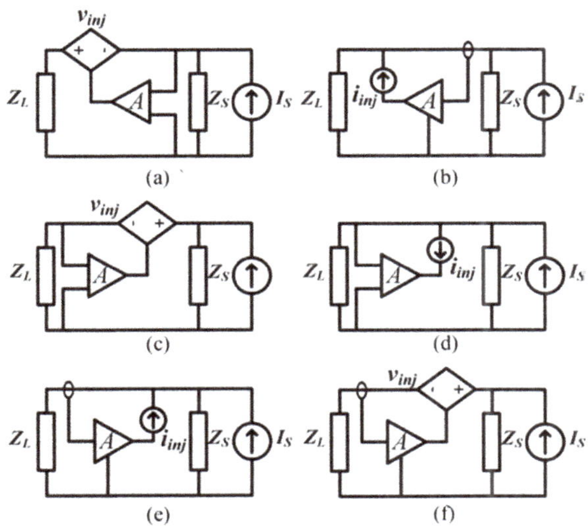

Figure 15. AEF topologies: (**a**) feedforward voltage-sense voltage-cancellation, (**b**) feedforward current-sense current-cancellation, (**c**) feedback voltage-sense voltage-cancellation, (**d**) feedback voltage-sense current-cancellation, (**e**) feedback current-sense current-cancellation, and (**f**) feedback current-sense voltage cancellation.

4.2.2. Passive Filtering Technology

Passive filtering technology is the most widely used EMI suppression technology in existing power electronic devices. It eliminates the resonance point of a converter by changing the impedance characteristics of the coupling path to suppress EMI conduction noise. The research on passive filters mainly focuses on three aspects: reducing filter sizes, increasing the number of DM inductors, and reducing the external leakage flux of the filters.

Fan et al. [92] proposed a method of adding a passive device absorption circuit. This method used ferrite magnetic beads and an improved gate driver to minimize EMI sources. Zhang et al. [4] proposed that the parasitic parameters of EMI filters seriously deteriorated the high-frequency performance of the filters. A method of introducing resistors into the CM coupling path to improve filter performance and reduce EMI was proposed. A conducted CM EMI suppression method based on an independent floating radiator was proposed. It suppressed high-frequency CM EMI by improving the electromagnetic source and coupling path in the central mode. Chu et al. [97] compared the EMI spectrum before and after adding the CM inductor. It is proposed here that adding CM inductance between LISN and the converter can effectively suppress the CM current. The implementation plan is shown in Figure 16. Dai et al. [98] proposed a new integrated multi-function CM choke based on an EMI filter. It combined a current transformer and a common choke into one element. This versatile CM choke eliminated the need for current transformers to optimize the topologies of hybrid EMI filters. Han et al. [9] paralleled the X-type capacitor on the input DC filter capacitor and added a pair of Y-type capacitors between the high-voltage positive and negative buses on the PCB board and the ground to suppress EMI. Their results showed that this method could effectively suppress the conducted EMI in the range of 150 kHz–108 MHz. Tanim et al. [99] suppressed CM EMI by adding a CM choke to the Wheatstone bridge. Their experimental results showed that this method had the characteristics of reducing EMI noise and having easy-to-realize soft switching. Dai et al. [100] proposed two methods to suppress CM EMI in AC-DC-AC converters. In the first, a CM choke was added. A three-phase CM choke could be installed on the grid side or on the load side, and a two-phase CM choke could be installed on the DC side. In the second method, the Y-type filter capacitor was added, and the filter capacitor $C_{ap}N$

connected the neutral points of the filter capacitor on the rectifier side and the inverter side. The Y-type filter capacitor was connected in series with the line-equivalent inductance to form a path to reduce the CM current and suppress EMI. Hedayati et al. [101] designed CM filters based on LCL filters. Filter inductors and filter capacitors were inserted at the AC three-phase input in the study. At the same time, capacitors C_{y1} and C_{y2} were inserted at the middle bus of the rectifier and the inverter of the AC-DC-AC converter for filtering. In addition, the connection point of the two capacitors in the CM filter was connected with the three-phase filter capacitor of the LCL filter. The capacitor C_{Mg} was introduced between the connection point and the ground to make it as close as possible to the ground potential, thereby improving the filtering efficiency.

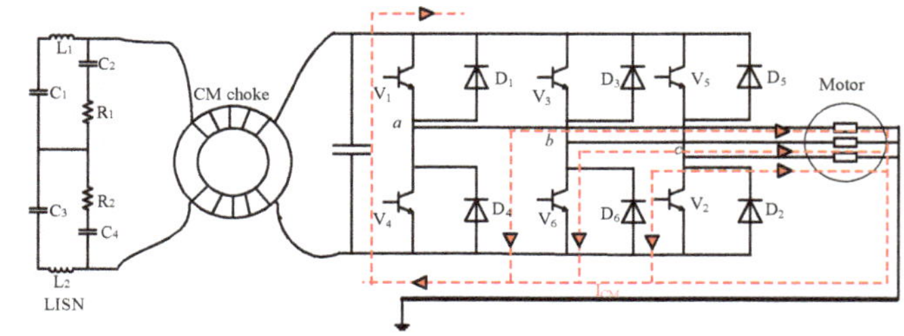

Figure 16. Schematic of EMI suppression with CM inductors.

4.2.3. Optimization of Package Layout

The package parasitic inductors have a great influence on the EMI sources in converters. Reducing the package parasitic inductors can not only suppress the switching voltage ringing, but also improve the switching speed of power electronic devices [102]. In addition, the parasitic CM capacitors of the package determine the impedance characteristics of the CM EMI coupling path. Reducing the parasitic CM capacitance of the package is an effective means by which to reduce system EMI. Through improved thermal management and packaging material processes, higher power densities and wider temperature ranges (i.e., operating temperatures above 200 °C) can be supported. Ultra-low inductance packaging can protect components from voltage transients and suppress EMI.

Bendicks et al. [103] proposed a hybrid half-bridge SiC power module and a hybrid half-bridge for TO-247 packaged SiC devices. Their proposal effectively reduced the total number of parasitic inductors and achieved the purpose of reducing EMI. Pahlevaninezhad et al. [93] proposed a high-frequency planar transformer for DAB converters. Its embedded EMI filter could effectively eliminate CM EMI in a DAB converter. Kumar et al. [104] proposed a low-inductance bus design method for discrete devices. This method could effectively reduce the coupling inductance in the loop, and in turn, high-frequency EMI could be better suppressed.

4.2.4. Bridge Balancing Technology

Bridge balancing technology uses the Wheatstone bridge balancing principle to construct the converter as a Wheatstone bridge. The impedance of the converter is matched to meet the bridge balance conditions, thereby eliminating the CM current of the converter and suppressing conducted EMI. Bridge balancing techniques require the adjusting of impedance parameters multiple times in order to achieve bridge balancing. In addition, for occasions in which a converter is required to be common ground, bridge balancing technology will destroy the common-ground characteristics of the converter. Additionally, its actual suppression effect is debatable.

Narayanasamy et al. [105] derived two CM noise sources and their characteristics and proposed a CM noise reduction technology with a large impedance ratio balanced bridge to achieve EMI suppression. Chen et al. [106] proposed a neutral-point clamp topology with bridge balancing technology that could achieve large EMI attenuation and reduce filter parameters to reduce size and weight. Yang et al. [107] proposed a CM equivalent circuit model of a multi-unit AC/DC traction system in which two series balanced capacitors C_B were incorporated at the input end, and the midpoint of the balanced capacitor was connected to the midpoint of the two DC buses to suppress EMI. Yang et al. [108] grounded the inverter casing with the inverter and the motor at the same time, and decoupled the inductor L_G in series to reduce the influence of the inverter on the CM noise of the rectifier, so as to reduce the CM noise of the system. Li et al. [109] changed the traditional three-phase hexagonal AC chopper bridge arm to multiple chopper AC voltage regulation circuits and inductors in series, as shown in Figure 17. A multistage AC hexagonal chopper, which reduced the dv/dt generated by the switching device, was formed so as to effectively reduce EMI.

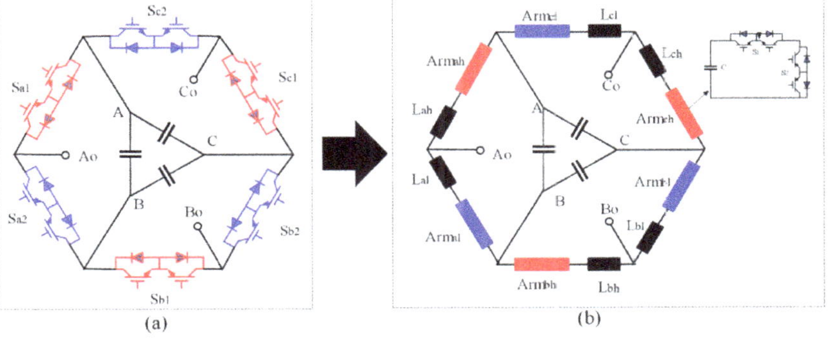

Figure 17. Schematic diagram of two AC choppers: (**a**) three-phase hexagonal AC chopper; (**b**) multistage AC hexagonal chopper.

Table 4 summarizes the advantages, disadvantages, and use cases of several EMI suppression methods.

Table 4. Comparison of EMI suppression strategies.

Classification	Suppression Strategy	Advantages	Disadvantages	Application Scenarios	Inhibition Effect	Cost
Suppress sources of EMI	Soft-switching technology	• It has a simple structure. • It has an easy design. • It has a low cost.	• Its EMI suppression is not good.	Scenarios where EMI suppression is low	It has an insufficient inhibition effect.	Low
	Optimization of the modulation method	• It improves power density. • It limits narrowband power. • It uses DM EMI suppression.	• It has limited effectiveness in limiting wide bands. • It is not applicable with resonant converters. • It is limited by specific topologies or control methods. • The effects of suppressing CM EMI are limited.	Scenarios where EMI suppression is low	Its inhibition effect is limited.	Relatively Low
	Optimization of the transformer structure	• It uses fundamental EMI suppression. • It is easy to implement.	• It needs complex modeling and calculation. • It is mainly used to suppress CM EMI.	Scenarios where EMI suppression is low	Its inhibition effect is significant.	Middle

Table 4. *Cont.*

Classification	Suppression Strategy	Advantages	Disadvantages	Application Scenarios	Inhibition Effect	Cost
Change the coupling path	Active filtering technology	• It has a low cost. • It uses fewer devices used. • Its operation is simple.	• Operation and calculation are complex. • Its suppression effect is not as good as that of passive filters. • New EMI will be introduced.	Scenarios where EMI suppression is low	It has an insufficient inhibition effect.	Low
	Passive filtering technology	• It has a simple structure • It has a good inhibition effect.	• The weight of the equipment is large. • It has a high cost. • It has a large size. • It has a limited ability to suppress high-frequency EMI.	Scenarios with high suppression standards	Its suppression effect is good.	High
	Optimized package placement	• It effectively suppresses high-frequency EMI.	• Its equipment structure is complex. • Its operation is difficult.	Scenarios with high suppression standards	Its inhibition effect is good.	Relatively High
	Bridge balancing technology	• It effectively suppresses CM noise.	• Its design is difficult. • Its operation is difficult. • It involves complex modeling and calculation.	Suppresses CM EMI	It has a good inhibition effect.	Relatively Low

5. Future Prospects

In this paper, the EMI generation mechanisms, coupling paths, modeling methods, and suppression strategies of power electronic converters using third-generation semiconductors devices have been investigated and analyzed. The direction of research on EMI in power electronic converters in the future is speculated upon. This paper aims to further promote the development and application of power electronics technology under new energy structures. The questions of how to fully optimize the layouts and selection of power electronic devices based on the requirements of power electronic converters, and how to design a power electronic converter that considers high efficiency, high frequency, high power density, and low EMI, pose serious challenges when applying third-generation semiconductor power electronic devices. Research on EMI in power electronics using third-generation semiconductors can be carried out from the following perspectives.

(1) Modeling method

EMI modeling is mostly based on the ideal circuit model of the converter. However, in practice, the control loop of the power electronic system, the layout of the PCB board, and the selection and placement of a power electronic device will change EMI. These factors will make the constructed model deviate greatly from the actual measurement results, and it is difficult to simulate its actual working conditions. Therefore, the question of how to construct accurate device and circuit models in complex backgrounds and obtain analytical waveforms with a high degree of fitting to actual measurement results is worthy of further study. The coupling path of EMI above 10 MHz is very complex. In practice, there are various non-ideal factors that pose challenges to EMI modeling in medium- and high-frequency bands. Therefore, by optimizing electromagnetic numerical simulation technologies and test schemes, one may overcome a technical bottleneck to establish a complete and accurate electromagnetic compatibility model including nonlinearity, time delay, broadband, and coupling sensitivity.

(2) Analysis of EMI model under non-ideal conditions

In the process of model idealization, some secondary factors are usually ignored, such as the relative smallness of some EMI sources and the influence of temperature on EMI model parameters. Although the distribution of EMI after the idealization of a model can be obtained relatively easily, the ignored factors will further lead to large errors between the model and the actual measurement results, resulting in inadequate simulation and experimental fitting. Therefore, it is of practical significance to study the EMI model under non-ideal conditions and discuss the direction of the influence of secondary factors on the EMI model. As the supporter of electronic components in the system, the PCB plays the role of electrically connecting electronic components. Due to the existence of parasitic effects, when the system is actually working, its PCB will show some non-ideal states. In particular, when a high-speed signal flows through the PCB trace, the PCB becomes the main path of high-frequency noise due to the aggravation of parasitic effects. The EMI characteristics of a PCB can be studied by following these three instructions: extract PCB parasitic parameters and establish a system EMI model, use this model to analyze the relationship between the structure and the EMI noise, and optimize the PCB structure according to the results of the analysis.

(3) Coupling path analysis and model simplification scheme

With the increasing demand for the miniaturization of power electronic devices, the operating frequencies of these devices is increasing, and the EMI caused by higher dv/dt is becoming more and more serious. EMI coupling path analysis and model simplification schemes under specific EMI sources will reduce the difficulty of EMI analysis and provide new ideas for EMI prediction. The interference model that has now been established has a large number of parasitic parameters. On the one hand, the values of these parasitic parameters are difficult to measure accurately. On the other hand, the influence of these parasitic parameters on interference is also very small. If all parasitic parameters are considered, the established model becomes too complicated. Therefore, model simplification is important. However, an oversimplified EMI model will affect the accuracy of the model. Therefore, conceptualizing an EMI coupling path simplification scheme with low complexity and high accuracy is the key issue of future EMI research.

(4) A comparative study of EMI between the converter using Si-based and third-generation semiconductors.

Third-generation semiconductor devices have the advantages of fast switching speed, low loss, and high voltage and current stress, which are increasingly favored by designers of power electronic devices. However, due to the high switching speed of third-generation semiconductor devices, their dv/dt will be higher than that of Si-based devices, and the resulting EMI will be more complicated. Therefore, addressing the question of how to flexibly use the frequency-domain method and the time-domain method to study the specific problems of third-generation semiconductor devices and compare them with Si-based devices to clarify the influencing factors and modes of action of EMI will be a direction for future EMI research.

(5) EMI research combined with the actual working conditions

The working states of power electronic devices are diverse. The runtime contains a variety of operating conditions. Studying the conducted noise under a certain working condition cannot explain the EMC characteristics of the drive system. It is necessary to study the conducted noise under various working conditions and establish its noise prediction simulation model. An EMI design method for the whole stage of hardware design should be developed. As for the design, the following measures are recommended: (a) the area of the power circuit should be as small as possible, and (b) the switching speed and loss in the power semiconductor device need to be balanced. For valve tower and system design, it is recommended to select the appropriate grounding, filtering, and shielding methods according to the operating conditions. It is recommended to combine the reduction of EMI sources with the weakening of EMI propagation paths. Therefore, it is of great significance in engineering to study EMI modeling and suppression methods in combination with actual operating conditions.

(6) EMI suppression strategy formulation

The traditional EMI suppression strategy mainly relies on two measures: suppressing the EMI source and improving the coupling path. It mainly includes absorption circuits, soft-switching technology, package optimization, and filter design, and relies mainly on previous experiences. New methods are needed to further improve the performance of EMI filters without the need for additional high-voltage capacitors to improve stability or to inject transformers into power lines. These new topologies can use all analog or hybrid analog-and-digital active circuits. In addition, there have been a few studies on auxiliary power supply, power loss, and protection of AEF, so as to encourage the wide adoption of low EMI modulation and low EMI topology. While meeting the power quality requirements, an EMI source should be made as small as possible by selecting the topology and modulation appropriately. The operation of simultaneously-switching large quantum SM should be avoided. For third-generation semiconductor devices, addressing the question of how to quantitatively derive EMI content according to the spectrum characteristic curves and the spectrum analysis results, and then specify suppression strategies, is another important issue for future EMI research.

Author Contributions: Article structure designation, editing, X.W.; writing-Sections 1 and 2, X.G.; review, J.W.; investigation, project administration, F.L.; supervision, J.D.; writing-Section 5, N.I.S.; writing-Section 3, Z.L.; writing-Section 4, S.D. and X.Z.; writing-Section 5, S.G. All authors have read and agreed to the published version of the manuscript.

Funding: This paper has been partially supported by the funding of Chengdu Guojia Electrical Engineering Co., Ltd. (No. NEEC-2022-B07).

Data Availability Statement: Not applicable.

Conflicts of Interest: The authors declare no conflict of interest.

References

1. Wen, F.; Tutuc, E. Strained SixGe1-x-Ge-Si core-double-shell nanowire heterostructures for simultaneous hole and electron mobility enhancement. *Appl. Phys. Lett.* **2018**, *113*, 113102. [CrossRef]
2. Zhang, R.; Zhang, Y. Power device breakdown mechanism and characterization: Review and perspective. *Jpn. J. Appl. Phys.* **2023**, *62*, SC0806. [CrossRef]
3. Wen, F.; Tutuc, E. Enhanced Electron Mobility in Nonplanar Tensile Strained Si Epitaxially Grown on SixGe1-x Nanowires. *Nano Lett.* **2018**, *18*, 94–100. [CrossRef]
4. Zhang, Z.; Hu, Y.; Chen, X. A Review on Conductive Common-Mode EMI Suppression Methods in Inverter Fed Motor Drives. *IEEE Access* **2021**, *9*, 18345–18360. [CrossRef]
5. Singh, S.; Chaudhary, T.; Khanna, G. Recent Advancements in Wide Band Semiconductors (SiC and GaN) Technology for Future Devices. *Silicon* **2022**, *14*, 5793–5800. [CrossRef]
6. Ma, C.T.; Gu, Z.H. Review on Driving Circuits for Wide-Bandgap Semiconductor Switching Devices for Mid- to High-Power Applications. *Micromachines* **2021**, *12*, 65. [CrossRef] [PubMed]
7. Zhao, S.; Zhao, X.; Wei, Y. A Review of Switching Slew Rate Control for Silicon Carbide Devices Using Active Gate Drivers. *IEEE J. Emerg. Sel. Top. Power Electron.* **2021**, *9*, 4096–4114. [CrossRef]
8. Zhang, B.; Wang, S. A Survey of EMI Research in Power Electronics Systems With Wide-Bandgap Semiconductor Devices. *IEEE J. Emerg. Sel. Top. Power Electron.* **2020**, *8*, 626–643. [CrossRef]
9. Han, D.; Li, S.; Wu, Y.; Choi, W.; Sarlioglu, B. Comparative Analysis on Conducted CM EMI Emission of Motor Drives: WBG Versus Si Devices. *IEEE Trans. Ind. Electron.* **2017**, *64*, 8353–8363. [CrossRef]
10. Dalal, D.N.; Christensen, N.; Jorgensen, A.B.; Jorgensen, J.K. Impact of Power Module Parasitic Capacitances on Medium-Voltage SiC MOSFETs Switching Transients. *IEEE J. Emerg. Sel. Top. Power Electron.* **2020**, *8*, 298–310. [CrossRef]
11. Wen, F.; Shabani, J.; Tutuc, E. Josephson Junction Field-Effect Transistors for Boolean Logic Cryogenic Applications. *IEEE Trans. Electron Devices* **2019**, *66*, 5367–5374. [CrossRef]
12. Xie, Y.; Chen, C.; Huang, Z.; Liu, T.; Kang, Y.; Luo, F. High Frequency Conducted EMI Investigation on Packaging and Modulation for a SiC-Based High Frequency Converter. *IEEE J. Emerg. Sel. Top. Power Electron.* **2019**, *7*, 1789–1804. [CrossRef]
13. Han, D.; Sarlioglu, B. Comprehensive Study of the Performance of SiC MOSFET-Based Automotive DC–DC Converter Under the Influence of Parasitic Inductance. *IEEE Trans. Ind. Appl.* **2016**, *52*, 5100–5111. [CrossRef]
14. Marlier, C.; Videt, A.; Idir, N.; Moussa, H.; Meuret, R. Modeling of switching transients for frequency-domain EMC analysis of power converters. In Proceedings of the 2012 15th International Power Electronics and Motion Control Conference (EPE/PEMC), Novi Sad, Serbia, 4–6 September 2012.

15. Jia, X.; Dong, B.; Wang, H.; Hu, C.; Xu, D. Characteristics of SiC inverter powertrains on common-mode EMI noise. *J. Power Electron.* **2021**, *21*, 354–363. [CrossRef]
16. Adamowicz, M. Non-isolated resonant quasi-Z-source network DC-DC converter. *Electron. Lett.* **2019**, *55*, 855–856. [CrossRef]
17. Li, K.R.; See, K.Y.; Bandara, R.M.S. Impact Analysis of Conducted Emission Measurement Without LISN. *IEEE Trans. Electromagn. Compat.* **2016**, *58*, 776–783. [CrossRef]
18. Kang Rong, L.; Kye Yak, S.; Xing Ming, L. Inductive Coupled In-Circuit Impedance Monitoring of Electrical System Using Two-Port *ABCD* Network Approach. *IEEE Trans. Instrum. Meas.* **2015**, *64*, 2489–2495. [CrossRef]
19. Aiello, O. Hall-Effect Current Sensors Susceptibility to EMI: Experimental Study. *Electronics* **2019**, *8*, 1310. [CrossRef]
20. Lemmon, A.N.; Cuzner, R.; Gafford, J.; Hosseini, R.; Brovont, A.D.; Mazzola, M.S. Methodology for Characterization of Common-Mode Conducted Electromagnetic Emissions in Wide-Bandgap Converters for Ungrounded Shipboard Applications. *IEEE J. Emerg. Sel. Top. Power Electron.* **2018**, *6*, 300–314. [CrossRef]
21. Zhu, R.; Lin, N.; Dinavahi, V.; Liang, G. An Accurate and Fast Method for Conducted EMI Modeling and Simulation of MMC-Based HVdc Converter Station. *IEEE Trans. Power Electron.* **2020**, *35*, 4689–4702. [CrossRef]
22. Lai, J.S.; Huang, X.; Pepa, E.; Chen, S.; Nehl, T.W. Inverter EMI modeling and simulation methodologies. *IEEE Trans. Ind. Electron.* **2006**, *53*, 736–744.
23. Rabkowski, J.; Peftitsis, D.; Nee, H. Silicon Carbide Power Transistors: A New Era in Power Electronics Is Initiated. *IEEE Ind. Electron. Mag.* **2012**, *6*, 17–26. [CrossRef]
24. Jones, E.A.; Wang, F.F.; Costinett, D. Review of Commercial GaN Power Devices and GaN-Based Converter Design Challenges. *IEEE J. Emerg. Sel. Top. Power Electron.* **2016**, *4*, 707–719. [CrossRef]
25. Millan, J.; Godignon, P.; Perpina, X.; Perez-Tomas, A.; Rebollo, J. A Survey of Wide Bandgap Power Semiconductor Devices. *IEEE Trans. Power Electron.* **2014**, *29*, 2155–2163. [CrossRef]
26. Ben Dhia, S.; Boyer, A.; Li, B.; Ndoye, A.C. Characterisation of electromagnetic compatibility drifts of nanoscale integrated circuit after accelerated life tests. *Electron. Lett.* **2010**, *46*, 278–280. [CrossRef]
27. Fernandez, R.; Berbel, N.; Gil, I.; Morata, M. Impact of NBTI on EMC behaviours of CMOS inverter. In Proceedings of the Asia-Pacific International Symposium on Electromagnetic Compatibility/Technical Exhibition on EMC RFIMicrowave Measurements and Instrumentation, Beijing, China, 12–16 April 2010.
28. Dienot, J.-m.; Batista, E.; Ramos, I. Thermal-electromagnetic susceptibility behaviors of PWM patterns used in control electronic circuit. In Proceedings of the International Workshop on the Electromagnetic Compatibility of Integrated Circuits, Edinburgh, Scotland, 10–13 November 2015.
29. Wang, D.W.; Zhao, W.S.; Gu, X.Q.; Chen, W.c.; Yin, W.Y. Wideband Modeling of Graphene-Based Structures at Different Temperatures Using Hybrid FDTD Method. *IEEE Trans. Nanotechnol.* **2015**, *14*, 250–258. [CrossRef]
30. Wang, Z.; Li, H.; Chu, Z.; Zhang, C.; Yang, Z.; Shao, T.; Hu, Y. A Review of EMI Research in Modular Multilevel Converter for HVDC Applications. *IEEE Trans. Power Electron.* **2022**, *37*, 14482–14498. [CrossRef]
31. Cheng, W.; Huang, Z.; Xu, S.; Sun, W. Novel Hybrid Analytical/Numerical Conducted EMI Model of a Flyback Converter. *IEEE Trans. Electromagn. Compat.* **2017**, *59*, 488–497. [CrossRef]
32. Wen, F.; Yuan, J.; Wickramasinghe, K.S.; Mayer, W.; Shabani, J.; Tutuc, E. Epitaxial Al-InAs Heterostructures as Platform for Josephson Junction Field-Effect Transistor Logic Devices. *IEEE Trans. Electron Devices* **2021**, *68*, 1524–1529. [CrossRef]
33. Willemen, J.; Andreini, A.; De Heyn, V.; Esmark, K.; Etherton, M. Characterization and modeling of transient device behavior under CDM ESD stress. *J. Electrost.* **2004**, *62*, 133–153. [CrossRef]
34. Hsu, W.; Wen, F.; Wang, X.; Wang, Y. Laser Spike Annealing for Shallow Junctions in Ge CMOS. *IEEE Trans. Electron Devices* **2017**, *64*, 346–352. [CrossRef]
35. Yuan, Y.; Qian, Z. An improved lumped-charge model and parameter extraction approach of PIN diodes. In Proceedings of the IEEE 33rd Annual IEEE Power Electronics Specialists Conference Proceedings, Cairns, QLD, Australia, 23–27 June 2002.
36. Roccaforte, F.; Fiorenza, P.; Greco, G. Emerging trends in wide band gap semiconductors (SiC and GaN) technology for power devices. *Microelectron. Eng.* **2018**, *187*, 66–77. [CrossRef]
37. Smith, B.; Fleming, G.; Parrish, K.D. Mean Free Path Suppression of Low-Frequency Phonons in SiGe Nanowires. *Nano Lett.* **2020**, *20*, 8384–8391. [CrossRef] [PubMed]
38. Merkert, A.; Krone, T.; Mertens, A. Characterization and Scalable Modeling of Power Semiconductors for Optimzied Design of Traction Inverters with Si- and SiC-Devices. In Proceedings of the IEEE Vehicle Power and Propulsion Conference (VPPC), Seoul, Republic of Korea, 9–12 October 2012.
39. Xie, Y.; Chen, C.; Yan, Y.; Huang, Z.; Kang, Y. Investigation on Ultralow Turn-off Losses Phenomenon for SiC MOSFETs With Improved Switching Model. *IEEE Trans. Power Electron.* **2021**, *36*, 9382–9397. [CrossRef]
40. Wang, J.; Jiang, X. Review and analysis of SiC MOSFETs' ruggedness and reliability. *IET Power Electron.* **2020**, *13*, 445–455. [CrossRef]
41. Johannesson, D.; Nawaz, M. Analytical PSpice model for SiC MOSFET based high power modules. *Microelectron. J.* **2016**, *53*, 167–176. [CrossRef]
42. Xin, Z.; Li, H.; Liu, Q.; Loh, P.C. A Review of Megahertz Current Sensors for Megahertz Power Converters. *IEEE Trans. Power Electron.* **2022**, *37*, 6720–6738. [CrossRef]

43. Wang, J.; Zhao, T.; Li, J. Characterization, modeling, and application of 10-kV SiC MOSFET. *IEEE Trans. Electron Devices* **2008**, *55*, 1798–1806. [CrossRef]
44. Sun, K.; Wu, H.; Lu, J.; Xing, Y.; Huang, L. Improved Modeling of Medium Voltage SiC MOSFET Within Wide Temperature Range. *IEEE Trans. Power Electron.* **2014**, *29*, 2229–2237. [CrossRef]
45. Wang, X.; Wang, Y.; Wu, T. The Review of Electromagnetic Field Modeling Methods for Permanent-Magnet Linear Motors. *Energies* **2022**, *15*, 3595. [CrossRef]
46. Huang, X.; Liu, Z.; Li, Q.; Lee, F.C. Evaluation and Application of 600V GaN HEMT in Cascode Structure. In Proceedings of the 28th Annual IEEE Applied Power Electronics Conference and Exposition (APEC), Long Beach, CA, USA, 17–21 March 2013.
47. Pajnic, M.; Pejovic, P.; Despotovic, Z.; Lazic, M.; Skender, M. Characterization and Gate Drive Design of High Voltage Cascode GaN HEMT. In Proceedings of the 19th International Symposium on Power Electronics (Ee), Novi Sad, Serbia, 19–21 October 2017.
48. Xie, R.; Wang, H.; Tang, G.; Yang, X.; Chen, K.J. An Analytical Model for False Turn-On Evaluation of High-Voltage Enhancement-Mode GaN Transistor in Bridge-Leg Configuration. *IEEE Trans. Power Electron.* **2017**, *32*, 6416–6433. [CrossRef]
49. Xie, L.; Ruan, X.; Ji, Q.; Ye, Z. Shielding-Cancelation Technique for Suppressing Common-Mode EMI in Isolated Power Converters. *IEEE Trans. Ind. Electron.* **2015**, *62*, 2814–2822. [CrossRef]
50. Carrasco, L.C.M.; Forsyth, A.J. Energy analysis and performance evaluation of GaN cascode switches in an inverter leg configuration. In Proceedings of the 2015 IEEE Applied Power Electronics Conference and Exposition (APEC), Charlotte, NC, USA, 15–19 March 2015.
51. Garsed, P.J.; McMahon, R.A. Understanding the cascode switching process. In Proceedings of the The 1st IEEE Workshop on Wide Bandgap Power Devices and Applications, Columbus, OH, USA, 27–29 October 2013.
52. Liu, Z.; Huang, X.; Zhang, W.; Lee, F.C.; Li, Q. Evaluation of High-Voltage Cascode GaN HEMT in Different Packages. In Proceedings of the 29th Annual IEEE Applied Power Electronics Conference and Exposition (APEC), Fort Worth, TX, USA, 16–20 March 2014.
53. Huang, X.; Li, Q.; Liu, Z.; Lee, F.C. Analytical loss model of high voltage GaN HEMT in cascode configuration. In Proceedings of the 2013 IEEE Energy Conversion Congress and Exposition, Denver, CO, USA, 15–19 September 2013.
54. Huang, X.; Li, Q.; Liu, Z.; Lee, F.C. Analytical Loss Model of High Voltage GaN HEMT in Cascode Configuration. *IEEE Trans. Power Electron.* **2014**, *29*, 2208–2219. [CrossRef]
55. Parikh, P.; Wu, Y.; Shen, L. Commercialization of high 600V GaN-on-silicon power HEMTs and diodes. In Proceedings of the 2013 IEEE Energytech, Miyazaki, Japan, 21–23 May 2013.
56. Chen, H.; Wu, J.; Zheng, X. Elimination of Common-Mode Choke Saturation Caused by Self-Resonance of the EMI Filter in a Variable-Frequency Drive System. *IEEE Trans. Electromagn. Compat.* **2019**, *61*, 1226–1233. [CrossRef]
57. Fu, K.; Chen, W. Evaluation Method of Flyback Converter Behaviors on Common-Mode Noise. *IEEE Access* **2019**, *7*, 28019–28030. [CrossRef]
58. Li, Y.; Zhu, H.; Wang, S. Investigating Switching Transformers for Common Mode EMI Reduction to Remove Common Mode EMI Filters and Y-Capacitors in Flyback Converters. *IEEE J. Emerg. Sel. Top. Power Electron.* **2018**, *6*, 2287–2301. [CrossRef]
59. Deng, L.; Shi, Q.; Jiang, F. F. Modeling and Analysis of Parasitic Capacitance of Secondary Winding in High-Frequency High-Voltage Transformer Using Finite-Element Method. *IEEE Trans. Appl. Supercond.* **2018**, *28*, 1–5. [CrossRef]
60. Zhang, H.; Yu, L.; Wang, S. Common-Mode EMI Noise Modeling and Reduction with Balance Technique for Three-Level Neutral Point Clamped Topolog. *IEEE Trans. Ind. Electron.* **2017**, *64*, 7563–7573. [CrossRef]
61. Duan, Z.; Fan, T.; Wen, X.; Zhang, D. Improved SiC Power MOSFET Model Considering Nonlinear Junction Capacitances. *IEEE Trans. Power Electron.* **2018**, *33*, 2509–2517. [CrossRef]
62. Dillen, D.C.; Wen, F.; Kim, K.; Tutuc, E. Coherently Strained Si-SixGe1-x Core-Shell Nanowire Heterostructures. *Nano Lett.* **2016**, *16*, 392–398. [CrossRef]
63. Duan, Z.; Fan, T.; Zhang, D.; Wen, X. Modeling and prediction of electromagnetic interference in whole SiC three phase inverters. *Adv. Technol. Electr. Eng. Energy* **2018**, *37*, 1–7.
64. Jia, X.; Hu, C.; Dong, B.; He, F.; Wang, H.; Xu, D. Influence of System Layout on CM EMI Noise of SiC Electric Vehicle Powertrains. *CPSS Trans. Power Electron. Appl.* **2021**, *6*, 298–309. [CrossRef]
65. Zhang, W.; Wei, Y.; Zhang, X.; Wang, S.; Yan, L. Wide-band Modeling and Conducted EMI Simulation of MMC-HVDC Station. *Saf. EMC* **2019**, *5*, 91–97. (In Chinese)
66. Revol, B.; Roudet, J.; Schanen, J.-L.; Loizelet, P. EMI Study of Three-Phase Inverter-Fed Motor Drives. *IEEE Trans. Ind. Appl.* **2011**, *47*, 223–231. [CrossRef]
67. Li, Y.; Chen, C.; Kang, Y. Using ultra-low parasitic parameter to reduce EMI noise of condition for SiC module. *Power Electron.* **2018**, *52*, 22–25.
68. Liu, Q.; Wang, F.; Boroyevich, D. Conducted-EMI Prediction for AC Converter Systems Using an Equivalent Modular–Terminal–Behavioral (MTB) Source Model. *IEEE Trans. Ind. Appl.* **2007**, *43*, 1360–1370. [CrossRef]
69. Duan, Z.; Fan, T.; Zhang, D.; Wen, X. Differential Mode Conducted EMI Prediction in Three Phase SiC Inverters. In Proceedings of the 2nd Asia Conference on Power and Electrical Engineering (ACPEE), Shanghai, China, 24–26 March 2017.
70. Zhou, W.; Pei, X.; Xiang, Y.; Kang, Y. A New EMI Modeling Method for Mixed-Mode Noise Analysis in Three-Phase Inverter System. *IEEE Access* **2020**, *8*, 71535–71547. [CrossRef]

71. Ohn, S.; Yu, J.; Rankin, P.; Sun, B. Three-Terminal Common-Mode EMI Model for EMI Generation, Propagation, and Mitigation in a Full-SiC Three-Phase UPS Module. *IEEE Trans. Power Electron.* **2019**, *34*, 8599–8612. [CrossRef]
72. Wang, K.; Lu, H.; Chen, C.; Xiong, Y. Modeling of System-Level Conducted EMI of the High-Voltage Electric Drive System in Electric Vehicles. *IEEE Trans. Electromagn. Compat.* **2022**, *64*, 741–749. [CrossRef]
73. Elrayyah, A.; Sozer, Y. An Effective Dithering Method for Electromagnetic interference (EMI) Reduction in Single-Phase DC/AC Inverters. *IEEE Trans. Power Electron.* **2014**, *29*, 2798–2806. [CrossRef]
74. Hedayati, M.H.; John, V. Filter Configuration and PWM Method For Single-Phase Inverters with Reduced Conducted EMI Noise. *IEEE Trans. Ind. Appl.* **2015**, *51*, 3236–3243. [CrossRef]
75. Tang, Y.; Ma, H. Dynamic Electrothermal Model of Paralleled IGBT Modules with Unbalanced Stray Parameters. *IEEE Trans. Power Electron.* **2017**, *32*, 1385–1399. [CrossRef]
76. Narayanasamy, B.; Luo, F. A Survey of Active EMI Filters for Conducted EMI Noise Reduction in Power Electronic Converters. *IEEE Trans. Electromagn. Compat.* **2019**, *61*, 2040–2049. [CrossRef]
77. Han, Y.; Lu, H.; Li, Z.; Li, Y.; Chai, J. Analysis and Suppression of Common Mode Voltage for SiC Inverters in Electric Vehicle Applications. In Proceedings of the 21st International Conference on Electrical Machines and Systems (ICEMS), Jeju, Republic of Korea, 7–10 October 2018.
78. Chen, Q.; Chen, Z.; Chen, Y. Electromagnetic Interference Analysis of Single-Ended Fly-Back Converter Based on Secondary-Resonance-Technology. *Trans. China Electrotech. Soc.* **2019**, *34*, 728–737. (In Chinese)
79. Kim, J. Design of Compact Active EMI Filters to Reduce the CM Conducted Emissions. In Proceedings of the IEEE Symposium on Electromagnetic Compatibility, Signal Integrity and Power Integrity (EMC, SI & PI), Long Beach, CA, USA, 30 July–3 August 2018.
80. Xiang, Y.; Pei, X.; Wang, M.; Yang, C.; Zhou, P.; Kang, Y. A Separate Floating Heatsink Based Suppression Method for Conducted Common-Mode EM. *IEEE Trans. Ind. Electron.* **2021**, *68*, 10436–10448. [CrossRef]
81. Xie, L.; Ruan, X.; Zhu, H.; Lo, Y.-K. Common-Mode Voltage Cancellation for Reducing the Common-Mode Noise in DC-DC Converters. *IEEE Trans. Ind. Electron.* **2021**, *68*, 3887–3897. [CrossRef]
82. Omar, S.; Bevilacqua, R. Guidance, navigation, and control solutions for spacecraft re-entry point targeting using aerodynamic drag. *Acta Astronaut.* **2019**, *155*, 389–405. [CrossRef]
83. Natarajan, S.; Padmavathi, P.; Kalvakurthi, J.R.; Babu, T.S.; Ramachandaramurthy, V.K.; Padmanaban, S. Conducted Electromagnetic Interference Spectral Peak Mitigation in Luo-Converter Using FPGA-Based Chaotic PWM Technique. *Electr. Power Compon. Syst.* **2019**, *47*, 838–848. [CrossRef]
84. Mihalic, F.; Kos, D. Reduced conductive EMI in switched-mode DC-DC power converters without EMI filters: PWM versus randomized PWM. *IEEE Trans. Power Electron.* **2006**, *21*, 1783–1794. [CrossRef]
85. Dove, A.; Naude, J.; Hofsajer, I. An Argument for the Relationship Between Spectral Spreading and Probability Spreading for EMI-Reduction in DC-DC Converter. *IEEE Trans. Power Electron.* **2020**, *35*, 1459–1472. [CrossRef]
86. U Hasan, S.; Town, G.E. An Aperiodic Modulation Method to Mitigate Electromagnetic Interference in Impedance Source DC-DC Converters. *IEEE Trans. Power Electron.* **2018**, *33*, 7601–7608. [CrossRef]
87. Videt, A.; Messaoudi, M.; Idir, N.; Boulharts, H.; Vang, H. PWM Strategy for the Cancellation of Common-Mode Voltage Generated by Three-Phase Back-to-Back Inverters. *IEEE Trans. Power Electron.* **2017**, *32*, 2675–2686. [CrossRef]
88. Kumar, S.; Voruganti, S.K.; Akin, B.; Gohil, G. Common-Mode Current Analysis and Cancellation Technique for Dual Active Bridge Converter Based DC System. *IEEE Trans. Ind. Appl.* **2022**, *58*, 4955–4966. [CrossRef]
89. Xie, L.; Yuan, X. Non-Isolated DC-DC Converters with Low Common-Mode Noise by Using Split-Winding Configuration. *IEEE Trans. Power Electron.* **2022**, *37*, 452–461. [CrossRef]
90. Li, Z.; Qiu, S.; Chen, Y. Experimental Study on the Suppressing Emi Level of DC-DC Converter with Chaotic Map. *Proc. CSEE* **2006**, *5*, 76–81. (In Chinese)
91. Zhang, Z.; Bazzi, A.M. A Virtual Impedance Enhancement Based Transformer-Less Active EMI Filter for Conducted EMI Suppression in Power Converters. *IEEE Trans. Power Electron.* **2022**, *37*, 11962–11973. [CrossRef]
92. Fan, J.W.T.; Chow, J.P.W.; Chan, W.T. Modeling and Experimental Assessment of the EMI Characteristics of Switching Converters With Power Semiconductor Filters. *IEEE Trans. Power Electron.* **2020**, *35*, 2519–2533. [CrossRef]
93. Bendicks, A.; Gerten, M.; Frei, S. Active Cancellation of Periodic CM EMI at the Input of a Motor Inverter by Injecting Synthesized and Synchronized Signals (S-3-AEF). *IEEE Trans. Power Electron.* **2022**, *37*, 11951–11961. [CrossRef]
94. Adapa, A.K.; John, V. Eigenvalue Analysis and Behavioral Common-Mode Equivalent Circuit for an Asymmetric Active Phase Converter. *IEEE Trans. Ind. Appl.* **2020**, *56*, 3966–3975.
95. Mueller, D.; Beltle, M.; Tenbohlen, S. EMI Suppression of a DC-DC Converter Using Predictive Pulsed Compensation. *IEEE Trans. Electromagn. Compat.* **2021**, *63*, 2134–2142. [CrossRef]
96. Roy, S.K.; Basu, K. Measurement of Circuit Parasitics of SiC MOSFET in a Half-Bridge Configuration. *IEEE Trans. Power Electron.* **2022**, *37*, 11911–11926. [CrossRef]
97. Chu, Y.; Wang, S.; Zhang, N.; Fu, D. A Common Mode Inductor with External Magnetic Field Immunity, Low-Magnetic Field Emission, and High-Differential Mode Inductance. *IEEE Trans. Power Electron.* **2015**, *30*, 6684–6694. [CrossRef]
98. Dai, L.; Chen, W.; Yang, X.; Zheng, M.; Yang, Y.; Wang, R. A Multi-Function Common Mode Choke Based on Active CM EMI Filters for AC/DC Power Converters. *IEEE Access* **2019**, *7*, 43534–43546. [CrossRef]
99. Tanim, T.R. A new analytical conducted EMI prediction method for SiC motor drive systems. *Etransportation* **2020**, *6*, 100047.

100. Dai, H.; Torres, R.A.; Jahns, T.M.; Sarlioglu, B. Analysis and Suppression of Conducted Common-Mode EMI in WBG-Based Current-Source Converter Systems. *IEEE Trans. Transp. Electrif.* **2022**, *8*, 2133–2148. [CrossRef]
101. Hedayati, M.H.; Acharya, A.B.; John, V. Common-Mode Filter Design for PWM Rectifier-Based Motor Drives. *IEEE Trans. Power Electron.* **2013**, *28*, 5364–5371. [CrossRef]
102. Hsu, W.; Wang, X.; Wen, F. High Phosphorus Dopant Activation in Germanium Using Laser Spike Annealing. *IEEE Electron Device Lett.* **2016**, *37*, 1088–1091. [CrossRef]
103. Pahlevaninezhad, M.; Hamza, D.; Jain, P.K. An Improved Layout Strategy for Common-Mode EMI Suppression Applicable to High-Frequency Planar Transformers in High-Power DC/DC Converters Used for Electric Vehicles. *IEEE Trans. Power Electron.* **2014**, *29*, 1211–1228. [CrossRef]
104. Kumar, S.; Akin, B.; Gohil, G. EMI Performance of Active Neutral Point Clamped Phase Leg for Dual Active Bridge DC-DC Converter. *IEEE Trans. Ind. Appl.* **2021**, *57*, 6093–6104. [CrossRef]
105. Narayanasamy, B.; Luo, F.; Chu, Y. Modeling and Stability Analysis of Voltage Sensing based Differential Mode Active EMI Filters for AC-DC Power Converters. In Proceedings of the IEEE Symposium on Electromagnetic Compatibility, Signal Integrity and Power Integrity (EMC, SI & PI), Long Beach, CA, USA, 30 July–3 August 2018.
106. Chen, Y.; Jiang, W.; Zheng, Y.; He, G. EMI Suppression of High-Frequency Isolated Quasi Z-Source Inverter Based on Multi-Scroll Chaotic PWM Modulation. *IEEE Access* **2019**, *7*, 146198–146208. [CrossRef]
107. Yang, L.; Zhao, H.; Wang, S.; Zhi, Y.; Zhu, B. Develop Common-Mode EMI Noise Models for AC-DC-AC Traction Systems. In Proceedings of the 43rd Annual Conference of the IEEE-Industrial-Electronics-Society (IECON), Beijing, China, 29 October–1 November 2017.
108. Yang, L.; Zhao, H.; Wang, S.; Zhi, Y. Common-Mode EMI Noise Analysis and Reduction for AC-DC-AC Systems with Paralleled Power Modules. *IEEE Trans. Power Electron.* **2020**, *35*, 6989–7000. [CrossRef]
109. Li, P.; Wang, Y.; Adam, G.P.; Holliday, D.; Williams, B.W. Three-Phase AC-AC Hexagonal Chopper System with Heterodyne Modulation for Power Flow Control Enhancement. *IEEE Trans. Power Electron.* **2015**, *30*, 5508–5521. [CrossRef]

MDPI
St. Alban-Anlage 66
4052 Basel
Switzerland
www.mdpi.com

Electronics Editorial Office
E-mail: electronics@mdpi.com
www.mdpi.com/journal/electronics

www.ingramcontent.com/pod-product-compliance
Lightning Source LLC
Chambersburg PA
CBHW040806250926
51850CB00007BA/1096
* 9 7 8 3 0 3 6 5 8 7 0 4 2 *